Makers of the Telegraph

NEW-YORK UNIVERSITY.

Makers of the Telegraph

Samuel Morse, Ezra Cornell and Joseph Henry

KENNETH B. LIFSHITZ

McFarland & Company, Inc., Publishers
Jefferson, North Carolina

Frontispiece: University of the City of New York, circa 1836, steel engraving by R. Hinshelwood (courtesy New York University Archives).

LIBRARY OF CONGRESS CATALOGUING-IN-PUBLICATION DATA

Names: Lifshitz, Kenneth B., 1951– author.
Title: Makers of the telegraph : Samuel Morse, Ezra Cornell and Joseph Henry / Kenneth B. Lifshitz.
Description: Jefferson, North Carolina : McFarland & Company, Inc., Publishers, 2017. | Includes bibliographical references and index.
Identifiers: LCCN 2016053230 | ISBN 9781476665597 (softcover : acid free paper) ∞
Subjects: LCSH: Inventors—United States—Biography. | Telegraph—United States—History. | Morse, Samuel Finley Breese, 1791–1872. | Cornell, Ezra, 1807–1874. | Henry, Joseph, 1797–1878.
Classification: LCC TK5241 .L54 2017 | DDC 621.383092/273 [B] —dc23
LC record available at https://lccn.loc.gov/2016053230

BRITISH LIBRARY CATALOGUING DATA ARE AVAILABLE

**ISBN (print) 978-1-4766-6559-7
ISBN (ebook) 978-1-4766-2681-9**

© 2017 Kenneth B. Lifshitz. All rights reserved

No part of this book may be reproduced or transmitted in any form or by any means, electronic or mechanical, including photocopying or recording, or by any information storage and retrieval system, without permission in writing from the publisher.

On the cover: *top to bottom* Samuel Morse, (Library of Congress); Ezra Cornell (Cornell University Digital Library Collections); Joseph Henry (Library of Congress); telegraph (© 2017 mashuk/iStock)

Printed in the United States of America

McFarland & Company, Inc., Publishers
  Box 611, Jefferson, North Carolina 28640
  www.mcfarlandpub.com

To Fred Epstein

Table of Contents

Acknowledgments   x
Notes on Abbreviations   xi
Preface   1
Introduction   6

Section I: Knickerbocker Tales

1. The American Experiment   17
2. SPAAM (The Society for the Promotion of Agricultural Arts and Manufactures)   33
3. T. Romeyn Beck   37
4. The Albany Academy   42
5. The Big Ditch   43
6. A Tale of Two Cities   45
7. Portrait Painter   48
8. The Man Who Sneezed So Singularly   52

Section II: Henry's Influence on Morse

9. The New, Old School   57
10. The Vision at Palmyra   61
11. Ghost Story   66
12. Wrap Artist   68
13. The American Achievement   72

14. Que Viva Mexico — 75
15. Endless Debate — 80
16. Quantity vs. Intensity — 86
17. The Barnaby Mooer Side-Hill Plow — 92
18. Catching Colt — 96
19. Out of the "Fog" of Invention — 103
20. The Plow in Maine — 107
21. Sins of the Father — 112

Section III: The Madman and the Telegrapher

22. A Federal No-Show Job — 117
23. Cable Problems — 123
24. Big Confab at Little Relay — 129
25. The Trouble with Fisher ... — 133
26. On the Third Floor of the Patent Office — 138
27. The Burden of Big Science — 145
28. Bartlett's Contract — 149
29. Cross-Cut! — 153
30. A Fight Over Pole Insulators — 156
31. Out of the Frying Pan — 158

Section IV: Relay Race

32. The Magnetic Telegraph Company — 169
33. A Red Herring — 173
34. The Mule Kicks Back — 177
35. The State Fair — 182
36. Raising Cash — 187
37. When the Going Gets Tough ... — 191
38. ... The Tough Go to Europe — 196
39. *Trompe l'œil* — 198
40. "The Telegraph for Dummies" — 203

41. O'Reilly	210
42. Saxton Faxton's Love-Hate Relationship	217
43. Organization Man	220
44. Crossing the Rubicon	225

Section V: Prodigal Son

45. Audubon's Laundry	233
46. Tit for Tat	238
47. An Indispensable Plague	241
48. The New York–Offing Line	244
49. Rebirth of a Notion	248
Coda: King Edward of Kalamazoo	256
Afterword	262
Appendix A: Morse's Deposition	267
Appendix B: Questions Prepared for Professor Henry by Morse, 1839	269
Chapter Notes	270
Bibliography	321
Index	325

Acknowledgments

Acknowledgments have always seemed a little self-serving to me—like those people who have six hundred friends on Facebook—mostly because the people who work in research and archiving do not seem to do it for credit. Nevertheless, I would like to thank certain individuals who in the course of this extended themselves generously (and somewhat unreasonably) on my behalf: in particular Nancy Lyon of Yale University Libraries; Tad Bennicoff, Assistant Archivist at the Smithsonian Institution Archives; John McClintock; Eisha Neely, Reproductions Coordinator at the Division of Rare and Manuscript Collections, Cornell University; Fred Bassett, Senior Librarian at the New York State Archives and curator of the van Rensselaer papers, who provided invaluable assistance with regard to van Rensselaer; and Marc Rothenberg of the National Science Foundation for graciously taking the time to read and point out some of the more egregious factual errors. Most importantly I would like to express my gratitude to the late Dr. Carleton Mabee, Professor Emeritus at SUNY New Paltz, an acclaimed Morse expert and Pulitzer Prize–winning biographer, who, for no good reason, invited a somewhat bedraggled virtual stranger into his home to sit and discuss the foibles of some long-dead, strange and remarkable inventor, and then further spent a considerable number of hours editing an earlier version of this manuscript and corresponding with me about it (despite being nearly 100 years old at the time).

I would like to thank Michelle Figliomeni of the Orange County Historical Society (of New York) who spent a good deal of time patiently listening to my sometimes uninformed ruminations on the subject of Morse, Cornell and the telegraph. I would also like to remember in passing my "blue" Chihuahua, Toby, who sat on my neck for much of the while as I wrote this book (and who has since, sadly, departed).

Notes on Abbreviations

POJH   Papers of Joseph Henry (Ed. Nathan Reingold and Marc Rothenberg), Smithsonian Press.
MPLOC   Morse Papers, Library of Congress (online), General Correspondence, 1793–1919.
ECPKL   Ezra Cornell Papers, Kroch Library, Rare Manuscripts, Cornell University.

Silliman's *American Journal of Science*, though published by several publishers, is uniformly referred to throughout as *AJS*.

Where sources appear in various places, I have generally opted to cite only the earliest.

Preface

My interest in Morse derives, in a rather convoluted manner, from a summer 2004 visit to a defunct insane asylum in the town of Ovid, on the shores of Seneca Lake, one of the larger Finger Lakes in upstate New York. "What," the reader may reasonably ask "does a defunct insane asylum have to do with the history of the telegraph, or with Samuel Morse for that matter?" Willard State Asylum for the Chronic and Indigent Insane was named after Dr. Sylvester David Willard, who, like that other influential figure in Ezra Cornell's life, Joseph Henry, was a protégé of a somewhat obscure Albany physician named T. Romeyn Beck. The institution was the realization of Beck's lifelong dream of a more humane repository for what was society's detritus: the indigent insane.

As the bronze plaque situated in the driveway attests, it was also here, on the site of Willard, that Ezra Cornell made a first attempt at creating an institution of higher learning open to the general public. This dream would eventually, largely thanks to his participation in the telegraph, come to fruition—albeit not here on Seneca Lake, but on Lake Cayuga (the next Finger Lake over), in the form of Cornell University. The first (failed) effort was called the Ovid Agricultural Academy, and it functioned for one year before the Civil War drained its ranks of both instructors and students. It shuttered its doors, reopening three years later as the Willard State Asylum following S.D. Willard's dramatic death on the floor of the New York legislature (but that is another story).

Having myself attended Cornell University in the late 1960s (and I am saying this neither to brag nor to explicate my interest in insanity, but to at least partially explain my dual interest in Cornell the man and the institution), I became fascinated by Cornell's part in the creation of the telegraph, which had, in some sense, ended in the creation of the very institution in which I then found myself (Cornell, not Willard). Just as Ovid was the author of the *Metamorphosis*,

Cornell supposed to be the author of *my* metamorphosis. As an (as yet) unshaped piece of sophomoric (and as it would turn out, unshapeable) clay, I had taken on the sheltered role of college student in that tumultuous decade, and thus first came to encounter the stern visage of a seated, bearded Ezra Cornell in the form of his statue on the Arts Quad. Not thinking myself quite insane to be on the wrong hill on the wrong lake, I nevertheless felt myself consumed by angst over a variety of social and personal issues of my own at the time, so I must confess I took little note of his disapproving glare, the stern silent injunction of this icon of "firmness and trueness" (as his son Alonzo would later characterize him) to bear down on life. In some far-crannied corner of my brain, however, I felt him bearing down on me (on me alone, seemingly), asking me what exactly I was doing there, and my thus far untransformed (perhaps untransformable) self began asking questions to which I did not have any ready answer and perhaps did not want an answer (and indeed felt somewhat annoyed that they were being asked at all).

Ezra Cornell, as I would come to find out, was an even less likable and approachable individual in person than when cast in bronze and placed squarely athwart the path of my future academic success (or lack of it). His earliest benefactor, Colonel Beebe, had called him "that hickory Quaker." Samuel Morse had called him "the plague" in print and likely other things even less complimentary in private. His first and most loyal friend amongst the cast of characters of the telegraph was the commonly accorded villain of the piece: the crude and rapacious Fog Smith.

My interest in Willard happily coincided with an exhibit at the New York State Museum in Albany that chronicled the history of Willard Asylum as shown through the investigation of inmates' suitcases that had been abandoned long ago and then retrieved from an attic in the summer of 1995 by a cleaning woman. As I learned more about the subject, I was surprised to find that some of the same individuals who assisted in the birth of Cornell University had also been instrumental to Cornell's prior career with the telegraph, and that all these traced back to a somewhat obscure Albany physician named T. Romeyn Beck. As we shall see, it was also due in part to a student of Beck's, Joseph Henry, that Cornell was so quickly elevated from the rank of "ditch digger" to eventually Superintendent of the Telegraph.

We look to Beck as the ultimate organization man. The scientific societies and schools in which Robert Livingston as well as the Old Patroon played a hand are shaped to a large degree by his vision. Consequently, the early chapters of this book are devoted largely to Beck's part in setting up the infrastructure and foundation for the intellectual revolution that produced the figure of Morse and in particular, the creation of what may be called "big science." And in a sense, this book is as much about the emergence what is called "big science" in America as it is about the telegraph (hence what may seem like an inordinate amount of time is spent on the early organizations like SEUM and SPUA).

Government-sponsored big science did not spring from a void. It was an offshoot of these prior efforts like SEUM and SPUA and efforts of those like Henry and Morse who attempted to merge the two approaches in a practical manner. As we shall see, it attracted not only those of refined erudition or indomitable will like Beck, or those of special genius like Henry, but naturally also some well-practiced in chicanery, graft and self-promotion like Fog Smith, and to some extent (it could be argued) Morse; but in reality it took in some degree of all of these working together for the efforts to bear fruit (not that I am condoning chicanery or graft, only noting its ineluctable presence).

So while we may be accused of exaggerating the influence of both Beck and Henry in the construct of Big Science and particularly American Big Sciences, it is important to realize that before Beck, this model did not exist in America even in theory (with the exception of Hamilton's experiment, SEUM), and did not exist at all in a practical sense until the advent of the telegraph. As first Secretary of the Smithsonian (and Beck's protégé), Henry was also a crucial figure when it came to shaping government policy with regard to science and scientific agencies, and while Morse's name is generally far better known to us today than either Henry or Beck's, without them, safe to say, a Samuel Morse could not have existed. The making of the telegraph therefore offers us this unique microcosm wherein we see for the first time the interplay of these forces emerging, and also provides a laboratory in which we can see these various types of individuals interacting and evolving.

Putting aside the incipient melodrama and metaphor for the moment, there remains perhaps the expectation of those coming to this book that all the major or salient events surrounding Morse's career and the telegraph will be dutifully recited, fairly and completely portrayed. They will not—the intent here is to provide a more complete portrait of a limited set of events than has been offered to date; to illuminate through the vignette; to collate a glossary of incidents that allow the reader to form a "higher resolution" picture of the individuals under the microscope of intense pressure of adverse circumstance. We will also focus on a device that became the central bone of contention between Morse and Cornell: the magnetic relay. We restrict ourselves to those periods in which we can clearly perceive the influence of two or three of these individuals concurrently at play: August 1823, Albany, during Morse's visit as a young aspiring artist; 1843–45, the period just before and after the success of the Test Telegraph; and lastly, 1860—when the battle over the renewal of the relay patent takes place and Cornell begins his educational project in

Ovid, N.Y. We stop short of wading into the myriad controversies and legal battles that emerged from the success and commercialization of the Test Telegraph and focus instead on the evolution of ideas. So there is no attempt at biography here in any conventional sense, but rather an attempt to investigate episodically in some greater detail the signal events (no pun intended) and periods in the birth of the telegraph and the careers of these three men, and avoid attempting to cover those other areas of more interest to biographers.

If we are speaking about metamorphosis, there is no apter subject than Samuel F.B. Morse. Yet, for all his capacity for self-transformation, Morse's vision of himself conversely seems oddly inalterable. It is well known that Morse started in life as an artist and sculptor. Eventually his main subject matter became himself, and his most monumental work, his own reputation. He rode the "hobby horse" of art well into adulthood and even long after distinguishing (or sometimes disgracing) himself in the far-flung fields of portraiture, engineering, politics and business; even well into middle age he persisted in regarding himself as an artist who had been temporarily waylaid by circumstances. In his valedictory speech given at a party in 1861 in his honor at Delmonico's Steak House in New York, he made a point of calling his fellow artists "brothers," implying that at heart he had never really abandoned them.[1] His talent and avocation exist not side by side but at loggerheads, locked in eternal combat in some empyrean realm known only to him, awaiting the kinder machinations of fate to provide some resolution; consequently, his life seems to remain on some unrealized cusp even in the midst of great achievement and success, stubbornly amorphous and incapable of full realization. It is because of this central conflict that we sense that Morse is really a tortured individual, and as is the case sometimes, this impression may be in part intentional.

I had acquired, through other means than my brief association with Cornell, a profound interest in 19th-century American science and scientific organizations (which had led me to produce a novel and several other works on the subject), and in this context, the images of Morse, Cornell and Joseph Henry inserted themselves. Their shared association with the telegraph began swirling in my head like some kind of a kaleidoscopic fantasy, a mélange of metaphysical/psychological forces and actual events that seemed to operate through some arcane logic rooted in an obscure biblical guilt, some crippling seminal sin, but one leading to great aims and great accomplishments in the end. When we introduce the figure of Beck into this mixture, this seems to precipitate this arcane fate; and through this lens, Cornell's association with the telegraph seems more than just the product of serendipity.

While Beck is someone neither Ezra Cornell nor Samuel Morse likely ever met in the flesh, his influence on both seems undeniable and profound, and not only through Henry alone, but through other individuals as well. In some sense, Henry, as Beck's intellectual heir, provides the nexus of this relationship between intellect and the public sector and he offers the key to understanding Morse's ascent to such a vast and unprecedented public stage. Thus the subtext of Beck's influence on both Cornell personally and the shape of American science in general is a theme that is central to the book.

It was just when Cornell's association with the telegraph was coming to a close that yet another of Beck's students, this time Albany physician Sylvester David Willard, inserted himself in a dramatic way into Cornell's life, paving the way for Ezra Cornell's single greatest accomplishment, Cornell University. This is the juncture from whence, we note, our story emerges (as did my signal discomfiture on the Arts Quad)—but that is yet in part another story.

All of these men were in very different ways a product of their age; scientific achievement on a cooperative national scale seemed linked to the preoccupations and expanding appetites of the nation as much then as now. But in terms of the telegraph, Morse's name generally stands out head and shoulders above the rest. For what reason? Whatever else may be said of him, Morse seems to me to be an impartial self-chronicler. While there are, however, clearly editorial excisions here as well (particularly when it comes to the relay), in general, whether through hubris or circumspection, we find many of his own flaws freely exposed in the correspondence and materials he himself conserved, and though he may have endeavored to hide them from his contemporaries, they are (for the most part), there for the historian and consequently for history as well to examine. Most importantly, like Cornell, his papers are well-organized and available online through the Library of Congress website, thus allowing the luxury of unhurried research and greatly expediting what might otherwise have been an impossible task of scholarship. Cornell's papers, like Morse's, are scrupulously preserved and available online as well, and while (probably self-edited to a greater degree and) not as voluminous as Morse's, they do provide an illuminating counterpoint as one tries to gain a more balanced perspective upon events.

Henry's papers, of course, are available in the remarkable collection published by the Smithsonian. These were the three main resources for this work, and the book therefore naturally finds its focus on the interplay between Cornell, Morse and Henry. Beck's papers are nowhere near as complete or voluminous as those Henry, Morse or Cornell. Most are held by the New York Public Library, where I did some of this research. Other materials were found in the van Rensselaer Papers at the NYS Archives at Albany, but some of these were damaged by a fire and are therefore delicate and not presently open to public inspection.

Introduction

It had been five years (five tumultuous years) for the inventor of the device on which had been posed the stirring, four-syllable question, "What hath God wrought?"—tapped out by the nervous index finger of that same convoluted and rather constantly tortured individual known to the press and public as "The Lightning Doctor" or "The American Leonardo," titles he would semi-humorously arrogate to himself.[1] These electrical impulses had seared their way from a hushed chamber inside the Supreme Court in Washington, D.C., setting a device, some forty miles away, at the far noisier, bustling Mount Clare Station on Pratt Street in Baltimore, to clattering. In that instant the world had changed (a fact not lost on the assembled crowd), and the event evoked a murmur of wonder and satisfaction that already seemed to unify them with unseen far-flung places. Samuel Morse's longtime assistant, Alfred Vail, had been the one appointed to receive and read it. The message as transmitted was in some sense a self-fulfilling prophecy. The changes wrought in the country were soon to be profound and indelible, and it would be a nation transformed, indeed, as if the very hand of God had swept across its face.

The Test Telegraph was as yet a device "in embrio," but clearly that would come to transform the nation. Its eventual incarnation as a fully realized system was one vastly different in both efficiency and complexity from this original incarnation. Whether for reasons good or bad, this was a source of profound embarrassment to Samuel Morse, who preferred the appearance of his genius to be "sui generis" and of whole cloth. He would therefore spend a good deal of effort and time trying to obscure this fact of incremental improvement. It is, however, the incremental nature and scope of that difference and improvement that concern us here and which forms the crux of this book. This is not to diminish the overall accomplishment by focusing on the details. Newspapers at the time spoke of "the annihilation of space and time." Safe to say, half a decade later, America was not just a nation transformed, but truly, for the first time, united and unified (albeit temporarily) in a way that could not have been conceived nor anticipated; yet ironically, this sudden and precipitate drawing together produced not a stronger bond but pointed the nation to the brink of its greatest catastrophe.

Let us look for a moment on the face of the country before the rumblings of a fratricidal war distorted it. For those of that great humorist Mark Twain's generation, the move west and the Gold Rush represented a watershed, a dividing line between an age of congealed and prescribed manners and starched formalism in communications and the brusque and explicit hand of business. In the case of the former, we find in operation a formalized discourse based on floridly polite correspondence that arrived with the deliberateness of a party of snails heading to tea at grandma's. With the telegraph came a new lightning bolt form of communication that dispensed with politesse in favor of economy and speed.

It was, however, the discovery of gold at Sutter's Mill that sucked intellect, resources and influence west and sent cargoes of bullion and also sent new freshets of talent east. Gentility and greed for the first time stood together on the same public stage, and to paraphrase Gordon Gecko, suddenly "greed was good." By 1861 news traveled between coasts in the blink of an eye. The finest orators of the age awoke to an expanded and more attentive audience to attend their florid ideals, as the tides of history all were rushing in as in a great flood.

If there was any figure representative of that age who seemed to be helplessly in tow and at the mercy of supernatural currents and forces, it was Samuel F.B. Morse. His life and relationships were a game of shadows; friends and colleagues transmuted instantaneously and magically into "fiends" and "plagues"; as in a medieval joust, former allies became instantly implacable opponents, some left vanquished on a littered field of legal and economic battles. As for his main claim to fame, the telegraph, he seemed to stand on this accomplishment as if atop quicksand—claims against his patent and his intellectual authorship surface and foment, slip from view and then re-surface again, and we are left dumbstruck as forces appear to conspire constantly. The more he struggled, the more they threatened to suck him under, into either ignominy or oblivion.

Disregarding for the moment the multifaceted arguments over intellectual ownership of the telegraph, that would rage almost from its inception, to first give credit where credit is due, if there is a single individual who can be credited with bringing American science onto a world stage, it is Samuel F.B. Morse. And though he much preferred to regard himself as a self-created "phenomenon," he too in fact stood not only atop the shoulders of giants but abreast of them as well. Rather than elevating them by proximity, Morse seemed to cast a shadow on his peers, and it seemed he did this not by accident but, as we shall see, purposefully. To put Morse's achievements in proper context, therefore, it is first necessary to trace the evolution of scientific organizations in general that preceded him and particularly those that influenced those that provided the foundation and context for his accomplishments and those that he systematically sought to obliterate.

Whether or not one credits the somewhat mythical accounts surrounding his single-handedly inventing the telegraph aboard the *Paquebot Sully* while crossing the Atlantic in a flash of inspiration, like Athena springing from Zeus's forehead, which image has long since imprinted tiself in the American psych, is unquestionably Morse, like Zeus flinging a lightning bolt, who destroys the pervading and crippling self-doubt that had lain like a shadow over the American scientific community. In large part it was he that allowed America to ascend unhindered to heights of technological, scientific and even artistic achievement that hitherto would have been unimaginable. So while one may question his accounts, and however questionable or suspect may be his accomplishments in regard to the telegraph itself (which is a subject we shall explore in great detail), for that alone he will indeed always be remembered: that through his invention he opened the floodgates of capital, both intellectual and economic, and set the stage for a vibrant economy based on technological innovation that would eventually become the envy of the world.

I have alluded in the preface to what I characterize in the book as a somewhat mystical influence that Joseph Henry exercised upon both Morse and Ezra Cornell. In particular, with respect to Morse, he appears almost a "Delphic Oracle," a mystical fount to which Morse resorts cautiously and infrequently to draw upon a profounder, unquestionably more venerable and unerring but also perhaps dangerous inspiration (that is, until Morse decides to storm Mt. Olympus himself). To Cornell, Henry is a *Deus ex machina*, the transmitter of forces that have their origins in some darker, more ancient wellspring.

The triumphs of technology are often regarded merely as the justifiable moral reward, the inevitable offspring, of unerring genius when they sometimes owe more in fact to trial and error and miscarriages of both ethics and experiment, but these latter are most often omitted from the sunny tales of progress that emerges and forms most often, in a phrase Kenneth Galbraith coined, "the collective wisdom" concerning these events. We suggest rather explore these darker threads running through the often mischaracterized "sunny" tales of progress were not attributable, as Morse's business partner "Fog" Smith would at one point crudely and racistly complain, to "the nigerfied conditions of Washington, D.C.,"[2] but are part of a complex fate that seemed to inalterably connect the destinies of these three men, Morse, Henry and Cornell, and do

Cornell, and do so in some fashion that bridges the logical and meta-logical domains and which provides a darker resonance to this erstwhile homily of inevitable progress.

And all this somehow will become intertwined with a theme of some seemingly inexplicable connection, some original sin—an indelible prior act that wove the threads that brought these particular individuals together. A familial murder that occurred in the Plymouth plantation in the 17th century, one which had left a dark stain on the Cornell genealogy. These strands of fate seem somehow to lead directly to Beck and therefore to Henry (and Willard) as well. They raise their ghostly visage at signal points in Ezra's career and entangle him just at the point he is poised to ascend to a greater stage. Elaine Forman Crane's[3] treatment of this theme is really the first to note this, as she is the one who first framed Cornell's life thus, as devolving from this heinous murder, and it seems even more apt than when viewed in the context of his career with the telegraph.

It was Beck's lifelong interest in the plight of the insane that would therefore tie him most closely to Ezra Cornell via its eruptions in his family. (And while on the subject of darker forces and influences), whether or not we, as supposedly enlightened modern individuals, believe mental illness to be hereditarily transmissible, clearly the vast majority of the population of the United States in the mid-(even late) 19th century believed it to be the case. As noted, instances, especially in prominent families, were whispered about with grim embarrassment, sometimes by the relatives of the afflicted (and stage-whispered with grim satisfaction by those who otherwise despised or envied them). Cornell himself was very much a man of his time, and he too was no doubt an unwilling subscriber to the prejudices of the day, so he keenly felt the embarrassment of his cousin Alvin's insanity and later that of his brother, Edward. Whether or not his wife MaryAnn shielded him for a time from these eruptions within his family, they would still come to play an unspoken but no doubt distracting part.

Joseph Henry, also having a history of mental illness in his family, also often took steps to further obscure his paternity, as his biographer Moyer notes, often giving out incorrect dates for his birth to muddy the connection with his alcoholic father. Aside from Henry, Beck also affected Cornell through several of his other protégés: Dr. Sylvester David Willard, who, as mentioned above, would come to play a pivotal role in the creation of Cornell University; also Edward van Deussen, who eventually took charge of Cornell's brother Edward when he was committed to the insane asylum in Kalamazoo, Michigan. By any standard, these strands that connect Beck directly to Cornell seem somewhat indistinct and tenuous and it is doubtful the two ever met in person, yet somehow they seem drawn together by events that are channelled through one or the other of Beck's disciples or in the case of Alvin, Beck himself.

As a medical doctor dedicated to the humane treatment of the insane, Beck became involved with the plight of Alvin Cornell, just at the time that Ezra Cornell embarked from Ithaca on the quest that would end in his finding fame in riches from the telegraph. Alvin had been accused of the murder of his wife near Jamestown, New York, and Beck's intercession would result in Alvin's life being spared. While this had really little on the surface to do with Cornell and his career—and in fact, we find no mention of it all in his correspondence—it was unquestionably, in a larger context, a profound event for him and for his family.

Later, somewhat more indirectly but in a strangely parallel thematic echo, there is the involvement of Edward van Deussen, another one of Beck's protégés, who was treating Ezra's brother Edward. Remarkably, these intrusions of insanity into an otherwise seemingly well-ordered, robust and productive life seem to neatly bookend Ezra's career with the telegraph, occurring as they do in 1843 and 1860. Alvin's insane act coincided with the beginning of Ezra Cornell's association with Smith and Morse, and Edward's final breakdown came right at the cusp of Ezra's disassociation from Western Union, (coinciding with his creation of Cornell University). In both instances we can sense the corporeal or incorporeal hand of Dr. Theodoric Romeyn Beck somehow hovering over and shaping these events.

So, though never overtly mentioned, the theme of madness clearly plays a role in this otherwise sunny tale of progress. Certainly Ezra himself at several points felt the cold hand of this specter on his own shoulder; heard the ghostly, disembodied voice; felt a plaintive tug at his breech-coat and heard the dark whisper whose ministrations (real or imagined) he had sought to banish through the purgatives of "clean living" and frequent manly exercise.[4] As events reveal just how far the red tide of madness had penetrated the bloodlines of his family, there can be little question that he lived in dread of arousing this specter in himself, and that no matter how much fresh air and exercise he obtained, he feared one day it might erupt from his own otherwise well-regulated Quaker psyche to wreak havoc on all he had so painstakingly built. This was the dark cloud of foreboding under which all his efforts and labor lay, and which in a sense form the backdrop of our rather more technical tale.

Unlike Morse, the urbane, well-traveled, Yale-educated scion of a well-respected New England dynasty, Ezra Cornell was an iconic self-made man. He possessed the unvarnished confidence of someone committed to his own interests and instincts. Unlike Morse, evidence of self-doubt was not displayed on the surface for all to see. He is certainly uneducated (at least in any formal sense), and as a New York Quaker, he was undoubtedly considered (particularly by Vail, at least) as something of a country bumpkin. The impression is understandable, given that his spelling and speech are rustic and improvised, and his more technical suggestions, at least early on, were presented brimming with the confidence of a "may the better man win" credo—a forthright naiveté that was rivaled only by Morse himself in his vaunted ideal of a "scientific meritocracy."

Despite his lack of a formal education, Cornell managed to absorb whatever was available around him like a sponge and translate it into direct and practical effect. He had, however, unlike Morse no talent whatsoever for self-promotion. Consequently, his part in creating the telegraph is usually grossly underestimated and misunderstood. Having been made rich beyond his wildest dreams through his association with Morse, and with far greater accomplishments to his credit, in old age Cornell willingly ceded Morse unearned intellectual sovereignty, intentionally shrouding his own contributions. Those contributions and claims that Cornell pursued so vigorously in his youth he denominated superfluous and trifling in old age, preferring to gloss over what was in effect a bare-knuckle crawl with the crepe paper of retrospective bonhomie.

As with Morse, while clinging to what was an outmoded idealism in some respects, Cornell was far from a *naïf*; however, when it proved convenient, Morse would adopt that as a kind of convenient slur on his employee. When threatened by Cornell's invention, the magnetic relay, he chooses the word "clumsification" to describe it, a term no doubt chosen purposefully to highlight Cornell's humble background and educational deficits, hoping in the process to render Cornell's claim to authorship of such a complex component all but ridiculous. In fact, Cornell was at the time Morse's employee, but he had taken steps to protect his intellectual rights thru a contract with Amos Kendall. Unfortunately, thanks to Morse's capacity for deviousness, these would fail. Perhaps because of the final recognition that in the end they were more alike than different, Morse's relationship with the "ditch-digger," Ezra Cornell, despite several intervening violent ruptures (mostly also due to Vail's rabid influence, and unlike Henry, who apparently never forgave Morse) seemed to end on a note of mutual respect and admiration, with Cornell serving as a pallbearer at Morse's funeral.[5]

On the other hand, the rather one-sided hero-worship relationship that evolved between Morse and Henry devolved into a bitter recrimination and personal attacks, had always been based on a thinly veiled jealousy.[6] Though Henry came from a family background that should have rendered his intellectual credentials perhaps as suspect as Cornell's, because of Beck's influence and shielding, Henry's background was rarely if ever called into question—certainly not by Morse.

Beck's influence was sporadic and seemed to erupt at key points throughout Cornell and Henry's careers. Beck's personal commitment to furthering the education of those qualified, not just by the possession of money and connections, but by native intellectual talent, was what had brought a young and impoverished Joseph Henry into his purview and Cornell would be deeply affected by this same attitude. It would inform the university he would come to shape. Samuel Morse was also one who seemed to favor, at least on the surface, the idea of an "intellectual meritocracy." He was always seeking to surround himself with those possessed of a native intellectual curiosity. His respect for Joseph Henry, at least early on, was in part a reflection of his somewhat accurate estimate of his own comparatively meager talents which were, in all fairness, somewhat anemic, and to paraphrase Einstein, based on "99% inspiration and 1% perspiration."

Beck first encountered Henry, as the story goes, in a production of *Hamlet* at the Green Street Theater in Albany. The story of the telegraph therefore also seems derived from *Hamlet* and the ghost of Romeyn Beck, Hamlet's father, stubbornly refuses to be exorcised. We can almost sense the sunny public face of progress and amassing of fantastic wealth being o'ershadowed by a private, darker drama being played out. These intricate patterns of guilt and interdependence all seem to display themselves in the context of a somewhat humorously stiff 19th-century formalism. So there emerged in this context this kind of mystical subtext of a connection between Beck and Cornell through Henry and Willard with the haunting theme of insanity lending the whole tableaux a compelling aura of mystery.

By the end of the Civil War, for various reasons, Cornell's Ovid Agricultural College on Seneca Lake was no more, and in its place stood a gleaming new granite limestone insane asylum financed by New York State legislators who had assumed the mortgage from the good citizens of Ovid. Near the foot of that other lake, Cayuga, stood Cornell University, a product of the Morrill Act and Cornell's vast fortune, derived mostly from his Western Union stock. It was really Cornell's dedication to agriculture that first introduced him to F.O.J. Smith and hence Morse. How exactly Ezra Cornell came to own such a large proportion of Western Union stock is not quite the subject of this book, but his journey from plow salesman to Superintendent of the Telegraph comprises a large portion of our story, and the reader can likely guess some of the rest.

Throughout his career, Morse would find himself having to fend off charges that his ideas were derivative, either of Henry's or others', but the lengths that he would go to in order to deny these were in a word "Herculean." Clearly the vituperation directed at Cornell over his magnetic relay was not on the level which Morse mounted in his later attacks on Henry, but clearly one was rehearsal for the other. The device known as the magnetic relay was something that clearly embodied Henry's ideas on "intensity" and "quantity" and it came to be at the crux of the fundamental struggle between Morse and Cornell, so we devote a perhaps inordinate amount of space to it and to Henry's part. Conversely, as something that Charles Grafton Page once called "the life and soul of the telegraph," it had an importance that is underestimated, and it deserves a great deal more attention. It is clearly central to much of the conflict between Morse and Cornell.

The point is not to compare the scientific acumen or accomplishment of these two (Henry and Cornell), but only to note that they both, at different points in time, put Morse's paramountcy with regard to the telegraph in great jeopardy, and thus inspired similarly devious and sometimes vicious responses. While Morse initially defended Cornell from attacks and jealousy, when he needed one Alfred Vail would become the hatchet man (so to speak) with regard to both Cornell and Henry. In Henry's case, it was through an ongoing propaganda effort that began with the (virtually) entire (and according to Henry, intentional) omission of Henry's contributions to electromagnetism and the telegraph in Vail's book. This would spark a decades-long and irreparably bitter feud, not just between Vail and Henry, but between Morse and Henry as well. Vail needed no prompting to undermine Cornell. Their mutual dislike was almost instantaneous, and we will find that, rather than encourage it, by 1845 Morse felt he positively had to rein in what had become increasingly rabid and public attacks on the man who by then assumed Morse's former title of "Superintendent of the Telegraph."

Unlike Cornell, Henry and Morse's reputation rests almost entirely on their scientific accomplishments, but in some respects Morse was somewhat closer in character and style to that other great showman of his age, P.T. Barnum, but the influence of both he and Henry on the progress of American science is undeniable. For those like Henry, who consciously resisted the pull of the marketplace, and for whom abstract endeavor was enough reward in and of itself, the parity of American intellectual endeavor with European efforts and the shape of American intellectualism became the issue of paramount concern. For Morse, for whom the furies of commerce, science and belief all were forged into a singular, remarkable (if somewhat morally questionable) juggernaut, his accomplishment would be forever paired with the physical device on which it was based.

Before Morse and Henry there may have been other American scientists, but there was really not yet such thing that could distinctly be called "American science." American artists and scientists existed, but were primarily regarded as an adjunct to European intellectuals and intellectualism. They were used to European condescension, and sought to make their marks more in the realm of "ingenuity" rather than that of "genius." Everyone recognized America's great potential, but the essentially political decision of where and how and to what end to expend her intellectual and physical capital, would become a central preoccupation of those who inherited the American Revolution and the Industrial Revolution as well, and who would become the intellectual heirs to European science and culture. The resulting split in American scientific endeavor (before there was really a culture of which to speak), between pure and applied science or technology, would thenceforth come to shape the character and destiny not only of the national scientific and artistic communities, but of the nation itself as a whole. In effect the most enduring incarnation and merger of the two would come only within the precincts of the great university where, as Cornell put it somewhat blandly but nevertheless powerfully, they would offer a place where "any person might find instruction in any study."

Section I
Knickerbocker Tales

"List …
if thou didst ever thy dear father love—"
(Old King Hamlet)

1

The American Experiment

It was, most of all, an act of faith when a world- and bone-weary George Washington relinquished the office of the presidency of the United States in 1796, voluntarily abandoning power that was his for the taking—an act virtually unheard-of in the history of nations. While the unwieldy young country still tottered, like a life-drunk colt on freshly found legs, his coterie of fractious political heirs argued over the shape of his inheritance, how and in what form his faith and trust might somehow survive his imminent physical retirement from public life. It was this simple faith, that a government of laws could survive and thrive without a single charismatic central leader, that came to define America, and while the legacy of order hovered dimly in the vacuum afterward, fragile and fraught with difficulties, yet it also seemed somehow remarkably and surprisingly durable.

That very vastness of the new landscape, not any emerging schism of ideals, thus far had been the main obstacle mitigating against eventual success of the grand endeavor of building a new and, more importantly, unified country. Vast would become something of an understatement. Perhaps Washington at the time suspected that within a mere score of years its size would be more than doubled; perhaps not. But what was certain was that with that growth, the trouble and faction would likewise multiply in every quarter and in every niche of policy. After more than a full decade of peace, one thing had become abundantly clear: the political and social earthquake that was the American Revolution had done very little to bring about a corresponding realignment in the intellectual order of things. American scientists and artists still suffered from an inferiority complex. War, it seemed, could change everything—everything except.

Despite the political resolution afforded by the end of the conflict, the strands of culture and scientific achievements of the mother country still hung like an impending shadow that swaddled the new nation, eclipsing its endeavors and dimming its aspirations. Though political subservience had been repudiated by force of arms, its continental intellectuals still regarded their colonial offspring as mere errant children, somehow derivative and inferior cognates. Americans themselves, even having successfully thrown off the yoke of economic dependency and oppression, still subscribed to this persistent sense of inferiority when it came to the arts and sciences.

Thus it was as the new century dawned. With communications feeble and resources limited, America's best scientific minds focused not on innovating, but on supplementing and serving more tractable endeavors going on apace in Europe. American artists still humbly made pilgrimage to the stations of European culture, and our scientists routinely sent back their specimens of New World flora and fauna to their continental colleagues to be dissected and classified in the better-equipped labs of their European counterparts.

The esteemed Albany jurist, Chancellor James Kent, had written in 1819, complaining to his friend, Yale professor Benjamin Silliman, of the relative states of American and European scientific endeavor:

> If we must regard it is a much less a national reproach, then it certainly is a national loss that most of the products of the Chemical art are imported into this country from Europe—We may be confidently assured that until intelligent chemists are *trained up at home* and induced to awaken to the introduction and extension of *the chemical art*, the U[nited] States will never attain to that pinnacle of national superiority which G[reat] Britain and France owe more to the *successful cultivation and application of natural science*, than to their military prowess and successes.

Benjamin Silliman. Portrait by Samuel Morse (courtesy Silliman Family Papers [MS 450], Manuscripts and Archives, Yale University Library).

So, while the political landscape had radically shifted, this misbegotten sense of intellectual inferiority persisted, and despite the efforts of top-notch artists like Trumbull and Peale and enlightened amateur scientists like Franklin, the aspirations of American culture were regarded mostly as interesting amusements. Even Franklin's well-received electrical demonstrations were seen as more the good luck of the amateur than reward for sustained and coordinated intellectual effort.[1]

Some efforts in the right direction were made. One of the earliest examples is the community that formed at New Harmony, Indiana, with William Maclure and others. While it may have focused mostly on the evils of organized religion, it was perhaps not until socialist reformer Robert Owen delivered his valedictory speech to his acolytes at New Harmony on July 4, 1826, his so-called "Declaration of Mental Independence," that the problem was sufficiently portrayed. This persistent dependency in the intellectual sphere was finally recognized as something with which the scientific community and the country as a whole needed to reckon.

America's most notable accomplishments to that point had seemed destined to be confined mostly to the social or political spheres, where her best minds (and her best scientific minds) had (sometimes unwillingly) migrated. Having imbibed—along with the mother's milk of European culture—this ingrained subservience to European technological and pedagogical superiority, American scientists in the late 18th and early 19th centuries, like most of the rest of the population, (not surprisingly) prided themselves more on delivering goods stamped with the imprimatur of practicality and ingenuity rather than innovation and novelty—on doing much with little and for not far overstepping the bounds of the necessity governing the common man. Inventor and artist both labored, sometimes willingly, under this diminished and derivative stature. They did not protest when their finest accomplishments were classed as "ingenuity" rather than "genius," whether that was framed as "Yankee ingenuity" or "American ingenuity," but as one American scientist who had studied in Europe put it, awe of European science had bred in this case "not self reliance but self-distrust and intellectual timidity."[2]

And there were those who firmly believed that American science should confine itself solely to the realm of practical problems. Those who hoped to focus on more abstruse subjects for their own sake—such as chemistry, mineralogy, biology and the infant science of electricity—styled themselves natural philosophers, and often adopted odd manners and habits and to cloak their serious ambitions under the guise of levity as protective coloration. Some (whom we shall come to know better later), like the Beck brothers of Albany, pursued their more abstract studies only at the risk of being labeled quacks or dilettantes behind their back. So even in exercising the knack of doing much with little, the American genius would eventually find its most profound expressions in two very different and disparate visions of America's scientific future: those of technology and of pure science.

Thus, while heirs to the Federalist/anti-Federalist debate wrangled on in

Washington, another more subtle rift was forming in the scientific sphere. There is something to be said for simply correlating Hamilton's comprehensive Federalist vision with the promotion of technology, and Jefferson's anti-Federalist one with that of pure science; and while it would make the author's chore far easier if the split in the scientific community hewed exactly along the lines of the corresponding political rifts, this would be a gross oversimplification. While there were social and economic dimensions to this argument of which figures in both camps were well aware, as another noted author put it, "A great deal of ink has been wasted by historians who assumed that the motivations for engaging in scientific work must be either purely scientific or purely social/economic."[3] However, there are parallels to be drawn, and while the political rift was most clearly personified (despite its complexity) by those two iconic statesmen, Jefferson and Hamilton, the parallel scientific split would eventually come to be represented even more starkly by those two equally seminal scientific figures of the 19th century, Joseph Henry and Samuel F.B. Morse.

Morse did not vault American science into world prominence single-handedly by any means. Without the firm foundation created by early scientific organizations like Franklin's APS, Alexander Hamilton's SEUM and Robert Livingston's SPUAM, and the work of Henry in particular, Morse's accomplishment would have been classed as one more oddity that was better left to the Europeans for serious investigation. So, though Morse himself often sought to claim status a true self-created "phenom," growing not from the ground of previous accomplishment but descending as if from an empyrean, arrogating to himself (only partially in jest) the title of "Lightning Doctor—or the American Leonardo," this was in part just an expression of the kind of pure showmanship at which Morse excelled.

As in the political sphere, the underlying scientific argument had been seeded by the clash of two great minds (and bitter moral antagonists), Hamilton and Jefferson. America was also possessed, by the beginning of the 19th century, for the first time simultaneously, with an expendable abundance of both monetary and intellectual capital. Perhaps the single point that these two men agreed on wholeheartedly was that both these forms of capital had to be somehow merged and conjoined for the nation to thrive and to successfully compete on a world stage. Their vision of how to achieve this end, however, was markedly different.

As Hamilton conceived it, if American science was to be finally wrenched from the hands of the enthusiastic amateur (amongst whose ranks he classed Jefferson), the first task at hand was to lay a firm economic context for future government-sponsored scientific endeavor. This required, in his view, a concentration of both types of capital within a confined area, a pressure cooker of innovation.

Hamilton's brainchild had been conceived as an incubator for scientific endeavor, a living example *ab reductio* of his larger and more ambitious plans for the country that hinged on creating a modern manufacturing base that would become the driving force of the economy. In March of 1791, Hamilton, along with Tench Coxe, Washington's former wartime aide-de-camp, conceived a plan to set this scientific/manufacturing dynamo into motion at the base of the great falls near what is today Paterson, New Jersey.[4] The resulting organization aimed at conjoining scientific achievement to economic imperatives and resources, and would be known as the Society for Establishing Useful Manufactures (SEUM).

SEUM, however, had not magically sprung forth in a vacuum. Hamilton had earlier been tasked by the Congress with establishing a foundation for American manufacturing. His report on the subject had been requested twenty-three months prior, and the request had been partly in response to Washington's first annual message to the joint session of Congress delivered on January 8, 1790, in which he had noted the need for industrial endeavors to ensure the safety of the country. Washington said the safety and interest of a free people "require that they should promote ... manufactories ... as a means to establish an industrial base capable of producing arms and munitions without foreign dependencies." SEUM thus represented Hamilton's vision of how best to fulfill Washington's injunction, but what he delivered was far broader and more ambitious than a mere munitions factory.

While Hamilton, like Morse, was clearly an advocate of the "lightning-bolt solution," the comprehensive politico-econo-scientific plan, Jefferson's approach (though he was often characterized as the more radical of the two, thanks to his Francophile tendencies) had been shaped more by his slow-paced Virginia upbringing. This background produced a natural inclination towards the circumspect steering and distribution of wealth toward incremental solutions rather than the accumulation of capital and its sudden mobilization. This attitude, coupled with an innate respect for moral absolutes, had made him a deep thinker, but also a man of deep moral contradictions, and what he saw in Hamilton's plan was the earthly embodiment of evil. To him and fellow Virginian Madison, Hamilton's proposed SEUM seemed nothing less than an ethical and economic abomination. Not only did it seek to diminish man's divine intellect to the status of an economic tool, a cog in a machine, even worse, they suspected it to be part of a nasty political trick designed to bolster support for Hamilton's Federalist agenda that would be embodied in his upcoming "Report on Manufactures." The most telling criticism of Hamilton thus far had been that he preferred to deal in abstractions. If SEUM succeeded, Hamilton would have a concrete example of his ideas by the time his report was due to be delivered to the Congress.[5]

Whether it was in response to Hamilton's experiment or not, Jefferson too in that year of 1791 felt impelled to lay the groundwork for a scientific infrastructure. Unlike the unrepentant utilitarian social engineer Hamilton, Jefferson harbored a sincere personal interest in and respect for the sciences independent of their ability to deliver social change. His vision was tinged by a desire to utilize science first and foremost to recapture a lost innocence, an agritopia wherein each man could profitably till his plot unmolested by tyrants (petty or otherwise). When Robert R. Livingston formed a scientific organization dedicated primarily to advancing agricultural innovation, Jefferson had seen in this an opportunity to weld an uneasy alliance, one that would support his Republican ideals and provide a counterweight to Hamilton's efforts in the heart of Federalist New York.

It is thus not altogether difficult or unrewarding to ferret out these diametric political underpinnings and tendencies shaping American scientific inquiry in the late 18th and early 19th centuries, but these had yet to be translated into any concrete policies. Aside from Washington's rather narrowly framed statement about the common defense in his farewell address, neither was there an overpowering urge among the electorate to do so. This was not yet a battle between pure science and applied technology; rather, it was a tiff over whether applied technology and innovation should mainly be directed to the manufacturing or the agricultural sector. There were, however, already shadings in the spectrum. Abstract endeavors clearly held greater potential benefits for manufacturing than for agriculture. Guided by a "show and tell" mentality, legislators found agriculture far better at showing, while more abstract endeavors such as Hamilton's were better at telling. With ninety-five percent of the working population engaged in the pursuit of agriculture, the dividends and benefits to them from relatively arcane endeavors like mineralogy or chemistry, or even Franklin's amusing electrical demonstrations, seemed to be in the end just too abstract to care much about.

Hamilton's idea of consolidating intellectual and economic capital in a single physical locale also potentially subjected it to the aims of the governing corporate entity. Aware of the dangers of parochialism, in addition to granting it special tax status, he sought to make its governing body exempt from designs of the state in which it resided. He hoped in this manner to provide the greatest benefit and mutual incentive to industry, capital and science to work together, if not in harmony, at least in concert. Some of the implications of this policy were not well thought out, nor were his choices for personnel to carry out this vision, and so some on SEUM's board took this special status as a license for abuse.

In desperate need of start-up capital for the venture, Hamilton had turned to several somewhat shady characters and speculators, associates from his long career at the helm of monetary power. One of these was William Duer, son-in-law of Lord Stirling[6] of Revolutionary War fame, who had served much of the war as the "bagman" for the Continental Congress, literally carrying bags of currency to officers in the field to pay the troops. Thanks largely to Hamilton's having created a nascent centralized economic infrastructure, Duer and those of his ilk, following the war, had become skilled currency and stock manipulators. They were quick to realize this new plan of Hamilton for a "manufacturing collective" could only benefit those activities. Indeed, part of Hamilton's vision was to redirect to a more socially useful purpose some of the feverish speculation that had emerged in the wake of the federal assumption of the individual states' war debts. Six of the initial thirteen members of SEUM's board were New Yorkers who had been involved with one or another of Duer's previous speculative schemes.[7] What they had seen in SEUM was a new means of "market making": of controlling the availability and price of key goods, and a mechanism to counter European superiority in manufacturing without resorting solely to mutually destructive tariffs.[8]

Hamilton's experiment, however, while suffering the pangs and uncertainties of a somewhat over-hasty birth, would eventually become the prototype for a new and distinctly American approach to the marriage of science and economics,[9] one encouraging intellectual and technical innovation, but only so long as it was firmly tethered to the needs of industry and the marketplace.[10] It would last in one form or another well into the twentieth century.

Though plagued by financial scandals early on, SEUM would eventually become a manufacturing dynamo that would drive the economic and intellectual progress of the early nation and (as Jefferson had correctly intuited) thereby tie its intellectual achievements to a centralized economic policy. The report, which included detailed recommendations for tariffs and fisheries as well as advocating government subsidies for projects like SEUM, bound all these aims together under a single umbrella and called for the establishment of a single governing board dedicated to promoting the arts, agriculture, manufactures, and commerce—a board funded through those same commercial tariffs that would protect the nascent manufacturing base. SEUM would end its life in a less grand manner, as a textile hub for the silk manufacturing industry.

While the ongoing American social experiment had been viewed with awe among certain elements of European society, when it came to the sciences and the arts, this awe was still tinged with amusement. Cultured Europeans laughed up their sleeves at their coon-skinned colleagues and their paltry efforts at erudition, and this attitude of intellectual condescension (as mentioned) persisted long after the establishment of the de facto political equality. As late as the middle of the century, "of three-hundred-eighty-two American books reprinted in England in 1833–43, only nine were scientific."[11] Well into the 19th century Americans still looked primarily toward Europe as the font of all that was most conducive to her physical happiness and comfort. While Americans now had the (dearly bought) right to compete with European manufacturing, it was not yet clear she had either the capability or the capacity to do so, nor the finesse in manufacturing to make such efforts worthwhile.

Hamilton, to his lasting credit, had realized this one singular fact early on; if the scales were to be finally and fairly adjusted, and if this was not to become a nation of second-rate minds and self-content, complacent farmers puttering about their fields as the juggernaut of industrial progress whooshed past them, something in the larger milieu had to radically change. As the industrial revolution gained momentum, it became increasingly clear that he was right. If America's manufacturers were to stand toe-to-toe with their continental cousins, this would occur only if her best and most enterprising minds, regardless of class or social background, had been kept abreast of the innovations in their respective fields. Embracing technology meant not only the timely acquisition of knowledge but secondly and equally importantly, its efficient and democratic distribution.[12] The lack of an existing scientific infrastructure, and again, the very vastness of the new country, had both mitigated against this end, and this in part was what SEUM had been conceived to ameliorate.

If, as Hamilton conceived it, capital was the fuel of the engine of innovation, expendable capital, at least in the post–Revolutionary era, still belonged mostly to a class of people that did not present themselves as the natural allies of innovation. Despite the proliferation of stock manipulators eagerly following his every move, there was as of yet no coterie of tycoons whose fortunes had been built on technology. There were instead cabals of large (intermittently) philanthropic landholders and merchants whose natural inclination was to favor endeavors of a less abstract stripe. In any case, it was clear to all that the exchange of ideas had to be better facilitated somehow if America was to become a first rate manufacturing power, and as Hamilton had recognized, under the conditions that then predominated, and this could only occur if manufacturing and intellectual endeavors were physically concentrated and conjoined in a fashion that violated existing demographic and social norms. To Jefferson it was also the means for Hamilton to create a political power base external to that of the perennially self-interested large landholders.

SEUM had been conceived in this same spirit of thrifty self-reliance, of creating a new kind of necessity that would in turn become mother to a new kind of American invention, and it relied on an "enlightened self-interest" of the monied classes. Therefore, to reduce Hamilton's plan to just another stock-jobbing scheme, as did Jefferson and other critics, does a grave disservice to his particular genius. Previous attempts to funnel intellectual and economic energies into a single channel were not lacking. Some, like Franklin's American Philosophical Society, had not worked, or at least not worked well, when it came to creating a real-world engine for economic progress. While providing a forum for novel ideas, they remained mostly the venue of well-connected dilettantes who presented their accomplishments to their well-heeled colleagues as a form of *noblesse oblige* and self-congratulation. It was in this important sense that an Albany physician named T. Romeyn Beck and his protégé Joseph Henry would come to define and represent an entirely new and iconic strain of socially engaged yet fiercely independent intellectualism in American culture (and to some extent Europe as well), motivated by something besides simply making money—that of the professional scientist. (And it could be said, if not for this singular fact, perhaps no one today might have heard much of Joseph Henry.)

Situated seven miles west of New York City at the base of the falls of the Passaic River, SEUM was from the start not just a fraternal organization like the APS, but one wherein the interchange of ideas could serve the far grittier aim of fostering real technological innovation within the context of a state-supported, pre-planned industrial and manufacturing complex.[13] It took shape as a tax-exempt, stock-issuing entity, a living laboratory wherein technical improvements could rapidly gestate, prove themselves or fail in a pressure-cooker stew of money, industry and intellect.

Before being appointed as SEUM's first governor in 1791, Duer had been tasked with raising the capital for the enterprise; to create the facility, which included the buildings; developing viable sources of power, which meant harnessing the Passaic Falls; and filling the canals for transportation and drayage—in short, rendering the raw, undeveloped site suitable for large-scale industrial production. Hamilton had exhibited remarkable foresight in this, if not in his choice of personnel, and Duer's main focus was to leverage these grandiose plans to underwrite his own less savory private ventures. When Duer was jailed in March of the following year for various swindles, the effects on the finances of SEUM were severe. A majority of the funds he had "borrowed" from SEUM to leverage his failing personal investments were from its start-up coffers. Hamilton had written the Board of Directors a somewhat unjustifiably upbeat assessment obviously referring to this fact: "Among the disastrous incidents of the present juncture, I have not been least affected by the temporary derangement of the affairs of your Society. If however no real misfortune shall have attended any considerable part of your funds, the mere delay will be no very serious evil."[14] Duer's duplicity, however, had set a bad example. Shortly thereafter, another of SEUM's trustees made off with $50,000 earmarked for purchase of machinery. (Incidentally, this is a pattern we will see echoed over and over as the reins of new scientific societies that emerged were awarded to men of commerce as opposed to erudition, and in particular with regard to Morse and his early association with Congressman F.O.J. Smith.)

Hamilton evidently had envisioned SEUM as not only creating a manufacturing powerhouse and a sponge for speculative capital but as part of a comprehensive solution to the federal debt crisis, adding a new dimension to the tariff battles.[15] But Jefferson's cynical assessment was at least in part correct: SEUM *did* have a pronounced political dimension apart from easing the need for tariffs. Hamilton's goal was to create a manufacturing base that would depend upon a strong central bank run by his political cronies.[16] Unfortunately, thanks to his questionable choices in management personnel, the project would be not only slow to start but rife with corruption and malfeasance, providing a poor model on which to base his grander political agenda. Congressional inaction would leave the project to founder for almost a decade. His "Report on Manufactures" would be delivered to a largely uninterested Congress that would—despite, or perhaps because of, the ongoing efforts in New Jersey—look upon it largely as merely another intellectual exercise gone awry. Apart from the early missteps, Hamilton's grand vision of industrial progress was just too comprehensive and far-reaching to be acted on as a whole, and it would be tabled by a Congress distracted by the looming tariff battle, to be acted on piecemeal at a later date.[17]

Ironically, it was Duer's malfeasance that would clear the decks and allow for the introduction of truly entrepreneurial capital. By the turn of the century, SEUM, to a large extent despite itself, was proving a successful and viable concept. Thanks largely to Samuel Colt, one of its greatest early successes, it would fall exactly in line with the goals initially set by George Washington in his farewell speech: providing domestic arms manufacturing. This, in any case, was the situation at the dawn of the American industrial age: while remarkable advances were being made by isolated individuals—the steamboat, the cotton gin, the sewing machine, the nautical screw propeller, and the McCormick Reaper—the winnowing of true achievement from charlatanism had as yet no dependable mechanism other than the marketplace.

Meanwhile, in the arts, the works of Gilbert Stuart, Ezra Ames, Rembrandt Peale and John Trumbull were beginning to attract loyal domestic audiences (which did little to alleviate the assumption that they were still inferior or derivative endeavors to their European counterparts). Philadelphia, the supposed cultural capital of the American enlightenment, was in fact the bastion of a conservative establishment in both science and art that embarrassed itself first with the rejection of Audubon by the American Academy of Sciences, and later with the pernicious investiture of "scientific racism"[18] of Samuel George Morton. Some of the better minds were led to flee what was becoming, much like Albany, an increasingly parochial and self-involved metropolis. Even those at the forefront of education and innovation could not shake off the shackles of an assumed inferiority. The speech given by T.R. Beck in 1835 to the New York legislature sounded almost like an *apologia* rather than a call to arms, referring once again to the vast landscape: "We differ in many respects from scientific Europe.... Our men of science are scattered over the surface of a large country and can scarcely pursue its progress except through the medium of journals and societies. That we have abundance of talent is every day's observation but we are hardly aware how much of it is wasted."[19]

In any case, in the eight bustling decades wedged between the tail-end of the Revolution and the beginning of the next major conflict of self-definition, the Civil War, these hungry demons of an assumed innate inferiority, whether of talent, intellect, training, resources or race, would all rear their ugly heads in one fashion or another. One thing had become abundantly clear: these all had to be faced, catalogued and reshaped or expunged in some fashion before the country could assume its rightful place among nations. Such were the underlying concerns that informed T.R. Beck's vision, and by the time his seminal speech was delivered, many of the grossest deficiencies of which Beck complained had in fact begun to be rectified by efforts of those like Benjamin Silliman of Yale and others dedicated to creating a vibrant dialog within American science. But while these provided the infrastructure, in a sense it was two inventions, the revolver and the telegraph, that would finally and irrevocably reshape public opinion towards the view that there was such a thing that could rightly be called "American science."

Attitudes in Europe had changed at an even more glacial pace than in America. European scientists routinely regarded their American counterparts as errand boys, only good enough to mimic their experiments or convey them raw materials like seeds and mineralogical specimens for further studies in their better-equipped labs. In 1739 an English botanist had written the American John Bartram, who had had the temerity to suggest that the botanical studies in which they were jointly engaged could be equally well performed in the U.S.: "To draw learned strangers to you requires salaries and good encouragement which can't at present by complied ... given the infancy of your colony."[20] The contempt is almost palpable.

Even one hundred years later, the esteemed Joseph Henry had fallen victim to this same snide and superior attitude. While presenting his discoveries to a scientific conference in Liverpool, he was stung by some impolite remark regarding steam navigation rendered by an attendee, and noted after, "It is rather a hazardous affair for an American to make a communication unless ... the communications be addressed to ... his friends...."[21] It was not Franklin's venerable APS, nor any of the well-endowed dabblers, but the alphabet-soup SEUM and SPAAM and those that followed, baptized in the liquor of independence and harsh economic realities, who would set the stage for a second portentous revolution that would elevate American science and the American scientist as a central figure on a world stage.

Despite the early stumbles and missteps, it was clear that SEUM would eventually succeed as a viable endeavor. If Hamilton's adversary, Thomas Jefferson, was to find an effective counter to Hamilton's vision of an industrial/intellectual juggernaut with a toehold outside the world of politics, it certainly was not going to be found in the polite utterances of the professional societies like the APS (of which Jefferson was a member). To this end, Jefferson sought the help of the powerful Livingstons. However, just as Hamilton had so painfully learned (thanks to Duer), Jefferson would also find that joint economic and scientific endeavors could easily fall prey to the calculations and depredations of unscrupulous manipulators, who, granted access to both the levers the economy and the means of manufacture and innovation, would inevitably steer them to their own ends.

This held even truer for those who sought to pursue unalloyed science. Even in the relative backwater of Albany, Beck repeatedly had to swallow the bitter pill of entrenched self-interest that sought to subtly (and sometimes not so subtly) sway his scientific interests and endeavors. Though himself a true visionary, Beck was forced to cultivate the friendship and largesse of men of far less impressive intellect, the hereditary landlords, only to find when he had insisted on true intellectual independence and focus, the cord of funding would be (genteelly) cut, or at least temporarily choked off.[22] The argument can be made that early scientific agencies, even while claiming their main goal was "to serve the common man," arose primarily to serve the interests of the well-heeled and were far from free of political taint.

And freedom from political taint was not something that organized science apparently even found desirable. SPAAM[23] was no exception when it came to making strange bedfellows. Though Livingston was ostensibly Jefferson's political antagonist, Livingston idolized Jefferson's agriculturally-focused scientific acumen. Consequently, SPAAM had been identified early on by Thomas Jefferson as the counterweight to Hamilton's SEUM and his Federalist agenda, the latter an organization Jefferson felt intentionally designed to strip the mantle of intellectual erudition and innovation from the genteel, southern aristocracy and award it to hardscrabble, practical northerners. In any case, the founders of those two organizations had little interest in separating them from the political imperatives that undergirded them and made no secret of that fact. In most quarters, political influence was viewed not primarily as an intrusion but rather an indispensable spur to technological achievement.

If a sustainable and truly independent scientific dialog was to be nurtured, clearly social attitudes and priorities had to change; but clearly, even well into the nineteenth century, there was still no inkling of commonality, no shared purpose or direction on which to build such a superstructure. There were those proponents of pure scientific endeavor who favored "natural philosophy" over technology, and those on the other end of the spectrum who seemed to think that science existed only to serve some overarching political or economic end. When SPAAM moved to Albany at the turn of the century, it was to allow it to remain close to the levers of power, and they had even been assigned offices in the State Capitol.[24] While the move would also introduce the influence of some of the most powerful and profound minds of the early nineteenth century, it would also put it squarely under the thumb of some of the most venal and self-interested.

Once again it was the vastness of the landscape itself that played the key role in shaping the tenor of the ensuing intellectual conversation. Even with valuable resources like Silliman's *Journal* providing a connective tissue, the scientific corpus remained frustratingly heterologous. Communications and discourse between semi-isolated pockets of knowledge and intellectual endeavor remained exasperatingly difficult and slow. The dispersal of complex ideas outside the lecture hall was a cumbersome and painfully slow process that precluded a true "give and take," and in some cases thwarted accurate attribution for innovations.[25] The nature of this invited charlatanism and false claims, and those who gave in to this tendency often had to pay the price later on. (This was why the Patent Office early on had assumed such a central and active role in innovation and serving as arbiter of our intellectual patrimony.) Morse, with his connection to former Yale classmate Henry Leavitt Ellsworth as Patent Commissioner, would use this to his advantage, and was even able to claim (almost believably for a time) entire ignorance of prior work in the field of electromagnetism and electrical circuits preceding his "Eureka moment" aboard the *Paquebot Sully*.

In any case, reputable scientific journals like Silliman's, the Lyceum system, public lectures such as those given at the Albany Institute and New York Athenaeum, first-rate colleges and scientific societies—all these would play some role in overcoming this tendency towards obfuscation of achievement and dilution of intellectual effort. But these were all still cumbersome, not to mention expensive to maintain, and therefore also prone in some cases, much like SEUM and SPAAM, to becoming mere tokens in the political struggles of their major patrons.

In the 19th-century American intellectual cosmos, the constellation of New Haven, Albany, and Philadelphia formed a literal and figurative conduit through which established social ties, money, and acquisitioned intellect could cut (in the case of the Erie Canal, literally) a channel to a bright, shared future. It was partly this fact that places Albany squarely in the sights of our initial investigation: as the locale for what would arise, under the guiding hand of T. Romeyn Beck and other intellectual and educational pioneers like Amos Eaton, as a haven for the semi-independent abstract pursuit of knowledge. In this context Joseph Henry's early investigations into the emergent field of electromagnetism would take place.

Arguably, it was there in Albany in the waning few weeks of the summer of 1823 (as we shall see) that the seeds of a great future technological achievement were first sown; where a young New Haven artist name Finley Morse would first be re-baptized as S.F.B. Morse, the budding scientist. Ironically, it would be under the roof of T.R. Beck's patron, that staunchest icon of the old guard, Stephen van Rensselaer III, the Old Patroon, who lorded over the vast hereditary holdings of the van Rensselaer family, not only as an old friend of Morse's father Jedidiah, but at the time of young Morse and Silliman's sojourn there, the household employer and personal benefactor of Joseph Henry.[26] So while it had been an intellectual hotbed for the first quarter of the 19th century, Albany was, thanks in part also to the Erie Canal, fast becoming the economic as well as the intellectual focal point for the region as well as the country.

Albany, however, was not really Robert Livingston's bailiwick. Two cocks could not crow there. While his politics, like that of most well-heeled New Yorkers, generally leaned toward the Federalist view, his family were the largest landholders in the Hudson Valley (and likely in New York State) but like SVR, and his main guarantee of continued wealth lay in continuing to collect steady rents. Focusing on improving methods of agriculture harmonized well with both his and Jefferson's goals for financial stability. While it seemed Jefferson had succeeded in finding an enthusiastic political ally in the Chancellor, he would prove, in matters of philosophical predisposition, at best a lukewarm supporter.[27] Jefferson would quickly learn that the goal of forming a strictly agricultural lobbying and research organization along a Democratic-Republican line in New York State could prove problematic for several reasons, not the least of which was Livingston's longstanding "frenemy" relationship with the Old Patroon.

Jefferson's intrusion into local politics using science as a pry-lever was, much

Letter of the Old Patroon, Stephen von Rensselaer III, to Amos Eaton, August 30, 1823 (courtesy New York State Library Manuscripts and Special Collections, Amos Eaton Papers).

like Hamilton's, naturally viewed with suspicion, particularly by Livingston's fellow landlords, who sensed that a deal was being quietly struck without their participation on behalf of Livingston's friend, Governor George Clinton, to ascend to a national stage.[28] Such petty concerns would, however, soon fall by the wayside, and Jefferson's agenda would fail to flower in New York State. By the time Jefferson left office in 1809, just as Hamilton had foreseen, convulsive change was already in the air, indeed permeating it. The aroma of intellectual ferment wafting in from the European continent already was too potent a brew to be entirely dissolved in syrupy, bucolic visions of a simpler life. Europe was ablaze with innovation. Steam engines and ingeniously designed pumps expelled geysers out of formerly abandoned mineshafts. Huge looms shuddered and leapt with frenetic energy as Scottish mills integrated mechanical innovations and interchangeable parts for assembly lines, allowing them to double and then triple their cloth production. Balloons wafted into the halcyon skies of France, rendering to men (like Livingston) a view of God and man's works never before possessed. In Paris, gas street lights had begun to appear, illuminating the new crop of belles decked out in their gay shawls woven on the mammoth and lumbering Jacquard looms.

Although the pace of American manufacturing had picked up considerably, once freed from unfair restrictions and imperatives, the relative value of the manufacturing sector of the economy as a whole had, for the early decades of the nineteenth century, been declining steadily. Europe had little need for American-made hoes, hats, colored cotton cloth and ribbons, breech-coats or silver plates. What they still needed most was raw materials. Just as European scientists wanted access to American specimens, the raw goods of science, without the clutter of inferior commentary by their benighted colleagues, what was wanted and needed in the stalls of continental commerce was cotton, tobacco and increasingly flour (not those goods finished in an "inferior" style by American manufactories).

Hamilton's organization had been promoted as staunchly anti-tariff and thus more palatable to the Eastern states that still housed the great commercial ports and harbors, but the economic axis of the country had long since begun to tilt inexorably westward. As long as American appetites favored West Indian sugar and European fabrics, little was to be gained by tariff battles with European producers. Fresh horizons beckoned, new fonts of profit were waiting to gush forth as from a biblical rock in the desert. Forward-looking men in New York, like Livingston and fellow landlord Steven van Rensselaer, had realized early on that the path to America's future ran almost literally through their backyard, and they had begun planning a great canal that would provide a route both for eagerly anticipated goods and riches flooding in from the frontier and a easy head start for those being shuttled west. The path to wealth no longer lay exclusively over the vast ocean but beckoned from this new frontier as well. And for New Englanders at least, that path, the bloodstream of future riches, led inevitably through New York State.

2

SPAAM (The Society for the Promotion of Agricultural Arts and Manufactures)

Shortly before the official chartering of SEUM as a state-sponsored industrial/scientific entity, Jefferson had decided to go on his "botany tour" of New York, ostensibly to study the Hessian Fly that was afflicting Long Island's wheat farmers.[1] The botany tour had no doubt been arranged as a pretext to give Jefferson the opportunity to do a little discreet political arm-twisting in that viper's nest of Federalism, New York. In May of 1791, accompanied by James Madison, Jefferson traveled throughout Long Island and the Hudson Valley, meeting en route with local power brokers like Chancellor Livingston and Aaron Burr. Livingston and Jefferson, along with Franklin, had been on the committee to frame the Declaration of Independence. More recently Livingston had been corresponding with Jefferson regarding one of his own more dubious inspirations—a way to lubricate grist mills with mercury.[2] Livingston was unquestionably eager to pursue this discussion in person, having already asked Jefferson in December to present his sketches to the American Philosophical Society.[3]

It was also later, during this same botany tour, that Jefferson enlisted the poet and editor Philip Freneau to start up an anti-Federalist newspaper to counterbalance Hamilton's *Gazette of the United States*.[4] Jefferson's plan thus clearly was to leverage Livingston's respect for his scientific reputation and expertise into firm political support, but Livingston was no "babe-in-the-woods" either. The defect of the type of organizations such as Franklin had started early on in Philadelphia and of which Jefferson was now a member (and which tendency Franklin himself warned about), was the inclination to become too self-involved. Livingston's approach, by contrast, had been baldly political from the outset.

The exact nature of Jefferson's arm-twisting is not known, but the intent was to enlist both Livingston and Burr to set up an infrastructure in New York to counterbalance what Jefferson saw as the unchallenged Federalist propaganda machine in the state. As one observer noted, "There is every evidence of a passionate courtship between the Chancellor, Burr, Jefferson and Madison."[5] The "courtship" of Livingston, at least, had been going on for several years. Jefferson, who desperately needed a base of operations in New York, evidently thought Livingston might be ripe for plucking, being especially eager to find a counterbalance to Hamilton's scheme for a scientific/manufacturing complex which had begun to spring to life, albeit fitfully, just across the Hudson.[6] Though Livingston's long-held Federalist sympathies might have otherwise put them at odds, he and Jefferson both had learned that their natural interests and avocations were somewhat better aligned than their politics. Like Jefferson, Livingston fancied himself a modern Renaissance man, a philosopher king who presided over his vast agricultural realms by dint of intellectual superiority. So the political détente, with Livingston at least, was to take place on this bridge of supposed shared intellectual interests.

In February, Jefferson had introduced a bill in the Congress to "Promote the Progress of the Useful Arts." So it had likely been with Jefferson's cooperation and blessing that Livingston, along with Samuel Latham Mitchell, Ezra L'Hommedieu and Simeon DeWitt undertook to form the Society for the Promotion of Agriculture, Arts and Manufactures, or, as it was known, SPAAM, constituting themselves[7] only a few weeks after Jefferson's bill was introduced in the congress.[8] Based at first in New York City, SPAAM had a distinguished board comprised mostly of large landholders (with the exception of Mitchell), with Livingston as easily the largest among them, serving as its first (and as it would turn out, only) president. From the outset it was intended to have a political dimension and function as an agricultural lobbying organization under the guise of an intellectual one.

SPAAM had cultivated and maintained a broad grassroots following, a power base they would retain for the decade following. Though they initially held session only in New York City (and then only when the state legislature was in session), they maintained influence and lines of communication throughout the rural parts of the state through this network of county-based agricultural societies. By 1797 they were holding meetings both in New York City and Albany, and a year later, the organization officially relocated to Albany, following the legislature there.

By the turn of the century, SPAAM was the most entrenched powerful agricultural lobbying arm in the state.[9] In April of 1804 it reincorporated, officially changing their name to SPUA (dropping both the "manufactures" and "agricultural" from their title and adopting the more Republican "useful arts"). They now had their own rooms in the State Capitol as well as state funding and support. They were in effect state-sponsored agricultural lobbyists. While Robert Livingston was still president, it was largely honorary, thanks to his other duties as Jefferson's minister to France. With Livingston's absence, other influences within SPAAM began to come to the fore.

Thanks largely to Livingston's resources, SPAAM had been one of the few scholarly organizations (besides the APS) with the wherewithal to publish a respected scholarly journal. Not surprisingly, Livingston's own personal interests and avocations had dominated what would or would not appear in those pages. While he expressed a nominal lively interest in and engagement with a broad array of scientific subjects from paper-making to self-lubricating machinery (ranging later on as far afield as steam engines and navigation), it is not surprising, given his Scottish heritage, that Chancellor Livingston's most avid and longstanding passion veered inevitably towards sheep. While fancying himself a "natural philosopher" in the mold of another of SPUA's intellectual lights, Samuel Latham Mitchell, many of Livingston's early essays published in the SPUA journal focused solely on improving the methods of wool manufacture.[10] In 1807 the legislature published 1,000 copies of his *Essay on Sheep*. In 1808, again, at the encouragement of the society, the legislature passed an act "For the Encouragement of Manufacture of Woolen Cloth."[11] Having returned from France with a good breeding supply of Merino sheep, in the spring of 1810 he hosted a sheep-shearing contest at his estate in Clermont featuring his finest stock and a production of the "American Pastorals, wherein the shepherds vied with each other in singing the praises of the wool...."[12]

Slowly though, in his absence things had already begun to change. The first issue of the SPUA journal prominently featured then vice-president Ezra L'Hommedieu's treatise on manure, while the second issue (though still featuring Robert R. Livingston's obligatory oration on sheep) contained a lengthy treatise on a slightly more erudite subject: weights and measures (something in which Livingston also had an abiding interest), authored by Philip Schuyler.[13]

In Livingston's absence, there had also arisen a system of lyceums and private academies, most notable among these in New York being the Albany Academy and the West Point Lyceum. The Lyceum coagulated the cadre of professional scientists like Beck in Albany and John Torrey of West Point, men with unimpeachable credentials in chemistry, geology or mineralogy, into the semblance of a professional scientific corps, while the Academy sought to pass this knowledge on to mostly young men in an semi-organized fashion. Thus, while safely ensconced within the lyceum/academy context, these individuals were somewhat freer to set curricula and research priorities by their own lights, publishing in their journals articles of broader scientific interest.

Still, without the financial backing of a patron class for the parent organizations, the more costly and "frivolous" undertakings, like publishing a journal, were often the first to fall by the wayside. Thus, while they may have been freer to pursue their own research, these "professional scientists" increasingly began to resent having to beg funds of the well-to-do patrons to get their work published or to maintain up-to-date collections and laboratories. Benjamin Silliman at Yale had erected a notable exception to this pattern with his *American Journal of Science*, which offered an impartial and widely read venue, and most importantly, one without perceptible bias or agenda.

Despite the fact that Livingston's interests had broadened further afield from oviculture to the uses of steam power in navigation and other once (to him) seemingly esoteric areas, he was still bound by an intellectual corset into which he had laced himself—the original deal struck with Jefferson to focus on agriculture. SPAAM's charter had stated plainly its purview, "the objects of the investigations of the society shall be Agriculture, Manufactures and Arts,"[14] but its practical focus had remained on the first of these—agriculture. The world had changed, however, in the few short years since Livingston's return. Hamilton's ideas about manufacturing had been gaining traction, and by 1812, even Jefferson had more or less given up the ghost as far as hoping to influence the main direction for scientific studies in the country, if not to Hamilton, at least to the devil (though the difference to him apparently seemed negligible): "Our enemy has indeed the consolation of Satan ... from a peaceable and agricultural nation he makes us a military and manufacturing one."[15] Jefferson, however, would find an unlikely ally in T. Romeyn Beck in ensuring this vision did not expire in New York State.

3

T. Romeyn Beck

When Theodoric Romeyn Beck, a young, enterprising physician fresh from the Columbia College of Physicians and Surgeons, joined the SPUA in 1811, it would profoundly alter its future course and extend its role from being primarily an agriculturally focused society into a bastion of pure scientific inquiry. This in the end would result in the destruction of SPUA, but in the interim would create a new and more sustainable model for education and inquiry than had the direct patronage system, with the government increasingly taking on the role previously occupied by private donors. Beck's personality was a far cry from the smug van Rensselaer's, or that of effete, dabbling, coffee-table scientists like Livingston.[1] After graduating from Columbia at the age of nineteen as a fully licensed physician, he had never really practiced medicine but viewed himself primarily as a "natural philosopher" and educator in the style of that day, meaning that he was inclined by temperament to inquire into subjects that may have once been regarded as frivolous or inherently useless: mineralogy, chemistry and higher mathematics.[2]

His first professional position had been as instructor at the state medical college near Buffalo. The second was his role in Albany, where he focused on mineralogy and mathematics—building up the society's library as a cornerstone of its holdings. Many of the other medical and scientific lights of Albany soon added themselves to SPUA's rolls, including Dr. Jonathan Eights, Richard Varick DeWitt, and eventually Lewis Caleb Beck, T.R. Beck's brother, who was a noted geologist and probably the premier chemist of his day. Not only the constituency but the complexion of the society's journal also began to change as an ever larger percentage of articles between its covers dealt with nonagricultural topics such as mineralogy and chemistry.[3]

Livingston was growing old and feeble, and with the young lions at the gates, T.R. Beck, as head of the organization's newly formed Mineralogical Committee, made a presentation on the subject to the entire membership of

SPUA in February of 1813. Ironically, though well aware of the economic implications and change in direction he was suggesting, Beck stated in the preface to his remarks that the abandonment of agriculture in favor of manufacturing would be a grave mistake: "It cannot however be the wish of any true patriot, that the United States should become ... a manufacturing country. The disease, vice, and diversified forms of misery, that exist in those parts of England, from whence our hardware and cloths are obtained, are sufficient to make the most sanguine advocate for the encouragement of manufactures tremble."[4]

When Livingston passed away a few weeks later, Simeon DeWitt was appointed in his place. Though refusing to take the title of president (allowing Livingston to retain that sole honor), he became chairman officially in 1814, having been already *de facto* head of the organization for more than a year. Beck had been added to the board as recording secretary.[5] Under Dewitt, the young Turks had been given their head. Beck made sure that they were kept abreast of the advances in European science through the library, which he kept up to date with subscriptions to treatises and journals. Agricultural activities and interests (though still a central function of the organization) were relegated, if not to a subordinate position, at least to a reduced role. While the organization still maintained its hegemony over the state agricultural societies, members of the old guard, like Elkanah Watson, were encouraged to pursue their activities further afield, in the venue of county agricultural fairs rather than in the pages of the society's journal or in its lecture hall.

T.R. Beck was known as a vibrant speaker and sought after as a lecturer. His rock-star status in Albany and his high profile within SPUA lent the organization a new air of excitement and cachet that eclipsed the useful but dull image maintained under Livingston's tenure and seemed to guarantee the organization's continued political relevancy.[6] Beck's keynote address of 1813 had been a masterful and comprehensive presentation on the subject of mineralogy and had sparked a new interest in the subject not only among SPUA's constituency but among the general public. It had also brought him to the attention of Benjamin Silliman, at Yale, who was himself was an avid mineralogist and contributor to the *American Mineralogical Journal* and was about to bring out a publication of his own, the *American Journal of Science*.

The venerable Ezra L'Hommedieu had passed away two years earlier in 1811. This left only the aging Samuel Latham Mitchell and Simeon DeWitt of the original founding members of SPUA. DeWitt had always had a somewhat more catholic intellect than Livingston's, a fact which he had at times taken pains to conceal.[7] After graduating Queens College of New Jersey (now Rutgers), he had spent two years under Robert Erskine (a respected inventor and scientist in his own right and George Washington's map maker during the

Revolutionary War) as his assistant. As Surveyor General of New York State, DeWitt had presided over the distribution of military bounty tracts in western New York designed to compensate officers in the Revolution for the depreciation of their pay due to inflation (incidentally awarding himself in the process a prime parcel overlooking Cayuga Lake that would later become Cornell University).[8] In short, though like Livingston the scion of a prominent political family, DeWitt had an active and cultivated intellect that did not shy from more complex subjects. In 1825 he would present Joseph Henry, then an instructor at the Albany Academy, with a surprisingly erudite analysis of how variations in the earth's magnetic field might affect surveying operations.[9]

The shift under DeWitt had not been just one of emphasis but deeply structural, and in some sense organizationally suicidal. SPUA had long been the titular head of the network of local agricultural societies throughout the state. In their new pursuit of more abstract subject matter, they had let slip the reins of power, in the process becoming far less influential in the legislature and therefore less valuable to those landlords that controlled the board. Even as their focus shifted to more insular pursuits, they had broadened out from scientific inquiry into the field of the fine arts. SPUA was reorganized into several committees, with noted Albany portraitist Ezra Ames recruited to establish the new fine arts division. (Ames, along with Anson Dickinson, would help pioneer the "portrait miniature" concept of which Morse would make liberal use.)

In the meantime, with Beck and Dewitt firmly at the helm of this new accumulation and concentration of intellectual capital, the floodgates of progressivism opened even wider. In the months following Livingston's death, Stephen van Rensselaer III, along with other members of SPUA's board, incorporated an ancillary organization, the Albany Academy, apparently hoping to divert and provide an outlet for some of this more progressive spirit by educating the sons of Albany's well-to-do. But this did nothing to remediate the lack of focus in the parent organization. Intellectually, its prestige was never more lustrous, but as a lobbying organization, it was growing increasingly feeble.

Having seen the handwriting on the wall, SPUA's board had already resigned itself to paying the freight to sustain a wounded organization. They acknowledged this fact, issuing a public statement excusing themselves of any further intellectual leadership or responsibility: "While other bodies ... possess the patronage and favorable eye of the State, it cannot be expected that the exertions of the society will be either very extensive or conspicuous."[10] Not just new organizations, but new concerns had come to dominate the attentions of the patrician class. By 1816, three of SPUA's former trustees, Samuel Young,

DeWitt Clinton and Stephen van Rensselaer III, had been appointed as the commissioners for the new Erie Canal project.[11] Never before having taken much of an active interest in any of the organizations he nominally headed, van Rensselaer decided at this point to take a personal tour through the state in the company of another ex-board member, Chancellor James Kent,[12] ostensibly to survey the state's local agricultural fairs.[13] The true purpose was to drum up support for the highly controversial canal project.[14]

T. Romeyn Beck had been overseeing the school's day-to-day operations since its inception. On August 14, 1817, he was elected principal and professor of mathematics, which was a big enough event to make the Albany newspapers.[15] So while SPUA foundered like a leaky scow, its somewhat better-caulked and partly state-funded offspring,[16] the Albany Academy, thrived and expanded. And though SPUA's demise left the Beck brothers somewhat politically isolated, their role as educators in elevating the minds, morals and aspirations of the sons of Albany's better-heeled citizens was sufficient to keep them in a favorable light as far as van Rensselaer and the public were concerned.

SPUA's decline had been masked to some degree by the formation of the Albany Academy but even that had become a double-edged sword, diverting the limited pool of funds and interest away from the parent organization. But this new spirit of what van Rensselaer could only have viewed as "enervating progressivism" was finding expression in other areas as well. A year after Livingston's passing, the Albany Female Academy[17] would open just two blocks away. As the first female college preparatory school in the country, it was intended as an adjunct to the all boys Albany Academy. Yet while the scientists and educators continued to debate, SPUA's purse strings were still controlled by powerful men whose main social concern was the inevitable westward expansion.

Despite his evident genius and success as an educator, Beck's (relatively speaking) lightning ascension had essentially been the death knell for the organization. The *coup de grâce* came in the desultory response to the unusual and disastrous weather of the late spring of 1816. On June 11 of that year, a freak frost occurred in New York State. This had followed weeks of below-normal temperatures which had, on June 6, caused snow to fall on Albany, and which had a devastating effect on the entire agricultural economy of Northeast.

Rather than seek emergency assistance for their base constituency from the legislature as they had done in the past, SPUA's reaction seemed instead to be more akin to Nero's fiddling while Rome burned (or in this case froze). They focused instead on commissioning portraits of the society's founders by Fine Arts Chairman Ezra Ames,[18] an exercise in self-absorption and self-congratulation. It seemed to be, just as Franklin had warned, that any organization fundamentally based on fraternal and intellectual, rather than political bonds would inevitably turn inward, and this is what SPUA had done just when its influence might have proven critical. To add insult to injury, rather than paying Ames for his work (in an echo of what had occurred early on with SEUM) the treasurer of the society absconded with all the funds of the organization, leaving it essentially bankrupt.[19]

Suffering the dual embarrassments of disenfranchisement and financial misconduct, by 1818, SPUA was on the ropes. Having lost its state funding entirely and with membership declining, SPUA artificially inflated the membership rolls by electing members of the state legislature *ex officio* without their knowledge and then demanding payment of dues *post hoc*.[20] The implosion of SPUA had also led to a power vacuum as regards to the agricultural lobby in Albany. Without SPUA to provide direction, the numerous still-robust county agricultural societies found themselves suddenly and embarrassingly headless—like chickens headed for the broiler.

Various other organizations had adopted the term "agriculture" into their titles in hopes of becoming heir to the state funds formerly earmarked for SPUA. One of these was the newly formed "Board of Agriculture," which had been created by the legislature in the wake of the freak frost event to give the county-based agencies more responsive representation in Albany. Not surprisingly, it was the Old Patroon who "was one of a number of disinterested gentlemen who induced the legislature to pass this act" in 1819.[21] With an initial budget of thirty thousand dollars, the new State Board of Agriculture, with van Rensselaer at its helm, would become the vanguard of an expanded and powerful grassroots political network dedicated to realizing the Jeffersonian vision in New York State.[22]

4

The Albany Academy

The Albany Academy had first come into existence in March of 1813 when the Board of Regents granted a state charter for the school.[1] Stephen van Rensselaer's uncle, Philip S. van Rensselaer, then mayor of Albany, had laid the cornerstone on July 29, 1815.[2] It had opened a mere two months later under the supervision of the trustees, with Stephen van Rensselaer III as the chair. Four years later, on assuming his other role as secretary of the New York Board of Agriculture in 1819, van Rensselaer had decided finally to step down, encouraging the board to appoint T.R. Beck, who was by then principal,[3] in his place. The board had declined, instead choosing to appoint Beck principal lecturer in chemistry in addition to his current duties, for which position he was expected to raise his own salary by the sale of lecture tickets to the public (to which members of the legislature were invited gratis), as chemistry was apparently too frivolous a pursuit to merit state support.[4]

The board's reticence in following van Rensselaer's advice perhaps stemmed from lingering suspicions regarding the Patroon's motives and his part in the dissolution of the parent organization, SPUA. In executing what appeared to be a naked power grab to arrogate to himself all the goodwill and influence that SPUA still held with agricultural interests, even the Teflon landlord had, it seemed, stepped a bit far out on the limb of propriety. The new Board of Agriculture had proposed conducting an agricultural and geological survey of Albany County, but Van Rensselaer, seeking to deflect any incipient criticism resulting from the dissolution of SPUA, decided to jump in and personally underwrite "geological/mineralogical" surveys of Rensselaer and Albany Counties, initially appointing the Beck brothers to head it.[5] The results of the survey, however, were to be delivered not to the Board of Agriculture but directly to the Albany (County) Agricultural Society, headed by none other than the Old Patroon himself.[6]

5

The Big Ditch

Amos Eaton had arrived in Albany in 1818 while on the upstate/New England lecture circuit.[1] Portly, explosive and prone to fainting spells, he cut an odd figure indeed amongst the staid burghers of the old Dutch town.[2] As a naturalist he had studied under Benjamin Silliman and Eli Ives at Yale but he also had something of a shady past, having been convicted in 1811 for check-kiting and forgery in Catskill, New York.[3] He had lectured briefly at Williams College in Williamstown, but Albany certainly promised him both a wider audience and a more attentive (and sympathetic) one.[4] By the time he returned to New York, he desperately needed a fresh start, having been pardoned just the year before by Governor DeWitt Clinton. His easy-to-understand presentations on abstruse scientific subjects were well received, and he outdid even the "rock star" Beck in drawing the attendance of the general public. His flamboyant personal style filled the lecture halls with Albany's cultivated and powerful elite—drawing, however, from much the same audience as would have been otherwise attending the more staid lectures at the Albany Institute.

Eaton had basically been an assistant to the Beck brothers on the mineralogical survey for the first two years of its existence. Having participated in the 1820 survey, Eaton had nominal charge of the 1821 expedition. Most histories credit Eaton with taking over the survey in 1821: "Prof. Eaton, with the assistance of Drs. T. Romeyn and Lewis C. Beck, under the patronage of Hon. Stephen Rensselaer, conducted an agricultural and geological survey of Rensselaer and Albany."[5] But the fact that Eaton did not really take control until 1822 is evidenced by the fact that it was T.R. Beck and not Eaton who delivered that year's report to the State Agriculture Board.[6]

The savvy Eaton no doubt had by then realized that the new canal project, especially with van Rensselaer as one of the commissioners, represented a bonanza when it came to leveraging money for any future projects he might contemplate.[7] "In 1822, again under the patronage of Stephen Van Rensselaer,

Eaton undertook a geological and agricultural survey of the district adjoining the Erie Canal."[8] Eaton had proposed abandoning the county-by-county approach, focusing on the canal instead as a kind of discount, open-air classroom to replace the mineralogical survey, defraying the cost by charging a subscription for students to accompany him. Van Rensselaer eagerly responded in the tone of a pleased, prospective buyer: "I have long contemplated the examination you propose but was apprehensive it would exceed the means of an individual ... your calculation is so moderate that I willing[ly] will engage in the enterprise."[9]

So, after two years under the Beck brothers, van Rensselaer abandoned his support for their methodical county-by-county approach in favor of a new expedition running entirely along the route of the Erie Canal and with Eaton at its head.[10] While the expedition was confined mostly to the canal and its banks, to the physically rotund and less than hardy Eaton, this was no doubt a small sacrifice, since it represented a virtual ready-made field classroom with built-in transportation and without the need of an inordinate amount of exercise in the country.[11]

The timing could not have been better. Though it would not officially open until 1825, by 1822, the long middle section of the western canal had already been partially opened to navigation, and the still freshly excavated banks provided a ready-made window onto the geological strata of New York's Northern Tier. With the dredges plunging their jagged jaws daily into the viscous clay, gypsum and shale or the Northern Tier, certainly it was a rare opportunity to classify what they churned into daylight as they burrowed their way ever further westward. Boats were soon plying its full length, a distance of some two hundred and twenty miles.[12]

The gentlemanly falling out between van Rensselaer and the Beck brothers had been exacerbated further by the unfavorable comparison with Eaton's lively, entrepreneurial (and politically savvy) style. In 1824, probably while on that year's summer canal trip, Eaton conceived a plan to create a competing institution to the Albany Academy across the river in Troy, which he hoped to put into effect with van Rensselaer's aid.[13] The Albany Academy had been structured around the classical Latin day school with an emphasis on Greek and Latin studies, and this increasingly contrasted unfavorably, in the cosmopolitan air of Albany, with Eaton's proposed "American" or "Rensselaerean"[14] system of education. It was probably to combat this perception of stodginess that the Academy's board made a radical shift in their curriculum once the new Rensselaer school was in operation: "Latin and Greek were increasingly seen as ornamental. Such innovations were not evident at the Albany Academy until the curriculum's reorganization based on the [Benjamin] Franklin plan in the late 1820s."[15]

6

A Tale of Two Cities

With few options, in February 1823, a frustrated T.R. Beck had written van Rensselaer asking him to take over the direction of a new organization that would combine the ailing SPUA with a new Lyceum under the banner of an entirely new organization called the Albany Institute.¹ Van Rensselaer at first demurred, citing his new position in Congress, saying any such position would of necessity have to be purely honorary.² It was not as if the Patroon's participation in Beck's other projects had required any substantial "hands-on" participation, so, after some further cajoling on Beck's part, in March, van Rensselaer reluctantly acquiesced. Three days after receiving van Rensselaer's reply, Beck had taken up a private subscription to pay for the Lyceum's move into the extra rooms in the basement of the Albany Academy.³ The Albany Institute would eventually take over the rooms on the second floor, northwest. It was here that Joseph Henry would conduct most of his experiments with electromagnetism.⁴

The Albany Lyceum had been conceived initially mainly as a repository for Lewis Beck's extensive mineralogical and biological cabinet (and also as a place to house SPUA's valuable library).⁵ Both had, in the interim, been relegated to the somewhat unsuitably damp basement of the Albany Academy. L.C. Beck's mineral collection constituted a major scientific trove that would have ranked the Lyceum as one the major institutions of natural history in the country at the time. Professor James Freeman Dana had been impressed enough by it to donate his own samples of New Hampshire mineralogy.⁶ However, lacking van Rensselaer's substantial patronage, the efforts to create it an independent entity were doomed. A petition to the mayor of Albany, literally denominated "A License for Begging for the Lyceum," was clearly a deliberate attempt to embarrass van Rensselaer which evidently backfired.⁷ Though the lead signature was that of Richard Varick DeWitt (Simeon DeWitt's son), the Old Patroon certainly would have figured out that the Beck brothers were behind this. In the end, Beck, albeit with the Board's help, had prevailed and the Lyceum would survive, but it had been a Pyrrhic victory at the substantial cost of van Rensselaer's goodwill, and then only temporarily, as the Lyceum would only last a few months on its own. It would be folded under the second department of the new Albany Institute.

By May the conjoined organizations were recognized under the banner of the new Albany Institute with van Rensselaer as the somewhat diffident and detached president (elected and serving in absentia). Probably with more than a bit of *quid pro quo*, the somewhat recalcitrant Stephen van Rensselaer IV was enrolled in the Albany Academy where he would attend the next six months under the personal tutelage of one of Beck's new rising stars, Joseph Henry.[8] Even as the Becks' influence with the Patroon waned, young Henry's relationship with the van Rensselaer family, at least while he remained in Albany, would grow more cordial from this point onward, and he would receive a stipend and lodging to serve as personal tutor to the van Rensselaer brood.

As described (albeit somewhat reductively) by Horatio Gates Spafford, author of the then popular hymn "It Is Well With my Soul," the Albany Academy in 1823 "was a large and elegant pile of masonry ... faced with red sandstone ... the same as that used in the capitol."[9] In fact, its three stories housed five teachers and one hundred forty students and had cost the city ninety-two thousand dollars. Despite his having helped secure the funds for its construction through the legislature, the bad blood between van Rensselaer and the Beck brothers over the Lyceum fracas simmered on.[10]

When Eaton had planted the idea of a separate institution to be built across the river in Troy, one of a more practical bent, with Eaton himself as the main academic attraction, van Rensselaer saw it as an opportunity to free himself from the out-of-touch "theorists" across the river. Eaton had extended the (three-fingered) hand of collegiality by inviting the Becks to bring their better students to Troy to witness the various scientific expositions in chemistry and electricity that he was conducting. But even in these supposedly cordial invitations, Eaton could not help but contrast the effete theoretical approach of the Albany Academy's experimentalists (Henry among them by now) with his own more down-to-earth approach. "You will not expect to see the nice delicate experiments of the electrometer as Dr. B[rown] makes no pretensions to science but you will see all that is strong and forcible."[11]

Eaton's more pragmatic approach promised tangible benefits. To date, the only commercial mineralogical find of major economic consequence in New York State had been the discovery of iron ore in Orange County in 1736 by Cornelius Board and Timothy Ward.[12] Significant gypsum deposits were daily being uncovered by dredging crews working on the Erie Canal as well as in the salt springs at Onondaga near Syracuse.[13] Who knew what other valuable mineral deposits might lie in their trajectory? Perhaps gold, or perhaps even the long sought-after mythical Dutch/Indian copper mine would turn up. Though none of these ever materialized from the rather disappointingly prosaic strata of New York's Northern Tier, the new canal would provide plenty of riches to the interested parties in other forms.

7

Portrait Painter
Morse in Albany

"The portrait I am now painting
is Judge Moss Kent, brother of the Chancellor."[1]

In March of 1822, Stephen van Rensselaer III, the Old Patroon, had been tapped to fill the unexpired term of his cousin in Congress. Solomon van Rensselaer had decided to retire from Washington life to take up running the Albany Post Office, which had been vacated so the postmaster, Solomon Southwick, could run for his seat.[2] Stephen had beaten Southwick handily despite displaying an aristocratic disinterest, refusing to campaign, and complaining to cousin Solomon, "I am too old to engage in any active Electioneering business."[3] This began a rather unexceptional seven-year congressional career, but for the fact that, through an odd collation of circumstances, Stephen van Rensselaer III would virtually single-handedly come to decide the results of the 1824 presidential election.

Unlike his more sanguinary cousin, and despite his protestations to the contrary, van Rensselaer had quickly become enamored of the Washington horse-trading mentality, and found his long rehearsal as philosopher-king within the precincts of Albany served him well in his new environs, where he relished the role of behind-the-scenes-dealmaker/powerbroker. He took up this role, presiding over the table at Peck's Hotel in Georgetown,[4] where he was routinely referred to as "The General." The majority of his Albany activities were confined to the summer months, when Washington became just too bug-infested and muggy for his (or really anyone's) taste. It was on one of these summer breaks that he received a visit from a curious and passionate young man in a floppy hat, the artist Samuel F.B. Morse, or, as his wife Lucretia called him, "Finley."

7. Portrait Painter

Morse's entry into Albany society had been mostly due to the auspices of his father, Jedidiah. As a staunch and practical Calvinist, Jedidiah Morse's opinion of his son's career choice as an artist was not an especially elevated one. He had hoped his erratic and hot-tempered son would take an interest that veered more towards the sciences or religion.[5] As perhaps the nation's premier geographer and also a respected man of the cloth, whatever Jedidiah's views on the subject of art and artists, they had done precious little to sway his son towards another vocation. Jedidiah had made no attempt to hide his exasperation with his son's career choice: "His parents had designed him for a different profession ... but his inclination for the one he had chosen was so strong ... we thought it not proper to attempt to control his choice."[6]

It was not that art could absolutely not be a profitable profession. There were abundant examples to the contrary. Several of Morse's contemporaries had already demonstrated that being an artist, while possibly not respectable to a New England Calvinist, certainly could be a lucrative and rewarding pursuit. The dean of American artists, John Trumbull, had realized, if not a fortune, certainly a tidy sum by exhibiting his masterwork, *The Declaration of Independence*, throughout the country in 1818 prior to its being installed in the Capitol Rotunda. Ticket sales in Boston alone had amounted to a hefty three thousand dollars, a fact which could not have escaped Jedidiah's attention. Another prominent member of the American Academy, Rembrandt Peale, had just recently garnered an astonishing nine thousand dollars from showings of his luridly allegorical *Court of Death* in venues throughout the Northeast. All this had been more than enough to sustain the simmering fires of avarice and ambition in a young Finley and stave off the threats of financial excommunication from Jedidiah.

Following Finley's less than illustrious academic career at Yale, Jedidiah had, somewhat reluctantly, agreed to underwrite a jaunt in Europe so he could continue his artistic studies under the tutelage of expatriate painter and poet, Washington Allston. Returning to Boston in 1815, Finley had then traveled throughout the country seeking portrait work, ranging as far as South Carolina, where he made the acquaintance of soon-to-be Congressman Joel Roberts Poinsett. Defying his father's sanguine expectations, Morse was already making a tidy sum with his portrait work in the South,[7] and also served briefly as director of the South Carolina Academy of Fine Arts (with Poinsett as president). In 1821 they had together organized a massive auction and exhibition that included not only Peale's acclaimed *Court of Death*, but other notable works by Sully and Vanderlyn. It had been the highlight of the Charleston social season.[8]

Following his return north, Morse had embarked on what he hoped would be his neoclassical masterpiece, *The House of Representatives*. His preparations for the larger work had required that members of Congress sit for miniature portraits, and their cooperation had no doubt likewise been secured with his father's continued influence and support. For the four months, before returning to New Haven in February of 1822, he had cooped himself up in a dusty dressing room next to the House chamber in the Capitol where he corralled not only members of the Congress but the justices of the Supreme Court, as well as pages orderlies, and even newspaper editors to sit for their miniature portraits. The larger work in the end had included all these as well as his old instructor at Yale, Benjamin Silliman, his father, and for some reason an Indian chief standing in one of the galleries together. In the end it had taken fourteen months of grueling fourteen-hour and sometimes sixteen-hour days, but the result was a massive 8' × 11' canvas that he planned to put on exhibit throughout the Northeast just as Trumbull and Peale had displayed their works.[9] And like Trumbull's stirring portrayal of a pivotal moment in the nation's birth, the thrust of it was to catalog (though not to scale) a somewhat stiff gathering of white men looking on as a significant moment transpired. However, unlike Trumbull's work, for Morse the central event was not the signing of a pivotal document in the nation's history, but rather an elderly black man lighting the House whale-oil chandelier. Not only was the focal point weirdly anticlimactic, in some sense the main subject was the architecture itself. As Morse's biographer Silverman would put it, "It was a history painting without history."[10]

Not surprisingly the public evinced little appetite for the subject matter. The exhibitions were uniformly a flop. Showings arranged in Boston, New York and New Haven, even with the help of positive notices in the local papers from friendly editors,[11] had drawn dismal attendance, barely covering Morse's expenses. All in all, he had spent close to two years on the project and had little to show for it but some new Washington connections. Unwilling to give up entirely, Morse had first hired an art student named Henry Pratt, and then businessman Curtis Doolittle, to manage the exhibitions while he turned his hand back to portrait work and to perfecting an invention, a machine to replicate classical sculptures.[12]

The portrait work, however, had begun to dry up. Finally, almost destitute, late in the summer of 1823, with his wife Lucretia and infant son Charles both ill, he had decided on one last try in Albany, hoping to acquire some portrait work there, once again through exploiting his father's connections. He arrived in Albany on August 9 aboard a Hudson steamer, in the company of Benjamin Silliman and his wife,[13] having left New York somewhat abruptly to accompany them. His departure had been so spur-of-the-moment he had to have his brother Richard load his clothes trunk in New York onto another northbound steamer after his departure, writing, "I came away from New York without even a change of clothing leaving Richard to send my baggage."[14]

As if to add insult to injury, he was told on his arrival that he had timed his visit poorly. It was precisely the wrong time of the year for portrait work, as most of the more prominent occupants of the city had abandoned its humid haunts for more bucolic surroundings.[15] The whole project seemed ill-conceived. Local artist and SPUA member Ezra Ames had already garnered much of the portrait work for local political figures, and the commissions Morse had arranged in advance through Silliman, those of Chancellor Kent and his niece, proved ephemeral as Kent had left town somewhat abruptly with his niece in tow. Morse's exasperated letters to his wife in New Haven were chock full of complaints of having too little to do except read the paper and sit in his rented room on North Pearl.[16]

There was, however, one very prominent individual who was in town for the summer and who could offer him further entry to Albany society. With the stifling weather, maneuvering and infighting over the upcoming election making Washington even more unbearable than usual, newly minted Congressman Stephen van Rensselaer III[17] had returned home to spend the summer months overseeing improvements to the manor house.[18] The day of Morse's arrival, van Rensselaer had called on him at his humble lodgings, where the young artist was still licking his wounds over his recent failures as an artistic entrepreneur. In light of these setbacks, Morse probably immediately commenced to regale van Rensselaer with his economic woes and his views on the Philistinism of the American public. Although his home was undoubtedly in some state of controlled chaos, the Old Patroon extended an invitation to the morose young man, asking him to come visit and apparently to paint his portrait at the manor house.[19] A week hence, Morse wrote, "I am painting the Patroon."[20]

Somewhat embarrassingly for Finley, his father's social connections, had once again provided his main avenue of income and entertainment.[21] But this invitation may have been more than just a cordial nod to the relationship with Jedidiah. Van Rensselaer's son (IV) also had been an intimate acquaintance of Morse during his Yale days, and they met up in England following graduation.[22] The Young Patroon came religiously to watch him paint almost every day in the countryside just outside London.[23] In addition to renewing old school ties, the visit to the manor produced the expectation of a commission for a full-scale portrait of the Old Patroon.[24] Morse informed Lucretia that the Patroon would be shortly traveling back to New Haven in the company of the Sillimans to enroll his younger son at Yale, entreating her to extend the same hospitality to them that had been extended to him in Albany, "who by the way will be in N[ew] Haven at commencement."[25]

8

The Man Who Sneezed So Singularly

Despite the muttered cautions greeting his arrival and the unending litany of complaints to Lucretia, Albany that summer had provided a tilt-a-wheel, head spinning social schedule for Morse. Not only had he renewed ties with his former chemistry instructor from Yale, Benjamin Silliman and his wife, and his former Yale classmate, van Rensselaer the younger, he had made the acquaintance of the brother of the chancellor, Moss Kent, who had been left to convey James' apologies for his unexpected absence.[1] Having just stepped down from the bench of the State Supreme Court in July,[2] the chancellor had decided instead at the last minute to take a "nature tour" through New England with his young niece rather than sit for his portrait.[3]

Silliman had been in poor health all the previous year and had decided to take the waters at Ballston Spa, which were well known for their therapeutic properties. He and his wife likely had arrived at Albany on the same steamer as Morse, who had written his wife the following day, "I presume I shall see Mr. Silliman on his return here."[4] Silliman had not only been Morse's instructor back at New Haven, they had since spent several summers together tramping through the leafy Berkshires. Following his taking the cure at Ballston Spa, Silliman and his wife, as expected, had returned to Albany to pay a few social calls before returning to New Haven (via Brooklyn). Morse would also spend the next few days renewing ties with the Sillimans in the congenial but somewhat noisy setting of the van Rensselaer drawing room.

By the final week of August, prospects had brightened further for the aspiring portraitist. He found himself hard at work concurrently on two portraits, finishing up the one of the Old Patroon, and a new one of Moss Kent (also an old friend of his father's).[5] The letters to his wife Lucretia show him in a somewhat better frame of mind than a few days earlier. Perhaps to further

8. The Man Who Sneezed So Singularly 53

soften the blow of his brother's sudden departure, Moss Kent had arranged for Morse to paint yet another of their relatives who just happened to arrive in town, Elias Kent Kane.[6] Kane was a soon-to-be state senator from Illinois and presently Secretary of State of the Territory of Illinois.[7] There was yet another familial tie to Morse as Kane's law partner was, by coincidence, Morse's favorite cousin and his brother's namesake, Sidney Breese. Also, as it turned out, Kane was a cousin of the Old Patroon. So the attendance at the impromptu salon was steadily growing, plus there was still the prospect of the two commissions originally promised by Kent of his brother, the chancellor, and his daughter (the chancellor's niece) on their return, which was imminent. From these Morse expected finally to realize enough funds at least to cover his expenses for the preceding slack three months and the poorly attended showings of *The House of Representatives*.

Moss Kent, though, had apparently over-promised again, as the chancellor returned to Albany only long enough to attend the opening of the Erie Canal lock at Troy on September 10.[8] He set out again almost immediately for New York City to take up the chair offered to him by Columbia University — once again with his niece in tow (probably on the eponymously named steamer *James Kent*). Though Morse himself probably had already departed Albany, the vessel that would carry the pair south along the Hudson no doubt carried with it, at least metaphorically, his shredded hopes of breaking even. Nevertheless, he had graciously exhorted Lucretia to entertain the pair when they passed through New Haven on the way to New York.[9] The gesture would pay off, as Morse would complete a portrait of the chancellor and his niece the following year, while both were in New York.

Though once again despondent and ill-at-ease, Morse's plans for departure were interrupted by the arrival by steamer of his aunt and uncle, Arthur and Ann Breese, on their way back home to Utica from a visit to the relatives.[10] Nevertheless, with again nothing to occupy his mind but the plaintive letters from Lucretia regarding little Charles's ill health, he promised her that if things didn't pick up soon, he would, within the week, pack his bags and head for New York.[11]

The city of Albany could not possibly have been at the time as bereft of charm and entertainment as Morse had painted it in his missives to Lucretia. Albany was an important city in the early 19th century, and though it was far from what one would conceive of as a metropolis, for Morse it had provided the closest thing to a social hubbub. In those few weeks it drew an astounding number of congenial individuals, former friends, family and former acquaintances who had, either by design or serendipity, all grazed its precincts in the waning weeks of the summer of 1823. He could, it seemed, scarce emerge from the shadow of his doorstep without bumping into someone he knew, either from New Haven, South Carolina or Europe.[12] Into this already spicy late summer stew was introduced another intriguing jalapeño—Morse's former colleague and partner in the short-lived Academy of Fine Arts from South Carolina,[13] Congressman Joel R. Poinsett. Poinsett had just recently returned from Mexico, where he had spent the entire previous year acting as a kind of secret agent for President James Monroe—no doubt he had tales to tell.

Section II

Henry's Influence on Morse

9

The New, Old School

The age of Franklin, the brilliant amateur and his "philosophical toys," was fast fading from view. There was, however, as yet nothing to take its place. The role of a professional scientist was something fomenting in the halls of academia. Eaton realized he needed to staff his new school at Troy with men of solid scientific reputation and he did not have to look far afield, providing the petty rivalries could be set aside. T.R. Beck had quickly been offered the role of vice-president, with Lewis Beck named junior professor (the "junior" underlined personally by Stephen van Rensselaer IV). Amos Eaton as the sole full professor at the new school in Troy.[1] Considering their relative standings in the scientific and educational communities, the structure of these appointments could easily have been construed as "a slap in the face" to the Beck brothers (rather than the olive branch it was presented as); however, if they wished to continue to benefit in any degree from the Patroon's largesse, they had little choice but to accept this rather embarrassing arrangement.

Having presided over the demise of SPUA, if anyone was in a position to see the handwriting on the wall, it was T.R. Beck. With the school in Troy ramping up, the Albany Academy had, like its forerunner, been to some extent supplanted from its perch at the Rensselaerean teat. It was, however, far from suffering financially, as by 1830, the Academy was the recipient of more than half that year's state budget for education.[2] In any case, it is safe to say, even with van Rensselaer's diminishing support, they were on solid footing and no longer required their "non–Latin" instructors to raise their own salaries from lecture ticket sales as they had early on.

By the time the western portion of the canal was officially opened to navigation, the Rensselaer School was already a state-chartered institution with Eaton as senior professor, Lewis Beck as junior professor, and theologian Samuel Blatchford as president, and T.R. Beck (whose academic credentials far outshone Blatchford's) as vice-president. The new school's motto harked directly back to Robert Livingston's "pre–Beckian" vision of SPAAM as dedicated first and foremost to the purposes of instructing persons in "the application of science to the common purposes of life."[3] Even the appointment of the Reverend Blatchford as president was a backhanded rebuke to T.R. Beck's disposition toward abstract studies and a-religiosity.[4]

T.R. Beck's brother, Lewis, had achieved some national recognition in both botany and mineralogy, but in Troy, at least, he was eclipsed by the showy Eaton. While the Albany Academy maintained its cachet for the sons of the upstate elite, Eaton had already made inroads through the appeal of his annual grand outings on the canal, using it as an excuse to actively poach some of Beck's star pupils.[5] He had begun referring to the expedition as his "floating University," implying that it offered a more practical, advanced career path even for graduates (or current students) of the Academy.[6]

T. Romeyn Beck would have been the last to stand in the way of the prospects of any of his students. Instead of sulking about Eaton's predations, and despite the near certainty of sly overtures from Eaton for them to leave the Albany Academy to join Eaton's new school at Troy, he encouraged his star students to participate in the canal outings.[7] With van Rensselaer to back him, Eaton's enticements were nothing to laugh at. A budding artist in his own right, James Eights,[8] had accompanied Eaton on the canal jaunt of 1823. On their return he found that van Rensselaer had ordered up a thousand full-color folios of Eight's geological maps and picturesque drawings that he had completed while on the canal.[9]

Increasingly Albany seemed embroiled in these pointless petty academic rivalries, overshadowed by truly groundbreaking engineering projects. Beck realized the bulk of the intellectual interest (and hence money) following the opening of the Erie Canal would likely be channeled into the disciplines of mineralogy or engineering. Joseph Henry's interest in rocks was somewhat perfunctory, so, perhaps hoping his young protégé might yet ride the crest of that wave, Beck had used his considerable political influence to get his friend, Cherry Valley lawyer Jabez Delano Hammond,[10] to appoint Henry to the State Road project. On his finishing the survey work, Beck had written directly to Martin van Buren requesting Henry be given a position with the Army Topographical Engineers.[11] Henry's inclinations seemed to center still around Albany and teaching. He had proven somewhat more diffident than Eights in responding to Eaton's overtures, and despite Beck's gentle nudging to go further afield, his attachment to the Albany Academy remained, for the time being, unshakeable.

A large part of Eaton's appeal (as he was no doubt aware) derived from the fact that, unlike the Beck brothers, he was clearly not cut in the traditional mold of European scientist or educator. He proclaimed loudly and publicly that the new school at Troy would have a "purely American character," saying, "This school is not Fellenbergian or Lancasterian[12] but is purely Rensselaerean. The unwillingness to admit the possibility of an American improvement in the course of education which generally prevails and the universal homage paid to every thing European has caused much effort to trace the Rensselaerean plan to some supposed shade of it on the other side of the Atlantic."[13] In reality, like his contemporary and colleague, geologist William Maclure, Eaton himself was somewhat less eager to eschew the shackles of European pedagogy, and inspired partly probably by Maclure's successes at New Harmony, Indiana,[14] adopted parts of the innovative Swiss Pestalozzian approach for use at Troy as well.[15]

That being said, unlike Beck, Eaton seemed to have firmly grasped the idea that the assumed inferiority of American science was one whose time was either up, or nearly up. No one could accuse him of being "stuffy." As a staunch advocate of practice over theory and heir to the Livingstonian mantle of *useful* science, Eaton insisted classroom instruction be alternated with *practicums* wherever possible. Aside from the annual canal jaunt, he arranged demonstrations at the sites of various local manufacturing operations including tanning, brick-making and bleaching shops, as well as botanical field trips to Rensselaerwyck, the sprawling van Rensselaer estate.[16] His was to be, if not purely, at least primarily an American approach, with a "hands-on" vision of education and without the taint of European sophistry or rationalism.

The experience of the Beck brothers should have offered a cautionary tale when it came to putting too much stock in Eaton's ideas or his integrity. Lewis Beck was by this time an extensively published expert in both botany and chemistry with several major works to his credit.[17] Eaton had shown his bullying side when Lewis Beck had been assigned the embarrassingly subordinate role of junior professor.[18] Whether or not the "junior professor" designation had been a sop to van Rensselaer, Lewis had knuckled under, supplementing his salary at the Rensselaer School by taking over Eaton's former position at Vermont Medical College in Castleton, teaching botany and chemistry.[19]

Thus far, the "frenemy" relationship between the Beck brothers and Eaton had consisted mostly of semi-collegial, intellectual muscle-flexing. For van Rensselaer, whose intellectual life now centered around Washington, this most likely had been a source of both amusement and edification, but even van Rensselaer was perhaps beginning to suspect that Eaton's creative approach to pedagogy was accompanied by an equally creative attitude toward finances. By 1826 Eaton found himself in the same boat as had the Beck brothers earlier, begging funds from the Patroon. By the fall term, van Rensselaer had begun to slowly withdraw his financial support from the Rensselaer School in the same manner as he had three years earlier with the Albany Academy and the Lyceum,[20] this time turning his munificence towards a more well-established institution.[21] Thanks in part to this generosity, van Rensselaer had been appointed a trustee of Rutgers,[22] and perhaps to make amends for his earlier slights, the following year he would help Lewis Beck to obtain a full professorship there in chemistry and natural history.[23]

Though the fires of experimentation, it seems, already burned as bright in him as the molten iron he had seen emitting from the West Point foundry furnace, Henry still thought of himself first and foremost as an instructor.[24] When he returned to Albany in the fall after his trip to West Point, he assumed a heavier teaching load.[25] By then he had been appointed special assistant for chemistry, a role that afforded him a better salary as well as allowing him to utilize the resources of the Academy for his increasingly ambitious experiments, conducting them under the guise of classroom demonstrations. This also put him nominally under the auspices of Lewis C. Beck, who was increasingly fascinated by the direction of Henry's novel ideas about magnetism and force exerted at a distance.[26]

Henry had managed to garner a substantial portion of the Albany Academy's (and the Institute's) resources for his ambitious efforts, but in return he had put them firmly back into the firmament of the lecture calendar.[27] He, along with Lewis Beck, Richard Varick and other members of the Second Department[28] organized a series of well-attended talks given on the second floor of the Albany Academy, with Henry featuring demonstrations of his increasingly powerful electromagnets.[29] (If Eaton had been intent on providing the public with shows of electricity that were "strong and forcible," Henry was not to be outdone in that department.) In the meanwhile, Henry continued assisting Beck with the chemistry lectures and working with him to devise an improved scale of chemical logarithmic equivalents. He attempted briefly to market this device, but after finding it too difficult for local artisans to produce cheaply and accurately, he abandoned the project.[30] By 1829, feeling confident enough in his own research and feeling increasingly confined in Albany, Henry resigned as curator of the Second Department (what had been the Lyceum), allowing Ten Eyck to take his place (he shortly would at the Academy as well), and began searching for a more prestigious position in academia.[31]

10

The Vision at Palmyra

T.R. Beck's course in chemistry was probably the first class ever taken in science by a young Joseph Henry at the Albany Academy. The son of a perennially broke day-laborer/drunkard, Joseph Henry had a face that grudgingly combined the qualities of thespian and street brawler: bruised and angular, battered by blows from scores of would-be "sculptors" leaving their marks in the clay of a native sensitivity and nobility. His stunning, pale-blue Scottish eyes, chiseled chin and stern demeanor[1] had no doubt made him, despite his native shyness, somewhat of a matinee idol. It was there he had been introduced by a mutual friend, Philip Ten Eyck,[2] to Dr. T.R. Beck, while playing the role of Hamlet at Albany's Green Street Theater. Seeing great potential in the young man (not necessarily in the field of acting), Beck had somehow convinced him to give up on his career as a thespian and take up the study of science at the upscale Albany Academy, where, among the progeny of the rich and well-connected, he must have stuck out like a sore thumb.

In April of 1823, T. Romeyn Beck's wife Harriet died, leaving him with the care of their five-year-old daughter.[3] Relieved at least of the burden of the annual geological survey, Beck had spent the summer completing his comprehensive work *Medical Jurisprudence*, for which purposes he no doubt had made ample use of the absent Chancellor Kent's extensive law library when he needed a break from babysitting. The book had appeared in September to a great deal of fanfare. It was a groundbreaking effort and for the first time established a system of medical standards for the courts to use that were not entirely beholden to English common law.

So, amidst all the already noted various comings and goings in Albany in August of 1823 in which Morse was active as a shuttlecock—between the Kents' residence and the Van Rensselaer Manor; the painting of portraits; the arrival of ailing scientists bearing erudite journals, politicians spouting tales of undiscovered wealth in lands to the south, conveying secret political intrigues; amidst

Memoranda

July 6. 1814. The Trustees, solicited from the Common Council, the erection of a building, that should cost from 10,000 to 15,000 Dollars.

Jan'y 24. 1816. Monthly meeting Committee, ordered. (See also Oct' 10. 1823)

July 14. 1818. Academical Library founded. No donations to be refused, but the main object to be the collection of works in those sciences, &c that are taught in the Academy.
The Principal to be Librarian.
Dec' 8. 1826. M' Henry to be do

Feb'y 10. 1823. The Lyceum have the S. West Room in Second Story, for five Years, provided they finish it.

June 10. 1825. Silas Walker.

July 8. 1825. Expense of Large Room
& furniture)   1770.34
                168.00
            $ 1938.34

Above and opposite: **Handbook memoranda from the Albany Academy: Shows Henry as librarian (circa 1823) and five-year lease of room to the Albany Lyceum (courtesy Archives and Collections of the Albany Academies).**

> April 23. 1827. The Albany Institute allowed to occupy the 'North West' Room on the Second Story during the pleasure of the Board.

all the ongoing edification of the next generation of learned Albanians—where, we might reasonably ask, was young Joseph Henry?[4]

Since he was too young to be appointed to the faculty, on his graduation or shortly after, Beck had found Henry a part-time position as his classroom assistant and school librarian, no doubt both poorly paid positions. Beck augmented Henry's modest income by securing him a stipend and a comfortable posting in the van Rensselaer household. In return for three hours a day spent tutoring the van Rensselaer children in mathematics and other subjects, he would receive free room and board and living expenses from the Old Patroon.[5]

No doubt, as a trusted member of the van Rensselaer household, Henry would have had the opportunity and surely would have taken pains to meet two such esteemed guests as Benjamin Silliman and Samuel Morse, who were visiting van Rensselaer that summer. The fact that there is no record of such a meeting seems baffling. The difference in their social station may have played something of a role. There may have been little reason for a somewhat self-involved son of privilege like Morse, a graduate of Yale and scion of a prominent New England family, to take notice of a pasty-faced, square-jawed, live-in household servant, the son of an alcoholic roustabout. Morse's presence was nothing of a secret. His high-profile artistic endeavors were well known in Albany and the ongoing exhibition of the *House of Representatives* and his solicitations for portrait work had been heavily advertised in the Albany papers.[6] What is more, Morse's father Jedidiah's geography was a staple of almost every preparatory school in the country, including the Albany Academy, where Henry was now the librarian.[7]

The Sillimans were the other houseguests of van Rensselaer that summer, and were also not likely to pass through town unnoticed by either Henry or Beck. Benjamin Silliman had long been somewhat of a personal hero of Henry's, and as a longtime friend of James Kent, was well known in the Albany Academy social circles. Henry's acquaintance with Silliman's *Journal*, as legend has it, stemmed from his youth, when he had devoured copies of the periodical while lying before a roaring pine fire in his house.[8] So it is again baffling that we can place Henry in the van Rensselaer household at the same time as Silliman and Morse, yet, if we are to believe their own subsequent accounts, they managed to avoid meeting one another at the time.

With the summer drawing to a close, Albany was once again abuzz with activity. A notice in the Albany newspaper of September 10 noted the opening of the northern portion of the canal. Boats ferried people up to the lock for a fee of $1 and ten thousand people assembled to hear the inaugural address by Governor DeWitt Clinton.[9] Five days later, T.R. Beck's *Medical Jurisprudence* appeared, to the almost universal acclaim of the medical and juridical communities.[10] Three weeks later the major section of the westward portion of the canal was opened to transportation.[11]

Henry had quickly fallen under the wing of a new professor at the Albany Academy, A.B. Quimby, whom Beck had recruited from Columbia College,

10. The Vision at Palmyra

focusing his experiments on the uses and study of steam. Steam was a hot topic in Albany by 1826; a new steam railroad between Albany and Schenectady was being contemplated, and came up that spring before the legislature.[12] Quimby was the closest thing there was to a mechanical engineer in the state (and perhaps the country) at the time.[13] From Quimby, Henry would learn his habit of careful notation of experimental results and the application of higher mathematics to physical experiments, particularly with reference to the property of what was known as "work" in relation to steam.

While his loyalties remained clear, Joseph Henry's curiosity regarding the activities across the river had begun to overcome his diffidence.[14] Fresh from his thirsty work on the road survey, Henry had decided to accompany Eaton on the leisurely summer canal jaunt. Though Henry was by now a full professor at the Albany Academy, Eaton, true to form, extended him no courtesy, treating him essentially as any other student, asking him to pay for the privilege of accompanying the expedition.[15]

They had set out from Troy about noon on May 2, 1826. Word that a series of lectures were being offered by two respected professors, Dana and Torrey (among others),[16] concerning studies of the recently discovered and exciting force of electromagnetism had apparently reached Henry at Palmyra.[17] Here he was faced with a choice: continue on with the canal survey or head to West Point for a discussion of a new force about which little was known. According to the Mormons, Palmyra, New York, was the spot where Joseph Smith had his vision and found the golden tablets. Henry must have had a comparable awakening of his own.[18] Geology and botany were fine, Henry had decided, but he wanted to put his experimental skills to work in another direction—investigating of the fundamental forces of nature. The detour to West Point would prove a life-altering one for Henry, and for science in general.[19]

11

Ghost Story

With regard to whatever lively intellectual luncheons took place at the Rensselaer manor in late August of 1823, where we conjecture finding Morse and Silliman (and possibly Henry) in attendance certainly, we would be remiss in not noting one further, perhaps peripheral and obscure, but disconcerting note—one that prefigures in a weird and disturbing way, the fate of the electromagnetic telegraph—one linking Morse not with the ephemeral Henry but with Morse's future employee, Ezra Cornell. In it we may once again perhaps glimpse the metaphysical influence of T. Romeyn Beck, who shows a pattern of insinuating his presence at every significant turn in Ezra Cornell's career, in person or by proxy. We trace this present *ab incunabulis* event to a familial murder that occurred in the Plymouth Colony of New England in the year of 1673.[1]

We can set this imaginary scene where the convoluted strands of an underlying psychodrama play out in the sunny van Rensselaer dining room. We possibly find Harriet Elizabeth Bayard (the wife of Stephen van Rensselaer IV, Morse's classmate) serving lunch to a young Samuel F.B. Morse, who is on break from his work painting the portrait of the Old Patroon. Benjamin Silliman has just returned from Ballston Spa and is there as well, sitting in an overstuffed chair. Possibly, as we further conjecture, a somewhat evanescent Joseph Henry was flitting about the margins, corralling the less studious younger members of the household. We can conjure therefore the first chilling calling card of a perplexing and labyrinthine fate that was delivered that late summer afternoon by Harriet, alongside the smoked kippers: that of a familial murder.

Whatever the obscure faction of furies that was pursuing the uncomprehending Morse and Cornell as well throughout both their careers with the telegraph, it seems to spring mysteriously somehow from their indirect relationship with T.R. Beck. And whatever had thrust this unlikely duo together in the first place, and which persevered in binding them in a convoluted tango of mutual need, hatred and distrust to the end of both of their days, it seemed to stem in some metaphysical manner from some seminal, heinous crime. Harriet Bayard van Rensselaer, it turns out, was the great-great-great-granddaughter of the first of several Cornell women to be brutally murdered by a member of her immediate family, Rebecca Cornell.[2]

12

Wrap Artist

The stopover had been partly to renew old ties. Silliman had last visited Albany in 1819 *en route* to Canada, at which time he had been the guest both of Chancellor Kent and of the Old Patroon. He wrote on the earlier visit, "We ... reached Albany dined and spent most of a day with Judge Kent in whose fine library of between two and three thousand volumes *you* would *revel* ... I was more than ever delighted with the Judge. We were also at the Patroon's, probably the most like an ancient baronial establishment of anything in America, it is a princely place."[1] The following October, Silliman had forwarded a copy of his recently published *Tour of Quebec* to Kent as a "thank you" for his hospitality.[2]

The Sillimans' return to Albany in the last week of August of 1823 was something of an extension of what was turning out to be a stubborn convalescence.[3] Despite an avid interest in New York's geology, when Amos Eaton had invited him (through van Rensselaer) to join him on that year's canal expedition, Silliman (through van Rensselaer) politely and uncharacteristically declined.[4] There were several physical and mental factors contributing to Benjamin Silliman's condition that had occasioned the sojourn to Ballston Spa. Aside from various digestive ills, he was still extremely depressed over the death of his colleague and friend, Professor Alexander Metcalf Fisher,[5] who had perished in the wreck of the *Albion* off the coast of Ireland the previous spring.[6] The loss of the *Albion*, the first of her class of transatlantic packets to go down, had sent shock waves throughout Europe and America.[7] The artist Peter Grain had depicted the event on a 120'-long canvas mural for the residents of Charleston![8] No one, however, was more affected than Silliman himself over the loss of his dear colleague and friend. His ensuing poor health was evident to both friends and acquaintances.[9]

Whatever his physical or mental debilities at the time, the Sillimans' arrival back in Albany no doubt was a source of pleasure to Morse and most likely the occasion of some more erudite lunchtime conversations at the Rensselaer

12. Wrap Artist

Manor (to which we alluded above), for which Silliman readily provided the subject matter. Silliman, as was his habit, had with him several of the latest copies of his *American Journal of Science* to distribute amongst his friends: van Rensselaer, Beck certainly (for the Albany Academy's Library), and one for the absent James Kent for his private library.[10]

As we have stated, whether Henry personally encountered Silliman at the time of his return from Ballston Spa is a point of conjecture, but it is certain he would have found the contents of that summer's *Journal* rather stimulating, in particular a brief (one-page) article by one of Silliman's colleagues and close friends, Professor James Freeman Dana of Dartmouth College. It concerned a novel experimental apparatus Dana had devised involving an electromagnet.[11] So, whether or not it was from Silliman's own hand, Henry, being the librarian for the Academy, acquired a copy of the 1823 *AJS* rather soon after its publication.

The experiment was relatively simple, at least on the surface. A cork was set floating in what was essentially a voltaic cell, created from a glass beaker of acidic solution. A wire coil, with dissimilar metal leads was immersed in the solution, and then wound around a needle sitting on the cork floating in the solution. It was, in essence, a self-magnetizing compass; although preceding Sturgeon's London demonstrations by a year, certainly this was something that had been amply demonstrated in Europe before. Since a current was being passed constantly through the coil, this created an electromagnet, so the needle always pointed north. What would have rendered this rather prosaic effort of more than casual interest to Henry is the fact that, in creating the electromagnet, Dana

Alexander Metcalf Fisher. Portrait by Samuel Morse (courtesy Yale University Art Gallery and Yale University).

had employed an over-wound insulated rather than an uninsulated coil.[12] This at the time was novel. As Henry was no doubt aware, all of Sturgeon's efforts to date in England had involved uninsulated, single-wound coils.

This "new" phenomenon of electromagnetism was not in fact all that new, having by 1823 received serious notice in Europe and particularly in England and Denmark, where the theory of electromagnetic circuits had been discussed and investigated in detail by Danish scientist Hans Christian Oersted.[13] Oersted had first noticed the phenomenon of deflection when a charged wire was accidentally laid on top of a compass. Johann Schweigger,[14] building on Oersted's discovery, had utilized a wire coil insulated with either silk or wax mounted on a wood frame to create his galvanometer. English physicist William Sturgeon,[15] following on Oersted's observations and discoveries, had coiled a bare wire attached to a galvanic cell loosely around an iron bar, creating an electromagnet that could initially lift a weight of five pounds or more. The bar had to be varnished to prevent contact, as Sturgeon's coil was made of *uninsulated* bare wire. Having begun his experiments in 1822, Sturgeon had really not made the device public until 1824 when he presented it to the Royal Society.[16] Using an uninsulated loose coil of single thickness, the apparatus remained limited to lifting about nine pounds.[17]

Dana's modest one-page article thus contained this critical improvement that previous experimenters apparently overlooked, one that under Henry's hand would eventually allow for creations of vastly stronger overwound magnets. While neither the use of uninsulated wire nor the use of a coil to induce magnetism was novel in itself, the use of a *tightly wound insulated* coil certainly was. Dana's use of insulated wire had opened up new possibilities for overwound electromagnets that it would fall to Henry to exploit.

The *AJS* article had apparently lain dormant in Henry's brain for several years until the trip to West Point in 1826, where he had a chance to meet Dana in person. This apparently had reawakened him, and by 1827 Henry was experimenting with both a single insulated wire coiled around itself and several insulated wires with multiple galvanic sources[18] wrapped tightly around an iron bar.[19] Using these techniques, within the span of a year Henry would be creating electromagnets that could easily lift thirty-times the weight of Sturgeon's.[20] For some reason Dana's improvement was destined to go largely unnoted, and though his work was seminal, use of an overwound insulated coil would later be attributed solely to Henry (something which Henry unfortunately made very little effort to correct).[21]

Henry was not the only one who had been deeply impressed by Dana's little overwinding trick.[22] Morse, too, when he moved back to New York from Albany, had specifically sought out Dana as an intellectual companion,[23] and later became an avid attendee of his lectures on electromagnetism given at the New York Athenaeum.[24] While Morse would later rabidly deny Henry's influence on the final form of the telegraph, he would freely acknowledge his debt to Dana and chide Henry for his corresponding failure to do so, pointing out explicitly, "Prof Henry gives in his deposition the time of his first acquaintance with the science of electro magnetism and it will be seen that it was subsequent to the delivery of these lectures of Professor Dana."[25] Indeed, when faced with charges of intellectual plagiary by Charles Jackson (in the context of Henry O'Reilly's infringement suit), Morse would produce Dana's actual original apparatus to show its similarity to Morse's later drawings.[26] So, while neither Henry nor Morse explicitly acknowledge reading, let alone being inspired by Dana's 1823 article, it likely had made a profound impression on both. Whether or not it was percolating in the rarefied but humid air of Albany in the summer of 1823 as a topic of lunchtime conversation at the Rensselaer Manor is a different question and a matter of speculation. But while Henry's interest in electromagnetism is usually said to stem from his visit to West Point in the early summer of 1826, it is a virtual certainty that for him as well as for Morse the germ of this obsession can be traced to Dana's 1823 monograph on the subject.

13

The American Achievement

Most count themselves lucky if they have half an idea in their lives. Einstein was very lucky—he had two: special and general relativity. And while it is not the intent here to make a case for Joseph Henry as inventor of the telegraph, as have others,[1] he did have at least one-and-a-half ideas arguably his own[2] which would come to have a profound influence on science in general, and undeniably on Morse's telegraph as well (despite the loud protestations of the latter). The first half idea, (if we are to give due credit to Dana) was, as mentioned, the idea of overwound coils, and second was his distinction of "quantity" and "intensity" circuits. Barlow and Sturgeon in Europe had consistently failed in their attempts to reach more than two-hundred feet with an electrical signal.[3] Aside from his main focus on electromagnets, Henry's other novel insight was that the manipulation of resistance in an electrical circuit could enhance the power of an electromagnet to send signals *over a theoretically unlimited distance.*

The difference was that Sturgeon and Barlow were still laboring under the misapprehension that electricity was essentially a fluid for which the wire was the conduit, and thus electrical flow would behave in general in a similar fashion to water flowing through a pipe. This was incorrect, and Henry, grasping that fact, had set about proving it. Working from his third-floor classroom, Henry had recast the entire problem from one of wire diameter and battery strength to one of overall circuit configuration and internal resistance within the coil and the battery. This was a leap not even Henry's idol, Faraday, had yet made.[4] Thus, ironically, Henry's first major contribution to science consisted in challenging his idol Faraday's concept of the nature of electrical currents by demonstrating internal circuit resistance in electromagnets.

Based on the weak performance of the zinc-copper galvanic batteries of the day, the future of electromagnets, even with Henry's improvements, was not promising. It was overshadowed by the suppositions of the European

13. The American Achievement

experimenters that a strong enough initial current could not be generated to make what was merely an interesting phenomenon into something commercially viable or useful. Henry remained bent on demonstrating his thesis: that through adjusting the circuit configuration, plus using Schweigger's galvanometer (which he called his "galvanic multiplier"),[5] and also insulating the overwound wires surrounding the bar[6] (in the same manner he had seen in Dana's article three years prior), distance should become a negligible consideration. It was a bold theory and one contradicted by every European experiment thus far.

By the middle of the fall term at the Academy, Henry was conducting various experiments of his own in electromagnetism, with Lewis Beck and Philip Ten Eyck as avid onlookers. The former made copious notes—even encouraging his visiting nieces to lend a hand by helping insulate the long copper wires with silk cloth. As biographer I.B. Sebring (rather blandly) put it, "[Lewis] Beck collaborated in 1827 with Professor Joseph Henry who later developed the electromagnetic principles used in the telegraph ... having apparently become ... interested in the problems ... on which Professor Henry was working."[7] Philip Ten Eyck pulled his weight by scrounging bulk supplies like wire and iron from local merchants,[8] and "borrowing" a thousand-plus feet of copper wire from a local hardware store,[9] which he strung throughout the building and down to the basement.[10]

Since silk, the material initially used for the purpose, was relatively expensive, this represented a significant outlay for the still somewhat destitute young assistant professor (as would have wire, had not Ten Eyck nicked it). Lewis Beck had come up with the suggestion of wrapping the wires in gum-saturated cotton instead of silk, a far cheaper solution.[11] There is some evidence, however, Henry preferred using silk, as Beck's niece recalls:

> When Mr. Henry was perfecting his magnet he required many yards of wire covered with sewing silk; whether because it was the cheapest I do not know, but he selected pink sewing silk, and he persuaded my sister Helen and me [then children] to cover the wire. This we did, with the assistance of our grandmother's small spinning-wheel. Sometimes he had to resort to bribery, but all the same we helped to make the great magnet that brought him fame and fortune. My uncle, Dr. Lewis C. Beck, and Dr. Philip Ten Eyck were indefatigable in working with him.[12]

So evidently Beck and Ten Eyck were both eagerly supplying materials and helpful suggestions, running commentary, and digging up the necessary wire, insulation and other materials required for creating these ever larger and more powerful electromagnets.[13] Within a year, Henry, working from his room in the southwest corner of the third floor, would be creating magnets that could lift several hundred pounds. This had begun attracting a great deal of notice, not only in Albany, but in New Haven, from Silliman, and undoubtedly in Europe as well.

As an indigent,[14] the son of an abusive alcoholic father, Henry was understandably cautious of exposing himself to ridicule. His continued employment depended in a large degree on his reputation, and this had rendered him oversensitive to any potential criticism. This had also made him reluctant to publish, but by 1831, the former dock worker and actor was finally ready to publish in a respected scientific journal under his own name. He had by then already been bitten once by reticence when he had delayed publishing his theories on internal resistance in batteries long enough for German physicist Georg Simon Ohm to publish similar conclusions and taking the credit.[15] If not for this one serendipitous event, Henry might well have become primarily known as a theorist rather than an experimenter in the field of electricity.

Though his article for the 1831 issue of the *American Journal of Science*[16] had been submitted officially too late for publication, Silliman had obligingly bent the rules and got it added at the last minute as an appendix. In the same issue was a similar article on electromagnetism by Moll[17] that contrasted poorly with Henry's obviously more impressive accomplishment. In the rush to publication, however, Henry made a political *faux pas* that would have lasting repercussions on his relationship with the Beck brothers. In the article he had credited Ten Eyck as a collaborator, saying, "A series of experiments were instituted jointly by Dr. Philip Ten Eyck and myself,"[18] but he consigned Lewis Beck to a footnote, crediting him, almost insultingly, solely with cost-saving suggestions.[19]

In response to a subsequent, purely innocent query by Silliman regarding the extent of Ten Eyck's contribution, Henry poured fuel on the fire by overgenerously suggesting Silliman award Ten Eyck a co-authorship credit.[20] It was an innocent enough oversight, the mistake of a novice author who, thinking Silliman's comment a suggestion, had unthinkingly acquiesced. This had offended Lewis Beck sufficiently that he complained to his brother about the presumptuousness of his fair-haired boy. The error was then compounded and the presumption of ingratitude reinforced when T.R. Beck learned soon after that Henry had been receiving overtures from Princeton University, and had informed Beck only several weeks after the offer of employment had been accepted.[21] It seemed Henry had finally moved on from *Hamlet*. T.R. Beck may have lamented of his former fair-haired boy: "Blow, blow, thou winter wind. Thou art not so unkind as man's ingratitude."

14

Que Viva Mexico

To return to our year of 1823, Albany was, at the time, the seventh largest city in the nation, and thanks to the presence there of such scientific lights as the Beck brothers and Amos Eaton, one of the most active intellectual crossroads. But its location rendered it somewhat a social backwater, and thus it presented neither a likely setting for political intrigue nor an especially profitable venue for an aspiring artist. Morse's constant complaints to Lucretia about having nothing to do, however, do seem partly concocted to set her mind at ease. Morse may have rationalized this marital fib on the basis that their son Charles was ill at the time and he did not wish to distress her further unnecessarily.

There was, in fact, no lack of entertainment or stimulation. Famed Albany portraitist and former head of SPUA's Fine Arts Department, Ezra Ames, had his studio two blocks away from Morse's rooming house on Pearl[1]; John Cook's reading room was a few hundred yards in the other direction along Market Street. The brand-new Albany Lyceum boasted not only Lewis Beck's extensive mineralogical and biological cabinet and the priceless SPUA library but one of the finest collections of fine art and sculpture in the Northeast. Nevertheless, when the two Kent family portraits evaporated, we find Morse once again complaining to his wife that he has nothing much to occupy his time there except to go to the newspaper office and catch up on his reading.[2]

Despite the continuing failure of the *House of Representatives*[3] painting to garner an audience, by late August, things seemed to have picked up a bit for the budding artist. Through pure serendipity, his visit to Albany had coincided not only with that of his old instructor from Yale, Silliman, but also that of yet another prominent former associate, his partner in creating the South Carolina Academy of the Arts,[4] Congressman Joel Poinsett,[5] who had recently returned from Mexico, and was likely in Albany on a secret mission for Henry Clay concerning the upcoming presidential election. Anticipating that the election might be thrown into the House of Representatives, he had come to sway freshman Congressman Stephen van Rensselaer III's vote, a fact that, given Morse's express interest in that body, he may or may not have divulged to Morse at the time.

Morse and Poinsett (relatively speaking) went way back.[6] The two had first become acquainted in Charleston, where they, together with several other architects and artists,[7] had established the short-lived South Carolina Academy of Fine Arts. While in South Carolina, Poinsett had proven a lucrative conduit of portrait work for him,[8] so Morse, in his present reduced circumstances, was eager to renew ties and garner news of the South. Poinsett, for his part, had plenty to discuss with Morse as well, mostly concerning his recent travels in exotic lands to the south, but the real focus of his visit was a conversation he was intent on having with the honorable van Rensselaer. Poinsett's unannounced[9] visit to Rensselaer Manor that summer likely had little to do with artistic endeavors and more to do with the upcoming four-way national presidential contest in which the Patroon would come to play a pivotal (and by most accounts, not entirely savory) role.

To elaborate on the political subplot: the machinations surrounding the upcoming 1824 presidential election were already in full swing. It was looking to be a standoff, with four candidates running, and there was abundant speculation that a deal could be arranged in which the New York delegation, led by van Rensselaer, could play a key role in deciding the eventual outcome. John Quincy Adams had maintained strong support throughout the Northeast with the exception of New York State, where the electoral delegation was split between him and William Crawford. Andrew Jackson, the Democrat, had the backing of most of the South, including South Carolina, which was pledged to him, but if the first vote did not result in a winner, then, by common agreement, all bets were off. What came to be called "the corrupt bargain" began with Adams reaching out to the Henry Clay supporters and the former boosters of New York Governor DeWitt Clinton (who had by now withdrawn) to gain their votes in the event that the election was thrown into the House of Representatives. Poinsett had been acting as a sort of "secret agent" for James Monroe in Mexico. Adams, with Clay's assistance, chose to take advantage of this talent for subterfuge to line up support for himself among Governor Clinton's former supporters.[10]

Van Rensselaer, at the time, after having taken over his cousin Solomon van Rensselaer's seat in Congress, was living in Martin van Buren's house in D.C. while the Congress was in session.[11] Once Clinton had withdrawn, van Rensselaer had quickly come out strongly for the candidate van Buren favored, this being Georgia Congressman William Crawford. If New York actually went for

14. Que Viva Mexico

Crawford, that would have set the stage for a three-way contest between Crawford, Jackson and Adams. Adams's best bet, should the election be thrown into the House, which seemed increasingly likely, was a two-way head-to-head contest with Jackson, and for this to occur, Crawford had to be eliminated on the first ballot.

As a dyed-in-the-wool Southerner, Poinsett had publicly been quite vocal in his support of Jackson and of the newly created Democratic Party. Given Poinsett's own upcoming bid for reelection, to oppose Jackson would have amounted to political suicide. As a realist, however, Poinsett knew that his man might well lose once the contest was thrown to the House. The great orator Henry Clay, who was also running and who despised the backwoods "Indian fighter" Jackson vehemently, had decided (as a backup plan) to organize a kind of coup, to forge an alliance whereby John Quincy Adams would be put into the White House should it come down to a two-man race in the House. It appears that Poinsett had been dispatched by Clay to Albany on the sly to work out this secret deal hatched by Clay to get van Rensselaer to throw his support over to Adams on the House floor, should it come to a vote there.[12] It did play out just as expected, and (most likely) as a result of Poinsett's visit to Albany in August 1823, van Rensselaer would come to play a crucial role in what would come to be called the dirtiest presidential election in American history.[13]

Poinsett and Clay both had ambitions for prominent roles in the new administration: Poinsett hoped to be appointed minister plenipotentiary to Mexico, and Clay had set his sights on the plum Secretary of State position or the vice-presidency. In return for securing van Rensselaer's vote if it went to the House floor, Adams seems to have agreed to grant both Poinsett and Clay what they wanted. While this theory of a deal engineered by Clay involving van Rensselaer is undocumented, the idea is certainly not exclusive to this account, and indeed seems logical from the way things played out. As Andrew Jackson would point out as evidence of a deal, Poinsett and Clay, both Adams's avowed political enemies, were soon after the election appointed to the respective positions they desired. So, though actual documentary evidence for what is called "the corrupt bargain," a term coined by Jackson to describe these events, is pretty much nonexistent, it is also, amongst historians, pretty much accepted fact. It had led to an ugly and fractious atmosphere, and van Rensselaer had taken some pains to appoint himself above the fray, in somewhat haughty tones that still would not sound out of place in the mouth of a more modern politician: "Party mixes in every question.... I absent myself when the battle commences ... it is too disgusting for my Ear as I have ever kept good company."[14]

After a discreet arm-twisting session with Henry Clay and Daniel Webster, van Rensselaer had thrown the election to Adams with a one-vote majority in the New York delegation.[15]

There are other theories extant as to why van Rensselaer changed his vote at the last minute, handing Adams the presidency, that don't involve the auspices of Joel Poinsett and Henry Clay, but all reek more or less of wishful thinking. The most interesting and imaginative of these by far (the one offered by van Rensselaer himself) had him responding to propitiously timed divine intervention. He claimed a ballot for Adams wafted providentially under his seat just when he was about to cast his vote for Crawford. Taking this as a sign from God, he had switched his vote at the last minute.[16] The only other version of events that does not appear patently self-serving or the result of divine intervention was the assertion made by a Delaware congressman that his turnabout had likely been prompted by the interference of Mrs. van Rensselaer (which at least has the benefit of being logical).

Whatever the actual case, to return to our own tale, we find the well-tanned and alertly nervous Poinsett arriving in the van Rensselaer's drawing rooms in Albany in late August of 1823 to find a somewhat dejected and pale young Morse dabbing away at his portrait of the Old Patroon.[17] Poinsett was brimming with tales of old Mexico with which he regaled Morse, at some point suggesting Morse accompany him on a planned return trip.

Having served as unofficial agent of the United States government in Mexico, we can assume Poinsett expected (provided his candidate triumphed in the presidential election) to be appointed as head of an official legation. He had invited Morse along as cultural *attaché* of sorts, leading the then poverty stricken Morse to believe that he had discovered, in his recent travels through Mexico, many of Europe's old masterworks secreted away in haciendas or churches, and that due to the present uncertain economic circumstances and political turmoil[18] surrounding the coronation of Agustin Iturbide as emperor, these could be acquired relatively on the cheap.[19]

According to his letters to Lucretia, Morse had been half-contemplating going south for the winter months anyway, possibly to New Orleans,[20] but by this time he had decided on the far less risky step of setting up shop in New York City as a portrait artist hoping to replace his old friend and competitor, Trumbull. He wrote to Lucretia, "Colonel T[rumbull] is growing old and there is no artist ... to take his place.... I may be possibly be promoted."[21] With no other real prospects in hand, Morse had leapt at the opportunity Poinsett presented. It seemed to reinstall a sense of purpose, offering the prospect not only of a grand adventure away from the dismal New England winter, but the possibility of getting rich quick, both of which no doubt had immense appeal for him at the moment.

14. Que Viva Mexico

Never having been that far south before, over the next week Morse evidently had begun to have second thoughts, peppering poor Poinsett with all manner of questions about the weather, health risks, acceptable forms and species of graft, etc. The questions involved not only health and safety issues, but also evolving a solid business plan, to which end he had involved van Rensselaer, who was shortly to leave Albany to return to Washington. He began his letter introducing his new plan by complaining of the general lack of interest in his work and the plight of artists in the United States in general: "The failure of my picture of the House of Representatives to produce me [sic] anything but empty reputation has led me for some time past to think of some enterprise to retrieve my lost property."[22]

Poinsett would quickly tire of this nervous assault, asking him in the future to write down his questions and pass them along through van Rensselaer. Morse obligingly did so, delivering a set of numbered scholastic interrogatories that must have harked him back to his Yale undergraduate days, along with which he noted Poinsett's replies: "How soon is it perfectly safe for a stranger to the southern climate to be at Tampico on his way to Mexico?" Reply: "Last of October or beginning of November." "What detention will he meet with at Tampico from custom house forms and other government restrictions?" Reply: "None, a trifling present smooths the difficulty."[23]

On the larger public stage, things had devolved pretty much as Clay had foreseen. True to his word, one of Adams's first acts after being elected president was to appoint his staunch political adversary, Joel Poinsett, minister plenipotentiary to Mexico, and his other former enemy, Clay, as Secretary of State, thereby inciting Andrew Jackson to coin the phrase, "corrupt bargain." Morse had already begun making plans for his departure and even shipped his baggage ahead to New Orleans[24] when everything came to a screeching halt. Seemingly out of the blue, Poinsett had been replaced by Ninian Edwards of Illinois.[25] The plan unraveled entirely, leaving Morse to fall on his backup plan of becoming a prominent New York City artist, hoping once situated there, on Trumbull's death or retirement, to be asked to take the helm of the prestigious American Academy of Fine Arts.

15

Endless Debate

Morse at one time would claim, "Nothing has been invented heretofore like my telegraphic system. No improvement has been made in it by any other person since I have had it in operation."[1] The timing and sources of Morse's inspiration for the telegraph have long been at issue, and Joseph Henry has always figured centrally in that controversy, which has persisted (albeit increasingly anemically), down unto the present day.[2] There is this question and also the secondary question of their first in-person meeting, which arose later in the context of a heated patent litigation and has bearing on the actual extent of Henry's intellectual influence.

We know that Henry was a member of the van Rensselaer household at the time of Morse's visit to Albany. (According to Nathan Reingold,[3] his employment there may have started as early as the academic year of 1822, the year he graduated).[4] He was clearly augmenting his modest income around this time by alternating with Beck as librarian for the Albany Academy library.[5] In any case, evidence indicates he was well ensconced under van Rensselaer's wing at least until the beginning of 1824, when one of his friends who was a composer prevailed on him to interpolate one of his compositions into the van Rensselaer Christmas musical for 1823.[6]

The date of Morse and Henry's first in-person meeting arises first in a public context in the course of a contentious patent infringement case that was heard before the Supreme Court in 1853.[7] What becomes clear is that Morse would often go to absurd lengths to deny Henry's influence. So, whether they first met in 1823, 1830,[8] 1837 (as Henry claimed), or 1839 (as Morse claimed),[9] clearly Morse felt strongly that it was in his best interests to portray any such meeting as late as possible.

The court case had been brought by Henry O'Reilly, one of the former contractors for the national network; O'Reilly, like Joseph Henry, by this time had become a serious thorn in Morse's side. In a deposition that had been

15. Endless Debate

elicited on O'Reilly's behalf, Henry asserted having first met Morse in New York in 1837, to which Morse rather pointedly replied, "It will be perceived that Prof. Henry has misrecollected the time and circumstances of our first acquaintance, making the date 1837 in New York, instead of 1839 in Princeton, and he leaves the impression that the time of giving me his certificate letter was in 1837 instead of 1842 the true date as the letter itself shows."[10] So though we have certainly placed Morse and Silliman, and almost certainly Henry, in the Rensselaer Manor in August of 1823, if we are to believe Henry and Morse's own words, apparently no meeting between the two men occurred at the time.

Addressing Henry's possible contribution to the telegraph, Morse had written:

> 1st I certainly shall show that I have not only manifested every disposition to give due credit to Professor Henry but under the hasty impression that he deserved credit for discoveries in science bearing upon the telegraph. I did actually give him a degree of credit not only beyond what he had received at that time from the scientific world but a degree of credit to which subsequent research has proved him not to be entitled. 2d I shall show that I am not indebted to him for any discovery in science bearing on the telegraph and that all discoveries of principles having this bearing were made not by Professor Henry but by others and prior to any experiments of Professor Henry in the science of electro magnetism. 3d I shall further show that the claim set up for Professor Henry to the invention of an important part of my telegraph system has no validity in fact.[11]

There can be no doubt that when Morse first embarked in earnest on the telegraph project, he looked on Henry as a mystical, almost oracular authority[12] (an attitude that, while well deserved, Henry seemed to encourage and even find amusing). On the other hand, he had found Henry personally (as perhaps befits an oracular being) somewhat aloof and distant, saying somewhat disappointedly of him (on the occasion they met in July 1842), "He is not of an enthusiastic temperament but exceedingly cautious in giving an opinion on scientific inventions."[13] Leonard Gale of NYU was also an admirer of Henry's. Up until the rupture caused by Vail's book, Gale had relied on Henry time and again as the final arbiter of any important scientific question that Morse had brought to him,[14] so for him to take sides against Henry must have been emotionally wrenching.

Morse's admiration for Henry was clearly not shared by Alfred Vail. Vail's animus toward Henry stemmed from an incident in mid–February of 1844. The underground Test Telegraph attempt had clearly failed by December of 1843, and Morse was reevaluating whether to abandon it and go to an overhead pole system. If he did that, he would be faced with a new problem, that of keeping the wires from grounding out at the pole. Having recently promoted Cornell to mechanical assistant, a position equivalent (at least in theory and in pay) to Vail's, Morse had stirred further competition, tasking both his assistants with coming up with a design for pole insulators—invoking Henry as the final arbiter as to whose solution would be better.

Going to an overhead system required a different way of thinking about the manufactured problems that had plagued the underground cable. If they went to an uninsulated, suspended bare wire system, this would still require some kind of barrier to insulate the wires where they met the pole. Morse had visited Henry in February at Princeton to collect his opinion on the subject (and a few other matters), carrying with him the two versions of a pole insulator—one designed by Vail and the other by Cornell. Cornell's solution employed glass knobs from the bureau in Fletcher's rooming house in D.C., and Vail's was some insulator embedded in an asphaltum contraption. Morse had preferred Vail's and to this point fully intended to use it, but when Henry expressed a clear preference for Cornell's, this had caused Morse to change his mind.[15] A perhaps somewhat apocryphal account has Henry shaking his head while holding up Vail's device saying, "This will not do, you will meet the same difficulty you had in the pipes."[16]

Henry's preference and Morse's eventual adoption of Cornell's "bureau-knob" insulator would result in a vendetta against both Cornell and Henry that manifested itself, in the case of Henry, in Vail's virtual total omission of Henry's contributions from his 1845 book on the history of the telegraph. Henry had ascended by then to the position of secretary of the Smithsonian and could have brushed it off (since, scientifically speaking, Vail was a nobody), but it had infuriated him nevertheless.

As Morse portrayed it,[17] he had only acceded to Henry's opinion under duress. In all likelihood Morse was simply seeking a way to employ Cornell's solution without offending Vail. Not only was it more elegant, it was cheaper and did not require entering into any new contracts with suppliers to

Joseph Henry, first Smithsonian secretary. Photograph card portrait by Titian Ramsay Peale (courtesy Smithsonian Institution Archives. MAH-10603).

produce it since the required parts were commercially readily available, all of which Morse would have immediately recognized. Vail's revenge for this perceived slight would be taken also in somewhat of a devious manner (leaving Henry almost entirely out of his 1845 history of the telegraph). This in turn would be the root of the public rupture between Morse and Henry that resulted in vituperative exchanges in the 1850s.

So Morse, the portrait artist, had gone to great lengths to paint Henry out of the picture, first through Vail's purported history, which, though Morse claimed to have had no hand in it, was clearly a case of intentionally revisionist history. Secondly, and more directly, in the context of the patent battle cases of the 1850s: Henry's appearance on behalf of O'Reilly in Boston, even by deposition, had been so odious to Morse, that Morse felt impelled to attack him personally, saying, among other things, that Henry's poor memory was to blame for this and several other mistakes in his testimony. Morse scrupulously traced his intellectual paternity, as far as the telegraph and even electromagnetism were concerned, in a manner specifically calculated to exclude Henry,[18] pointing instead directly to Dartmouth Professor James Freeman Dana and specifically to the lectures at the Athenaeum. Morse says, "He [Henry] says I commenced the study of electro magnetism in 1827 ... whatever of information respecting the science of electro magnetism is contained in these lectures could not possibly have reached me first through Prof Henry. They first came to my knowledge through Prof. Dana whose lectures I attended in 1827."[19] Morse had summed up his true opinions on their comparative contributions in a letter to Horatio Hubbell: "A claim for the original barren thought, however brilliant, is comparatively of little account in the eyes of the world. It is he *who first combines facts*, plans and means to carry out a brilliant thought to a successful result who in the judgment of the world is most likely to receive the greatest credit."[20]

So, while Morse's paranoia and propaganda was not limited to Henry alone, it is worthwhile to take a further look at the veracity of Morse's claims with regard to Henry for a moment. According to most biographers (and Morse himself), it was on his return trip from England aboard the *Paquebot Sully*, the autumn following the appearance of Henry's article in Silliman's journal, that the idea for the telegraph sprang from his brain, "Minerva-like,"[21] following a brief conversation on deck with Charles T. Jackson (Jackson would later attempt to claim credit for the idea himself). In fact Henry's ground-breaking article in Silliman's *Journal*[22] had appeared over a year-and-a-half prior to Morse's more dramatic, oft-repeated stroke of inspiration at sea.[23] It had a detailed description of his experiments in deflecting a compass with a magnetic pulse at a distance of ⅕ mile that Henry and Ten Eyck had conducted on the third floor of the at the Albany Academy starting in 1827.[24] When confronted with the coincidental timing, Morse denied any knowledge of Henry or of the article based on his being in Europe at the time. The possibility that this article could have conceivably influenced the telegraph is only logical, but Morse, in an ostrich-like argument, claimed it had nothing whatsoever to do with his insight. He said: "Let me for a moment grant, for argument's sake, that it is ... suggestive of my Electro Magnetic Telegraph,[25] what are the *probabilities* that it could reach me in Italy or France, where I was then residing, so as to influence my mind in the conception and construction of the Telegraph on my return voyage in 1832! The probabilities, it will be seen, are certainly very small. The *fact* is, it did not come to my knowledge until *five years after* my return, (in 1837)."[26]

Morse's "asked and answered" set of interrogatories are designed to support his supposedly ironclad claim of ignorance of both Henry and Henry's article (based on his being away in Europe at the time), appear disingenuous at best. Silliman's journal was distributed widely throughout America and Europe at the time, and it had brought Henry a degree of national and international attention of which no doubt Morse, given his avid interest in the subject of electromagnetism evidenced by his transcripts of Dana's lectures and his close relationship with Silliman, would have become aware well before his ocean voyage back to the United States.[27] Henry's colleague and friend from NYU, Leonard Gale (as was his habit throughout both their long associations), would in this case play the willing foil for Morse's absurd disclaimers,[28] asserting he had introduced Morse to Henry's article at a much later date. According to Gale, when he first showed Morse Henry's article, Morse had exclaimed with "great surprise" at the idea of overwinding.[29] Morse was evidently counting on these attempts to project a total and conveniently all-enfolding ignorance, one that was difficult if not impossible to contradict.

As for various reasons Morse's claims of personal or professional ignorance of Henry were tissue-paper thin, they were still more substantial than Henry's disclaimers of earlier acquaintance with Morse. Henry was no paragon when it came to being truthful about his personal circumstances. During his early days in Albany it was his habit to obscure his age, circumstances and familial relationships.[30] He was so ashamed and embarrassed of his humble upbringing that he often gave the incorrect year for his own birth to deflect any connection to his father (a notable alcoholic).[31] Perhaps the role of household servant/tutor for van Rensselaer embarrassed him as well, and so he preferred it remain entirely unmentioned. (In any case, it certainly was not a posting the esteemed secretary of the Smithsonian Institution would care to have underlined in his past.) Certainly Morse would not have made reference to it first and lacking any testimony of the protagonists, we can only speculate on the possibilities.

15. Endless Debate

While Henry usually did not deign to stoop (in public at least) to public retaliation, reserving his skepticism for snide comments delivered in the classroom to his Princeton students, Morse's unprecedented attack in *Shaffner's Telegraphic Journal* of 1855, a transparently self-serving exercise, required a response from some quarter, and so he had delivered to the regents of the Smithsonian a measured rebuttal of Morse's charges.[32] The regents subsequently convened a special committee to review Henry's communication and delivered an opinion unqualifiedly critical of Morse: "The first thing which strikes the reader of this article is that ... it is simply an assault upon Professor Henry, an attempt to disparage his character to deprive him of his honors as a scientific discoverer, to impeach his credibility as a witness and his integrity as a man. It is a disingenuous piece of sophistical argument such as an unscrupulous advocate might employ."[33]

Besides Henry and Gale, Dr. Charles Grafton Page of the Patent Office were the only other individuals with sufficient scientific credentials to pose any continuing real and credible threat to Morse's claim to sole authorship of the telegraph. Page, as it turned out, could be bought, and relatively on the cheap.[34] Page would baldly promise Morse his "warmest advocacy whenever the occasion may require."[35] As chief patent examiner, he had knuckled under to Morse, granting him a patent for a device, the magnetic relay, for which Cornell had already made a reasonable claim, again making no pretenses about his favoritism. When Page came to regret those decisions, publicly opposing Morse in the 1860 relay renewal battle, it was already far too late.

Ezra Cornell, "the Hickory Quaker,"[36] on the other hand, was not one who could be bought or intimidated in any fashion, but neither was he really credible as a scientific inventor or innovator. The way Morse treated Cornell when he felt threatened in this area allows us a front-row seat to all of Morse's deviousness and serves as a virtual primer on how to maintain ownership of intellectual property through personal influence, intimidation and manipulation of the patent and legal systems. That being said, technically at least, Cornell was Morse's employee, which neither Henry nor Page ever was (though Morse did at one point lend Henry a ten-mile spool of wire for his Princeton experiments, which debt he brought up on several occasions). As for Cornell, Morse needed his expertise, and this became in a sense the basis of a classic love-hate relationship. In later years, once he had become rich from the telegraph and stood atop on his own major achievement, that of founding a great university, Cornell would conveniently forget the intellectual larceny, and abjectly bow to even the most egregious of Morse's claims.

16

Quantity vs. Intensity

As Morse would later (somewhat accurately) claim, "In 1824 the voice of science in Europe had declared the idea of an Electric Telegraph to be CHIMERICAL."[1] By 1825, the periodic, feverish attempts in Europe to create a workable electromagnetic telegraph by Ampere, Sturgeon and Barlow had largely fallen by the wayside. The reason was the diminution in strength of the current over distance, a problem for which there seemed no remedy. English researchers Peter Barlow and William Sturgeon had realized some measure of success by incorporating Johann Schweigger's galvanic coil (sometimes called a "doubler") into their tests, but disappointingly found that even with Schweigger's new device, after traveling only 200 feet, the magnetic force diminished to the point where it was virtually imperceptible. Barlow would moan, "In a very early stage of electro-magnetic experiments it had been suggested that an instantaneous telegraph might be established by means of conducting wires and compasses.... I found such a sensible diminution with only 200 feet of wire as at once to convince me of the impracticability of the scheme."[2]

The root of the problem was that the experimenters of the day still conceived of electricity essentially as a fluid similar to water,[3] and therefore governed by the same Newtonian laws as governed any liquid relating force to mass and acceleration, and in the case of flow, particularly pipe diameter and volume. The next major European experimenter in the field of telegraphy following Barlow, William Cooke, remained captive to this false analogy, attempting all kinds of methods to push more "liquid" through the "pipe" (the wire) on the assumption that if increasing the diameter of a pipe increased the speed of water, likewise, increasing the diameter of a conductor should enhance the strength of the electrical current. When Cooke proposed this method to enhance the electromagnetic effect at a distance,[4] even the genius Faraday (who would himself be the one to eventually prove Coulomb's "two-liquid" theory

16. Quantity vs. Intensity 87

wrong) did not see fit to correct him, but rather encouraged Cooke in what was a rather misguided endeavor.[5]

What neither Cooke nor Barlow nor Faraday realized at the time was that, unlike physical objects, electromagnetic force had two qualitatively distinct parameters that did not follow the easy paradigm of a liquid in a closed pipe, not even of a bi-directional fluid as Faraday had proposed. Henry would recognize the utility of this distinction and designate the two parameters "quantity" and "intensity" (roughly what we know today as amperage and voltage), and while he was not really the one who invented the concepts, he was the one who figured out how to best exploit them.[6]

Henry had no doubt attended Barlow and Sturgeon's repeated failures with increasing interest through the various scientific journals acquired by the Albany Academy library.[7] Interest in electromagnetism had not abated, but work specifically on a telegraph had stagnated for several years until two intervening events made such attempts perhaps again practicable. The first was Henry's own improvement to Schweigger's coil,[8] which he called "The Galvanic Multiplier," and the second was the groundbreaking work of Georg Simon Ohm.[9] Ohm had provided a better understanding of the relation between voltage, amperage and resistance in a circuit, and this had profound implications for the renewed viability of the telegraph. It also allowed for vastly increasing the strength of electromagnets, which was something only Henry seemed to realize initially. Ohm's Law stated that the strength of a current at a given endpoint would be proportional not only to the initial amount of current and the distance traveled, but also to the internal resistance of the generating source (battery) in relation to the load. Clearly this was different from how water behaved in a pipe.

Sturgeon's earlier efforts had employed a single battery with just two compartments and two large plates. Based on Ohm's theories, Henry realized that, with the use of a trough (series or Cruikshank) battery with more (in the initial case, twenty-four) compartments and smaller plates and hooked up to Schweigger's galvanometer, the effective distance of a signal could be extended greatly—in fact, almost indefinitely![10]

Thus Ohm's work had provided essentially an entirely new tool for understanding electromagnetism, recasting the problem from one of battery power to one of overall circuit configuration. What Henry realized was that, by dividing the measure of strength of a current into those two parameters of quantity and intensity, one could thereby manipulate the strength of the current merely by adjusting internal resistance (voltaic potential) at the initial point of differential (the battery). Though Henry's trough battery possessed roughly the same total reactive surface area as Sturgeon's, the strength of current it produced was therefore quantitatively different.[11] Calling Sturgeon's configuration a weak "quantity" battery, he named this new one an "intensity" battery.[12]

Less talented investigators might have stopped here to garner plaudits, but Henry, intrigued by the ideas he was uncovering, plowed on, realizing also in regards to his own work (once again through the application of Ohm's principle), that a "Quantity *battery*" would work best with what he now called a "Quantity *magnet*," meaning a magnet wound with several coils in parallel (producing the lowest resistance), while an "Intensity electromagnet" would work best with an "Intensity battery" (one whose magnet coils were wrapped in series around the iron bar, thereby producing the highest resistance for the length of coil used).

Henry lectured[13] and created demonstrations of these principles (with Beck and Ten Eyck's assistance) for his students at the Albany Academy by running a strand of wire through his third-floor classroom and eventually around the chapel and around the basement of the school, a distance of about a mile and a half.[14] In addition to substituting the "Intensity Magnet" at the sending end, he augmented the motive power by incorporating a "Quantity Magnet" at the receiving end. He found he could thereby project a signal strong enough to ring a bell over that entire distance, demonstrating at a stroke an effect that shattered the limitations that thus far had hobbled the European investigators.

There was yet another development that upended the discouraging assumptions that to that point had gone unchallenged. Barlow and Sturgeon both had estimated the fall-off in current followed the mathematical inverse square rule. Henry and Philip Ten Eyck (another instructor at the Academy) contrived experiments showing Barlow's supposition about the rate of fall-off was wrong; that it was actually directly proportional to the distance traveled, not the square of that distance. So, by the end of 1827, Henry, with some help from Ten Eyck and Lewis Beck, had destroyed virtually all the assumptions mitigating against the development of a telegraph as a practical commercial device—and then he dropped it, turning his attentions instead to creating stronger and stronger electromagnets for Silliman at Yale, and then to investigating the problem of electrical self-induction. Having proven such a signaling device was workable *in principle*, pursuing the obvious commercial implications to their end seemed not to have interested him.[15]

To suggest that Henry was entirely above trying to profit from any of his ideas is false—rather, it seems he would become so all-consumed in whatever intellectual problem he was attacking at the moment, he had little energy to spare for considering the commercial implications. Once the underlying problem was solved, he had already moved on to the next intellectual challenge.

In 1827 he had attempted to market a logarithmic scale of chemical equivalents, a copper slide rule for deriving the weight of elements that he had invented jointly with Lewis Beck. The reason he dropped this project seems to lie with his difficulties of finding a reliable manufacturer (or perhaps he feared embarrassment regarding errors in the device).[16] He also attempted to extract payment for the use of his electromagnetic ore extractor by several mining operators.[17] However, when he tried to collect royalties, his tone was almost embarrassed: "I have heretofore been perfectly free in giving to the public any Knowledge I might possess of use and ... considered it almost below the dignity of science to ask pay for my Knowledge."[18] It may have been that by then, having absorbed the influence of Charles Babbage, he was just not inclined to seek profit from his endeavors beyond what his position as a researcher afforded.[19] Or perhaps he regarded discoveries of fundamental scientific principles and natural forces simply not eligible for patenting (whereas the chemical logarithmic scale was an ingenious device embodying a previously known principle—Wollaston's.) Perhaps he misconstrued the use of patents as pertaining only to novel physical devices and not to the discovery of the utility of abstract scientific principles in a novel context. Whatever the case, he would leave the practical exploitation of his ideas to others who could better navigate the treacherous waters of commerce and politics.[20]

Though tricked out with his port-rule and scribing apparatus rather than Henry's simple bell, Morse's early telegraph, though truly more a tele*graph*, was in a sense nothing more than a version of Henry's remote bell-ringer. To avoid having his invention traced back to Henry, Morse had to invent, in addition to the apparatus itself, what was essentially a fairytale to go along with it. Morse succeeded where Barlow and Sturgeon had failed only because a colleague at NYU, Professor Leonard Gale, introduced him to Henry's "Intensity battery" to use in place of the weaker "Quantity batteries" employed by his English counterparts. All of a sudden, virtually overnight, the telegraph was a practical project.

Over time, the fairytale was extended to encompass not just Henry's contribution, but that of others as well. But the most significant of these was the quantity/intensity distinction. Morse was fundamentally a visual, not an analytic intellect, and the quantity/intensity distinction, while known to him, was analytically complex. It was thus left to the unlikely figure of Ezra Cornell, who at the time was Morse's mechanical assistant, and whose main contribution to that point had been digging the ten miles of ditch for the Test Telegraph line, to translate the Quantity/Intensity distinction for Morse in the form of a sketch of an actual relay device, and it really was not until Morse saw this sketch that he fully grasped the utility of Henry's distinction in relation to the telegraph.

Just days after Morse's return from Europe in 1845, Cornell would patent and introduce his own version of a peripheral circuit called the "Cornell Magnetic Relay." The patent, however, had been so poorly and hastily executed that it was a relatively simple matter for Morse, once he returned, to circumvent it, thereby ensuring Cornell would never receive credit. Just as he had done previously when he felt his intellectual authorship challenged, Morse had resorted to acrobatic contortions to lay claim to the relay as his own. As Cornell by then suspected, Morse had planned this whole series of events, having brought back with him from Europe an improved version of the device created for him by European craftsmen to submit under his own patent. Morse had gone to great lengths not to share with Henry the public adulation he felt he had earned. Having come to Morse as a "ditch digger," if Cornell was attempting to ascend to the same stage as Morse as an "inventor," to share the spotlight, (as he would somewhat painfully learn), this was not something the "Lightning Doctor" would tolerate kindly, if at all.

Although Gale's advice to change over to an Intensity battery was what had made the telegraph feasible in the first place, there is little evidence that Morse had any real theoretical grasp of the advantages of an "intensity" over a "quantity" circuit. Thus, aside from the battery, he had failed to incorporate Henry's insight fully into his Test Telegraph until Ezra Cornell presented them to him in the form of a device.[21] Secondly, where Henry, even in his early experiments, had relegated the "receiving" chore to a secondary "quantity" battery, Morse (by his own later testimony), continued to rely on in-line current boosters to increase the power of the initial signal on the Test Telegraph at least until 1845 (and probably in actuality until 1846).[22] So when Morse referred to a "combined circuit" in his court testimony and in his later deposition of 1860 for the patent office, identifying it with the relay invention, he was actually referring to his clunky in-line repeaters that joined two sets of circuits at the pole, not the more sophisticated (relay) device (as conceived by Cornell) that sat solely at the receiving or sending end. Morse was obviously relying on the similarity in terminology and rather arcane distinction between the two to intentionally confuse those who were questioning him.

Clearly Morse had tried to obfuscate the omission by deliberately employing a vague term, "combined circuit," that could be construed in any number of ways, particularly by a non-scientist. He would then adduce the testimony of Leonard Gale to back him up on this. When he first asserted this, it was in rebuttal to Henry's deposition in the O'Reilly case (and then again when seeking a renewal of his patent for the electromagnetic relay in 1860). Both times he would claim to have incorporated the "Quantity/Intensity" distinction as early as 1837, but had chosen intentionally to leave it out of the original patent;

16. Quantity vs. Intensity

there is little but Morse's word to support this.[23] Furthermore, as patent commissioner Thomas pointed out in 1860, if in fact he *had* known of it and its application to the telegraph and chose to leave it out voluntarily from his original patent, this would have been grounds for revocation of the underlying patent.[24] Once again Gale had sworn to this version initially, and then when the implications became clear, the claim was quickly retracted.[25]

Thus the question of Henry's influence concerning the quantity/intensity question remained a hot potato, and one in which Morse was caught between the devil and the deep blue sea. Clearly Morse's claim that he had first conceived of peripheral circuits back in 1837,[26] and only left it out of his 1838 patent because it was unnecessary at the time, rings hollow.[27] Gale, in defending Morse's claim from Henry's possible influence, had gone to even more ridiculous lengths: "Prof Morse could not have derived this idea from Prof Henry for the following reasons viz. First. He did not become acquainted with him, as appears by Henry's evidence, until late in 1837 or early in 1838.... Secondly. He did not find a trace of it in Henry's article of 1831. On the contrary, that article ... must have tended to convince him that no such expedient was necessary."[28] To anyone who has read Henry's article, the second point could only be considered ludicrous, and the first, as pointed out elsewhere, is merely self-serving and sophistical.

The real question for the time being, however, was how could someone with Morse's inventive skills have effectively ignored such a critical distinction as the Quantity/Intensity one for so long? Morse was obviously acquainted with the concepts, at least by name, as it had been the basis one of his interrogatories prepared by Morse (or for him) for his meeting with Henry in July of 1842 (see Appendix).[29] Henry's answer had simply been, "Ohm has determined it."[30] Henry had always maintained the utility of his ideas for the telegraph should have been immediately obvious to anyone who understood the underlying principles of electromagnetism, which clearly included the Quantity/Intensity distinction; and though it may seem harsh to say, the nature of Morse's questions seem to indicate a lack of basic understanding of the by then well-established conceptual underpinnings of electrical circuits.[31] In actuality, as we shall see, it would be left to Morse's "mechanical assistant," Ezra Cornell, to realize and effectively introduce Henry's critical distinction into the telegraphic apparatus.

17

The Barnaby Mooer Side-Hill Plow

Ezra Cornell's benefactor and employer, after his family relocated to Ithaca from the Quaker enclave of DeRuyter, New York, had been Colonel Jeremiah Beebe, the gruff, curmudgeonly patriarch of Fall Creek. Cornell had moved quickly from handyman for Beebe's mills to supervisor for all of his varied enterprises. When Beebe suffered financial reverses in the panic of 1837, he announced he was retiring from business and closing the mill. Rather than pursuing one of the various trades he had learned while under Beebe's employ, Cornell decided (with the help of a loan from Beebe) to undertake a business venture of his own. New versions of agricultural plows with interchangeable steel parts were just coming onto the market, and the young, enterprising Cornell evidently saw his future reflected in the bright gleam of the carbon-steel blade of the Barnaby Mooer side-hill and flat-land plow.[1]

Though the parting from Beebe had been less than cordial, with a growing family to feed, Cornell needed additional backers in a hurry. Despite somewhat ambivalent feelings, he turned again to the now poorer but still well-connected Beebe, who put him in touch with a slippery ex-congressman from Portland, Maine, by the name of Francis O.J. Smith. With Beebe and Smith's financial underwriting, Cornell had purchased the patent rights on the plow for Maine and Georgia, and in late January of 1843, armed with a "golden list" of standing orders (courtesy of Mooer), some advertising leaflets, a letter of introduction from Beebe to two representatives in the Congress,[2] and a few dollars in his pocket, Cornell set out on a sales trip that eventually would cover over 4,000 miles, up and down the East Coast, much of it by foot.

Traveling first from Ithaca to Maine, he had laid out his prospective plan for acquiring (by ex–Quaker standards) immoderately quick wealth to his new underwriter, ex-congressman and then new editor of the *Maine Farmer*,

17. The Barnaby Mooer Side-Hill Plow

F.O.J. Smith.[3] From Portland, he had headed south, first to New York and then by ferry to Philadelphia, arriving there on the evening of February 3. From Philadelphia he headed toward Baltimore, where he took another steamer, the *Pocahontas*, to Richmond, Virginia,[4] arranging there to have a plow shipped to him in Plymouth, North Carolina, care of his cousin, John Cornell.[5] He arrived in Plymouth on the 8th, having in the interim already acquired an intense dislike for smoky railroad cars that he likened to "a cage," and began walking the countryside, Barnaby and Mooer leaflets in hand.[6] While the environs were pleasant enough, there was little money to be made. On the 26th he arrived in Charleston, South Carolina, once again by steamer.[7] Having apparently engaged in a prodigious amount of vomiting while at sea,[8] once safely back on land he joked that he felt obliged to weigh himself to "see how much I have fallen away."[9] With a thinning purse and thinning frame, he had determined to make most of the rest of the trip through the South by foot.[10]

Sales of the B&M plow were slow to nonexistent. The three he had sold in Charleston had been almost at his cost.[11] The reputation among the Southern dirt farmers and skeptical, stubborn sharecroppers regarding Yankee innovations and smooth-talking salesmen had made Mooer's golden customer list relatively worthless, and the money he had saved on railway tickets had been expended on the additional room and board occasioned by his choice of slower modes of travel. Despite all this adversity, the letters to his wife MaryAnn remained chatty and upbeat, evincing convincingly good spirits, describing his tour of the local Navy Yard in Charleston for his wife and children in detail.[12]

In the meantime, on February 22 a gruesome event had occurred back in western New York, one affecting the entire Cornell family. It was the brutal murder committed by Ezra's cousin, Alvin Cornell, at the Quaker Day School located near Jamestown, New York. The victim had been his wife Deborah.

The Quaker school was run by Mary Osborne Cornell, and Alvin had been teaching there on and off, in between bouts of crippling depression. He had just returned from visiting the Michigan branch of the Cornell family, where he had discussed his intentions, when he murdered his wife. Quick thinking and swift action had prevented him from also taking the life of Mary Osborne as she locked herself in the kitchen while Alvin raged on. He then had attempted to kill one of his deaf and dumb daughters, who somehow managed to escape, and then himself, slashing his own throat ear to ear.[13] Having failed, however, to sever an artery, he was attended by a local physician named Gilbert Hazeltine and so survived to stand trial for his misdeeds.[14]

All the gruesome details emerged at trial and were subsequently collated into a detailed assessment of Alvin's precarious sanity by an Albany physician named T. Romeyn Beck. Beck's report would be used to determine if Alvin should be hanged, as he had been sentenced, or serve out the rest of his years in a facility for the criminally insane. MaryAnn had been Ezra's sole source of news and gossip about the family during his present sojourn to the Southland. Shortly following the Jamestown murder, all correspondence from her had abruptly ceased. Whether the sudden drought of information was related to the shock over the events in Mayville or Ezra Cornell's own frequent changes of address, it appears Ezra, for the time being, remained blissfully ignorant of the recent blot on the family name.[15] When the correspondence did not resume after several weeks, Cornell had been reduced to begging, writing MaryAnn in an uncharacteristic tone of chiding disapproval, "I had been in expectancy of a letter from you for some time. I feared [you] might some of you be unwell."[16] There was still no mention of Alvin or events in Chautauqua, though possibly he had heard of them through other sources by then, but they were omitted from MaryAnn's eventual reply.[17]

In the meantime other events were transpiring which were soon to have a profound effect on Ezra Cornell's future endeavors. The day after the murder in Chautauqua, on February 23, the House had passed Morse's appropriation for the telegraph, 89 to 80.[18] Once the bill had passed the full Congress, Morse began enlisting personnel, lining up suppliers and obtaining the rights of way for the telegraph project. Logic would have seemed to dictate he attempt to create an overhead wire system on overhead poles, as had been utilized in England, but encouraged by Colt's successes at the Castle Garden demonstration a year earlier, Morse had opted for a buried cable. If a line could be made to function underwater, it could certainly be made to function underground. Cooke and Wheatstone's telegraph in England had gone to overhead wires only after having failed with the buried cable approach,[19] but to Morse, the perfectionist, an overhead cable seemed to present too many uncontrollable variables. Aside from vandalism and weather, possibly the artist in Morse abhorred contemplating the aesthetic effects it would inevitably wreak on the landscape of the country if he succeeded: an army of naked poles marching across previously pristine vistas of the country. Plus, this offered the pleasant prospect of rubbing his success in the face of his rival, Wheatstone. So, even before the success of the appropriation, Morse had determined on an underground solution and needing someone to trench out the line bed. For this he had turned to his former partner now in Maine, F.O.J. Smith.[20]

His reluctance to use an overhead solution stemmed from something that Joseph Henry had mentioned somewhat offhandedly to Leonard Gale. Aside from the weather and problems with possible line breakages due to vandalism, Henry had warned of the possible effects of what he called "Celestial Magnetism." "He has been experimenting making discoveries on Celestial electricity, and he says that Wheatstone's and Steinheil's telegraph must be so influenced in a highly electrical state of the atmosphere as, at times, to be useless."[21] Morse had exulted in this manner to Smith, at Henry's confirming his wisdom in not following Wheatstone's example.

Heeding Henry's warnings in this regard may have seemed eminently practical, but in reality the conclusions offered in this case were highly speculative ones.[22] Henry was certainly correct about the phenomenon of atmospheric electromagnetic induction,[23] but at this point it was far too early to determine its actual effects on any proposed telegraph lines, nor had he begun any experiments on this score, either, but for the moment Henry's status as "the oracle of Princeton" was sufficient to convince Morse that adopting the overhead approach would result in his ruin and failure of the entire project.[24] In reality, the reverse would almost prove to be true.

18

Catching Colt

The panic of 1837 that had left the patriarch of Fall Creek, Jeremiah Beebe, in Ithaca teetering on bankruptcy had also gripped the country in a withering depression. By 1839, under the burden of heavy debt, Beebe would begin close his mills and Ezra would take up farming. Morse had been plunged into a depression of his own. Not having been picked to create one of the large canvases for the Capitol Rotunda, he seemed to be casting about for some sense of purpose, occupying himself by taking daily weather readings.[1] His old mentor, Washington Allston, who had been advising the committee selecting the Capitol Rotunda commissions, had written Morse in disgust, and with the clear suggestion that Morse follow suit, saying, "I have done building castles on [the] public ground."[2]

The depression seemed to have crippled all Morse's ambition, not just for painting, but also including that which he still harbored for the telegraph. When in February the Congress had called for proposals for a national telegraph, Morse had not responded until after an article appearing in the *Journal of Commerce* in August had noted his efforts with the telegraph.[3] The positive mention seemed to revive him,[4] prompting a flurry of activity. It was as if he were a dreamer having woken from a sleep; several replies went from Morse to the *Journal* regarding his "Morse Code"; an inquiry was sent to his former classmate, now Patent Commissioner Henry Leavitt Ellsworth,[5] about obtaining a patent caveat[6]; and Morse had, albeit belatedly, responded to the congressional interest. By the end of September a detailed reply on the history of the telegraph and its prospects had been sent to Levi Woodbury, then Secretary of the Treasury, who was the one who had sent Morse the notice regarding the congressional exploratory effort.[7]

Morse would ignore Allston's advice about building castles on the public ground. From this point forward he would also more or less abandon his aspirations as an artist and focus on seeking government support for his telegraph.

Still not confident in his design,[8] he had enlisted Alfred Vail, who offered the resources of his family ironworks in Speedwell, New Jersey, to help create a better working model; by November, Morse had been able to send a signal over a distance of five miles.[9] The two had first met earlier in September, when Vail, who was also rooming at the university, walked in on a demonstration Morse was giving of the apparatus at the university.[10] Morse must have quickly realized that Vail had money to burn. More importantly, he was young, impressionable and eager to make his mark on the world, and the two "patricians" had hit it off. By the time Morse received his caveat, on October 6,[11] he had convinced Vail to underwrite the whole effort of re-presenting the telegraph to the Congress in exchange for a quarter interest in the patent.[12]

Apparently Morse had been able to convince Catherine Pattison[13] of Troy to come down to the city and help with winding the miles of wire that would be needed for the demonstration, as Vail had made an inquiry about her future plans and its possible impact on the project.[14] Based on the extensive correspondence between the two at the time and the warm tone of the letters, Morse was obviously pursuing Catherine at this point, and there may have been some friendly romantic rivalry going on between him and Vail. Miss Pattison had retreated to Troy at the end of November and Morse had written again to Woodbury saying that he was planning to come to Washington to demonstrate his telegraph, inquiring what was the latest practicable date he could make his presentation.[15] Morse and Vail finally arrived in Washington early in February of 1838 armed only with the caveat issued by Ellsworth[16] and a supportive but lonely recommendation from the Franklin Institute in Philadelphia declaring the telegraph "worthy of the patronage of the government." They brought the now fully working model of the telegraph, which he and Vail had proceeded to set up in the Capitol. Among those viewing the apparatus would be Theodore Faxton Saxton and John Butterfield (who would later partner with Morse to create the New York-Buffalo line) and Maine Congressman F.O.J. Smith.[17]

Morse had initially sought just $26,000[18] for the project. Smith, then chair of the powerful House Commerce Committee, following several demonstrations of its efficacy before the committee, had passed Morse's telegraph project on to the floor of the House on April 6[19] with a slightly enhanced price tag of $30,000.[20] Smith added his gushing personal recommendation,[21] saying it would bring about a "revolution unsurpassed in moral grandeur by any discovery that has been made in the arts and sciences."[22]

If anyone was ill-qualified to make a statement concerning "moral grandeur," it was Congressman Smith. The nickname "Fog" had been awarded him by his fellow members of Congress, referring not only to the predominant weather in his home state, but also to the dense atmosphere of obfuscation surrounding his personal financial affairs. The editor of another Portland paper, the *Eastern Argus*, having borne the brunt of some of Smith's recent financial shenanigans, would write to Smith, "Whatever report you may send forth to the world will be received with a distrust which will be ripened into disbelief ... if you were my friend ... I should advise you to retire from political life."[23]

So this was the man Morse had enlisted as his main advocate, underwriter and friend (and incidentally, whom Cornell would choose in much the same capacity)—a man derided by his colleagues and business associates alike.[24] Congress finally rejected Morse's bid, and Smith obviously counted this as a personal slur on his prestige. So sincere, though, did Smith's belief in the future of the telegraph appear that he had agreed to forego his reelection bid so he could promote the device unencumbered by his congressional duties.[25]

True to his word, Smith did not seek reelection that year and in fact abandoned Washington and his duties in the middle of the session of Congress. He had instead returned to Maine to run the local paper called the *Maine Farmer*[26] and pursue his legal practice. He had not, in fact, retired entirely from politics, but simply thought he had found a potentially more enjoyable, lucrative and less demanding means of fleecing the government. He was always adept at finding ways to use other people's money to finance his schemes, and the telegraph was another such avenue.[27] He offered to accompany Morse to Europe to pursue a European patent and even pay Morse a salary during a trip. Unfortunately, the trip would turn out to be an extended exercise in futility,[28] and Smith would contribute little in the way of hard cash, paying Morse, as was his habit, mostly in promises.[29] In exchange for this rather diaphanous support, he would eventually receive the lion's share (after Morse) of interest in the patent.[30]

Following the twin failures in Washington and then Europe, Morse entered a period in which he found himself mired in genteel poverty and depression.[31] He was still president of the National Academy of Design, so his failure to gain one of the commissions for the Capitol Rotunda had been a major source of embarrassment to him. Those fellow artists who visited his private loft saw no evidence of artistic activity except a few crayons scattered about the dusty floor.[32] In response to this, his friends had taken up a private subscription for him to paint a large historical canvas of his own, *The Mayflower Compact*, but Morse had never followed through. Somewhat embarrassingly, a journal article detailing his laxity had appeared, asking him to return the subscription money in full or be accounted a fraud.[33] Under the title "Mr. Morse's—'Cabin of the Mayflower,'" the article in the prestigious weekly arts publication, the *Mirror*, read:

18. Catching Colt

> When the four pictures to be placed in the Rotunda of the Capitol were given to Weir, Inman, Vanderlyn and Chapman, much dissatisfaction was expressed by Mr. Morse's friends that he did not receive one of the commissions. A subscription was entered into to procure a picture from Mr. Morse's pencil, of the same size as those intended for the Rotunda. We ourselves were among the subscribers. Not having heard any thing of this picture for upwards of a year, it has just occurred to us to inquire what has become of it? "Is Mr. Morse engaged upon it?" and "when is the picture to be done?" We have frequently had these inquiries made of us and we know no way of answering the question except by putting it to Mr. Morse himself.[34]

Morse himself had begun to entertain grave doubts on the feasibility of the telegraph project and for a time even stopped pestering Smith on the subject. Perhaps to convince himself that all his efforts had not been in vain, in May of 1839, he had arranged for a visit to Henry at Princeton (on the pretext of lending him a spool of wire),[35] intending to question him on, among other topics, the quantity vs. intensity issue that was still gnawing at him.[36] Henry had been nothing but encouraging, and during this visit, he answered all Morse's interrogatories, albeit with the same disconcerting succinctness on which Morse would later remark.[37]

The NYU building, where Morse was renting rooms, was a marble-clad, gothic structure with cantilevered pediments. Modeled on the cloistered style of an Oxford or Cambridge, it lorded regally over Washington Square Park from the eastern edge.[38] At its center, running the length of the structure, was a chapel faced with tall metal-muntinned windows that refracted the long Manhattan sunsets into a series of elegant tracery patterns on the street and the venerable trees that shaded the *nouveau riche* of New York City who had moved to this newly fashionable neighborhood as they went out on their daily walks.[39] The grandeur of the building was undercut only by the collection of rather indigent-looking figures scurrying furtively in and out of its massive oaken doors with bundles of wire, artists' materials, and various other apparatus bulging under their long coats. The university, strapped for funds, had resorted to renting out rooms on the spacious upper wings, and these had become a sort of bohemian enclave combining living space, artists' lofts and scientific laboratories.

Morse had been among the first tenants, along with fellow artist Daniel Huntington; also Alfred Vail, who was a recent graduate, still casting about for some purpose in life.[40] Morse rented five rooms on the third and fourth floors in the northwest corner of the building at $325 per year. He himself occupied the fourth floor and this was where he had both his classes and his telegraph experiments. Apparently the walls had been improperly sealed; the rooms were drafty and virtually weeped when it rained, and this became a sore point of contention during his entire tenancy.[41]

By 1842, Morse was sharing his laboratory space with Samuel Colt, who, after closing his firearms factory at the former SEUM site in Paterson, New Jersey, had relocated to university digs. Eventually Colt would take his own rooms in the south wing.[42] The two had seemed to hit it off, and Colt's project of creating electrically detonated underwater explosives for the Navy meshed well with Morse's goal of creating an electrical signal to communicate over greater and greater distances.

Morse's sudden costume change could not have but stunned and puzzled those of his friends familiar with him only as the deacon of the New York art scene. Virtually the only encouragement he received from anyone during this period (besides Vail) had been thanks to his old friend Congressman Joel Poinsett, who, as one of the co-founders of the National Institute, was in a unique position to help Morse in his new vocation. This came in form of a notice of his appointment in October of 1841 as a corresponding member of the prestigious National Institution for the Promotion of Science,[43] an honor which he now shared with his idol, Joseph Henry. This had prompted a collegial letter from Henry a few months later saying, "My dear Sir I am pleased to learn that you have again petitioned Congress in reference to your telegraph and I most sincerely hope will succeed.... I have not the least doubt, if proper means be afforded, of the perfect success of the invention."[44] For Morse, coming from the oracle Henry, this was virtually a license to proceed. The letter, however, had contained one troubling disclaimer which should have alarmed even an overly-sanguine Morse: "Though ... little credit can be claimed ... since it [the telegraph] is one which would naturally arise in the mind of almost any person familiar with the phenomena of electricity."[45]

The notice of his appointment to the National Institute had arrived at his door almost simultaneously with the squib article in the *Mirror* regarding the subscription painting. Its content probably damped any mood of self-congratulation. After years of focusing solely on the telegraph, Morse had done nothing on the *Mayflower* painting. Morse doubted whether he even had the talent to paint anymore.[46] He maintained a few students, but theoretically this was for instruction in "the literature" of art, not painting per se.[47] The monies raised for the *Mayflower* Compact painting likely had gone to the purchase of the ten miles of copper wire he had kept in his rooms at NYU and which had been lent out to Henry in Princeton. Morse, now honor-bound to pay back his investors, in the midst of the stubborn recession, had been once again left virtually destitute.

His previous efforts, however, had not gone unregarded in Washington. Late in 1841 he had begun to receive several somewhat shady offers to "lobby" for a new telegraph appropriation from a Washington attorney.[48] Now mired in poverty, Morse had grasped onto this as a saving spar. He had written Henry requesting the return of the wire, in the same note begging Henry for a recommendation to be sent along to the Congress.[49] Henry had written back a month later, apologizing for the delayed response, promising to deliver both the wire and the requested recommendation with some alacrity.

The eight-plus miles of wire Henry had returned was quickly loaned out again for the use of Samuel Colt. That August, Colt had been awarded a congressional grant of $15,000 to pursue his submarine battery.[50] The now impoverished Morse probably viewed the loan of the wire as a means to avail himself of Colt's goodwill and also his rather more substantial cash resources. Having almost given up on the idea of the Test Telegraph,[51] he found himself encouraged by Colt's consistent successes in sending electromagnetic pulses over longer and longer distances. He described his flowering collaboration with Colt as follows:

> During the last few months I have availed myself of the means which Mr. Samuel Colt has had at his command in experimenting with wire circuits for testing his submarine batteries also to test some very important matters in relation to the Telegraph. I loaned him in the first instance my two reels of wire which by the by is reduced to eight and a quarter miles....[52] The experiments were highly satisfactory, the magnetism and the heating effects which latter Mr. Colt desired being apparently stronger when the wire was stretched out than when in coil. We also found that *when one wire was coated, the other might be naked and passed to any distance.*[53]

When Colt, with the help of government monies and resources had succeeded in detonating a torpedo with an electromagnetic pulse at a distance of close to ten miles, Morse's hopes had revived sufficiently for him to make another serious attempt at persuading the recalcitrant Congress regarding his own project.[54] This new sense of optimism had been expressed to Leonard Gale, his electrical amanuensis at NYU: "If I can succeed in working a magnet ten miles, I can go around the globe."[55]

Aside from these somewhat prematurely grandiose visions, what he proposed this time to the Congress was far less ambitious than what he had first proposed in 1838: a plan to test the telegraph only within the precincts of Washington, D.C., setting a line between the White House and the Capitol. Writing Representative W.W. Boardman, another former Yale classmate of his, on his proposal, he had asked for a vastly scaled down appropriation of $3500. Boardman had, however, quickly rejected the idea, saying it was far too late in the session to introduce such a bill.[56]

Despite the now three-year-old congressional study, thanks largely to the poor economic conditions throughout the country, American interest in the telegraph seemed to have waned. The anemically attended demonstrations of his working model at NYU must have recalled to Morse the failure of his other grand project, *The House of Representatives*. In July of 1842, following another successful demonstration by Colt,[57] Morse again had sought out "the Princeton Oracle," Henry, soliciting his advice and support, and Henry had provided it (again somewhat stingily). This time Morse had brought a working version of the device with him which apparently had impressed Henry sufficiently, (if not with its novelty, certainly with its ingenuity), that he on this occasion had expressed more than just mild enthusiasm for the project. Morse could not help bragging to Smith about the encounter: "Professor Henry invited me a day or two ago. He knew the principles of the telegraph but had never before seen it."[58]

19

Out of the "Fog" of Invention

Castle Gardens was a beautifully landscaped plot of land, a fairyland of gardens and baths, a retreat for city-dwellers within the city, just off the Battery. Its reflection flickered across the waters of New York Harbor, lingering like some kind of mirage, a flawless gemstone set in the pinky-toe of a far grittier Manhattan. It was also the site for that year's annual American Institute Fair, a month-long event showcasing the best that America had to offer in terms of art, technology and agriculture. This usually bucolic, manufactured scene was, however, about to be rudely interrupted.

On October 18, 1842, as part of the organized activities of the fair, Samuel Colt, with the assistance of Professor James Cogswell Fisher of NYU,[1] had detonated a huge underwater explosion via an electromagnetic signal sent along a tarred cable. It blasted the (aptly named) two-hundred-sixty-ton brig *Volta* (decked out for the occasion with festive pirate insignia replete with doomed figure lashed to the mainmast), demonstrating thereby simultaneously the effectiveness of his new "Submarine Battery" and the just deserts for bad nautical behavior. Colt unfortunately had run out of cable at the last minute and turned to another exhibitor at the fair that year to supply it—his roommate and collaborator "Professor" Morse.[2]

The electrical signal was originally supposed to have been sent from on board the *North Carolina*, which was anchored in full view of the assembled crowd onshore. The papers had reported it as Colt had given it out saying, "The case containing the combustibles was sunk under the hulk, and a wire conducted from it to the deck of the *North Carolina*, distant some two or three hundred yards. At the moment fixed, (1 o'clock) Mr. Colt, on the deck of the Carolina, applied the acid to his plates, and quicker than thought, the doomed hulk was thrown into the air."[3] But evidently neither Commodore Perry nor Colt were on board when the circuit was closed.[4] Nevertheless, despite the sleight of hand of personnel, the audience was duly edified at the explosion and the marvelous sight which Colt described in his own flowery words: "The great bulk seemed lifted by some unseen power, the bow and stern sunk heavily, and the whole was enveloped by a huge pile of dense mist, some two hundred feet in diameter and about eighty high."[5] The explosion had blasted the *Volta* into smithereens, leaving just pieces, none larger than a man's arm.[6]

Morse's own public demonstration for the American Institute had been scheduled for the following day between Governor's Island and Castle Garden. Using the same cable he had set up the day before for Colt, he reconnected it to a receiving apparatus on Governor's Island, where his NYU colleagues Leonard Gale and James Fisher had been ferried out to receive the incoming message.[7] This demonstration had not gone quite so well as Colt's. While Morse did manage to send a few characters[8] to Gale and Fisher, a passing merchant ship snagged the cable and then severed it intentionally so it could pass.[9] The crowd, disappointed by this rather tame and now abbreviated display, especially after the successful fireworks of the preceding day, had booed.[10] As far as Morse was concerned, it was a success. Despite the accident, the idea had been proven, and between Colt's and his demonstrations, the possibility of sending a telegraphic signal across a large body of water was now a reality. If an electromagnetic signal could be sent beneath New York Harbor, just as he had expressed it to Gale, one could be sent round the world.[11]

Despite the embarrassing jeers of the crowd, Morse had been awarded a gold medal by the Institute.[12] Colt's method of making cable watertight by inserting it inside a lead conduit pipe, coating it with a mixture of beeswax and resin, and wrapping it in cotton impregnated with tar, would be adopted by Morse. Whether or not he personally received the credit, a major physical obstacle to the viability of the magnetic telegraph had been overcome—he had his proof of concept, flawed though it might be—if a cable could be made to work underwater, it certainly could be buried in the ground.[13]

Apart from the glitch with the cable, the conclusion had virtually been foregone. Colt had succeeded previously in a dress rehearsal in virtually the same location, albeit with a shorter cable of just a few hundred yards, on July 4.[14] Morse, suffering from ill health at the time,[15] had not attended that earlier demonstration, but it had reinvigorated his own efforts. Just a few weeks later he had invited Smith to New York, saying he had found some mysterious new investor for their project, but that he needed a signed release from all the patent partners to speak and act on their behalf.[16] There was, in fact, no private investor.[17] Morse just wanted to let them know he was finally ready to forge ahead with his stagnating plans to get the government to fund his efforts. Having learned his lesson, this time he first lined up testimonials from all manner of renowned scientists and scientific organizations.[18] For securing these

19. Out of the "Fog" of Invention   105

endorsements he had depended heavily on his two colleagues at NYU, James Fisher and Leonard Gale, as well as Henry Leavitt Ellsworth.

Henry, by now a full professor of Natural Philosophy at Princeton, having witnessed a demonstration of the device earlier in 1842, had lent his (albeit tepid) public support to the project.[19] Morse's colleagues at NYU, Fisher and Gale, had not been idle in the interim either. The model they were working on now at NYU was considerably more sophisticated than the one shown to the Congress four years earlier.[20] Enlisting Vail once again to set up a model telegraph communicating between two rooms in the Capitol (one of them likely the same coatroom where two decades earlier, as a young artist, he had done the miniature studies for his failed masterpiece, *The House of Representatives*, and armed with better knowledge now of the internal workings of the Congress, Morse had mounted his new assault on Washington.

Morse's efforts had drawn immediate attention, this time from the proper parties.[21] On February 21, the House Committee on Commerce, after viewing several demonstrations of the new device, passed a bill on to the Congress[22] who, this time, would fill his outstretched hand with a substantial sum. On March 3, they passed the appropriation in the full amount requested: $30,000.[23] Morse had nearly bankrupted himself getting the appropriation passed.[24] If the bill had failed, Morse was done. He had written Vail on the 23rd saying, "If ... the bill should fail in the Senate, I will return to N[ew] York *with a fraction of a dollar in my pocket.*"[25]

Though Smith had already been awarded a share of the patent earnings, he knew how government projects worked and was well aware that this stake might not produce dividends for years, if ever. Not being a patient man when it came to lining his own pockets, he had immediately begun making plans to avail himself of a substantial portion of the appropriation. With Colt's waterproof method as his ace in the hole, Morse planned to bury a cable between Baltimore and Washington by the end of the year. The trenching part of the operation would no doubt be, besides the wire, the largest expenditure, and being labor intensive, it also offered the greatest opportunity for graft. Smith had leapt into the breach, suggesting that he himself would take on supervision of the trenching operations, thereby relieving Morse of one of the more mundane aspects of the project.[26]

Smith's qualifications in this area, having himself no more than a nodding acquaintance with hard physical labor, were laughable, and it became immediately apparent he had no clue as to how to go about sizing the task. This was properly the province of a qualified civil engineer, not a lawyer/newspaper editor/politician, but when one would offer his services, Smith would reject him.[27] Two months later, Smith provided Morse with what was a ludicrously inflated estimate for the trenching.[28] Morse, not happy, had written back, "I have examined the contract and ... I am not exactly pleased with the terms," going on to say he had spoken to a contractor who would do it for less than one-half what Smith was proposing.[29] Smith had begun looking more and more like a bad bet: not only were his inflated estimates a potential embarrassment, but, in a signal that Morse could not have missed, the Secretary of the Treasury had just recently turned down one of Smith's requisitions for traveling expenses and rather pointedly rescinded an approval for his hiring of a "mechanician." Also, Smith's purchase of seven barrels of alcohol (purportedly to make varnish to coat the wires) had raised some eyebrows.[30]

Morse may not have been aware the extent to which Smith was *persona non grata* in the Cabinet, but he was quick to catch on.[31] With no lack of enemies in Washington, Smith was no doubt also aware that his scheme for tapping the appropriation would eventually be uncovered. When the proposed arrangement with Smith was balked at by the Secretary of the Treasury,[32] realizing he needed a front man, Smith had encouraged his brother-in-law, Levi Bartlett, to put in a bid for the trenching.[33] So, while Morse and Smith continued haggling privately over costs, it was Bartlett's name that would appear on the contract presented to the Secretary of the Treasury.

Smith's approach of keeping things in the family, in this case, worked as a double-edged sword. Bartlett was not only an old acquaintance of Morse's father, Jedidiah, but of an uncle of Lucretia, Morse's deceased wife. Years earlier, while a merchant in Boston, he had married Morse's aunt Clarissa Walker.[34] So Bartlett represented an important familial link between Morse and Smith. More importantly, as far as the Treasury was concerned, Bartlett was all that Smith was not: a reputable and successful businessman who was loved and respected in his home state of New Hampshire.[35] Smith had in fact first clerked for Bartlett's brother, Ichabod, a prominent enough New England jurist for Morse to have painted.[36] As a gentleman farmer he had been a frequent contributor to several respected periodicals,[37] including the (somewhat optimistically named) *Granite Farmer*. He also had established himself as a lecturer on agricultural subjects,[38] and would eventually become an advisor to the Patent Office on agricultural affairs.[39] All in all, Bartlett was a solid citizen and one whose name Morse did not have to blush to put on a major contract to be delivered to those in the government auditing the enterprise.

By the end of May, Morse and Smith had reached an accord on the trenching costs. On June 10, Morse had entered into the contract with Bartlett for $6,000 for the entire forty-mile run, an arrangement that seemed to satisfy everyone concerned.[40] While the contract had been signed with Bartlett, Smith still was pulling the strings, and he and Morse had continued hammering out the details of the trenching well into July, when a discouraged Ezra Cornell, fresh from his failed sales trip, had shown up on the scene.[41]

20

The Plow in Maine

In June of 1843, a tired Ezra Cornell had turned back north briefly, toward Ithaca and home, but soon set out again for Portland, where his main underwriter for the sales trip, ex-congressman F.O.J. Smith, had been in the interim consumed by other matters besides the failing plow venture.[1] By the time Cornell arrived in Portland, it was a blazing hot mid–July. He no doubt expected to immediately be taken to task by his underwriter, Smith, for his poor sales performance, but imagine Cornell's surprise when he entered the editorial offices of the *Maine Farmer* to find Smith scribbling furiously on the wood floor with a fat piece of white chalk—apparently ecstatic to see him.[2] As Cornell put it, "I found Smith on his knees in the middle of his office floor with a piece of chalk in his hand, the mold-board of a plow lying by his side ... various chalk marks on the floor before him…. On my entrance Smith arose, and ... said, Cornell you are the very man I wanted to see. I have been trying to explain to Robertson[3] a machine that I want made, but I cannot make him understand it."[4]

The failure of Cornell's sales trip[5] had overlapped with Smith's machinations over the trenching for the telegraph. Smith's more down-to-earth brother-in-law, Bartlett, had by now no doubt given Smith a better idea of the cost of hand-digging a forty-mile-long trench, and Smith realized, constrained by Morse's demands, he himself stood to make very little. When a lean, sunburned, thoroughly dejected Ezra Cornell had walked into the offices of the *Maine Farmer*, an avenue leading directly back to the broad trough of federal money seemed to have magically opened up before him. If a machine could be devised to lay the cable, Smith might yet be able to pocket the savings. Turning to Morse's pet hobby-horse, "invention," Smith realized he might be able to convert the apparently unsellable plow into an implement needed by the government and in the process skim a good deal more from the Bartlett contract. He set Cornell the task of coming up with this invention.

If anyone knew how to economize, it was Ezra Cornell. His use of the

Barnaby and Mooer sales flyers for his letters home to MaryAnn visually demonstrated his obsessive frugality, and in case that was not sufficient, he had described for MaryAnn, sometimes in excruciating detail, all his daily expenses for food and lodging.[6] As if to further underline the fact, he squeezed the chatty text into every square inch of blank space, back and front, between illustrations and advertising verbiage, in the margins as if to graphically demonstrate his intent to conserve resources. The cramped hand, like his goals in life thus far, had been symbolically reduced to capitalizing on the margins of failure, but all that was about to change.

Ezra Cornell (courtesy Division of Rare and Manuscript Collections, Cornell University Library).

Just as Cornell was setting out for the Southland on what would be a failed sales trip, in February 1843, the Mutual Life Insurance Company opened an office in New York at 44 Wall Street.[7] The cost of a policy in those uncertain times was $100 for a single life for one year.[8] Among trustees of the company were two former directors of SPUA, Theodoric Romeyn Beck and Richard Varick DeWitt, former mayor of Albany and one of the founders of the Albany Academy; Philip S. van Rensselaer; and Ezra Cornell's father Elijah's business partner, director of the Farmer's Loan and Trust Company, Robert Comfort Cornell.[9] On April 12, 1842, the Mutual Life Insurance Company of New York was incorporated by an act of the New York State Legislature.[10]

Despite the hefty price tag for a policy, Mutual's first customers were not among the very wealthy (nor were they drawn from the very poor). Of the seven hundred ninety-six life policies issued in their first year and a half of operation, three hundred ninety-six were to clerks and agents, thirty-seven to brokers, and twenty-five to manufacturers.[11] Despite the fact that he could barely afford food at the time, Ezra Cornell seemed to have suddenly become acutely aware of his own mortality and possessed of an urgent need to purchase life insurance. Lacking the funds, he had instructed MaryAnn to borrow the necessary monies from her father.[12] This stark departure from his customary

20. The Plow in Maine

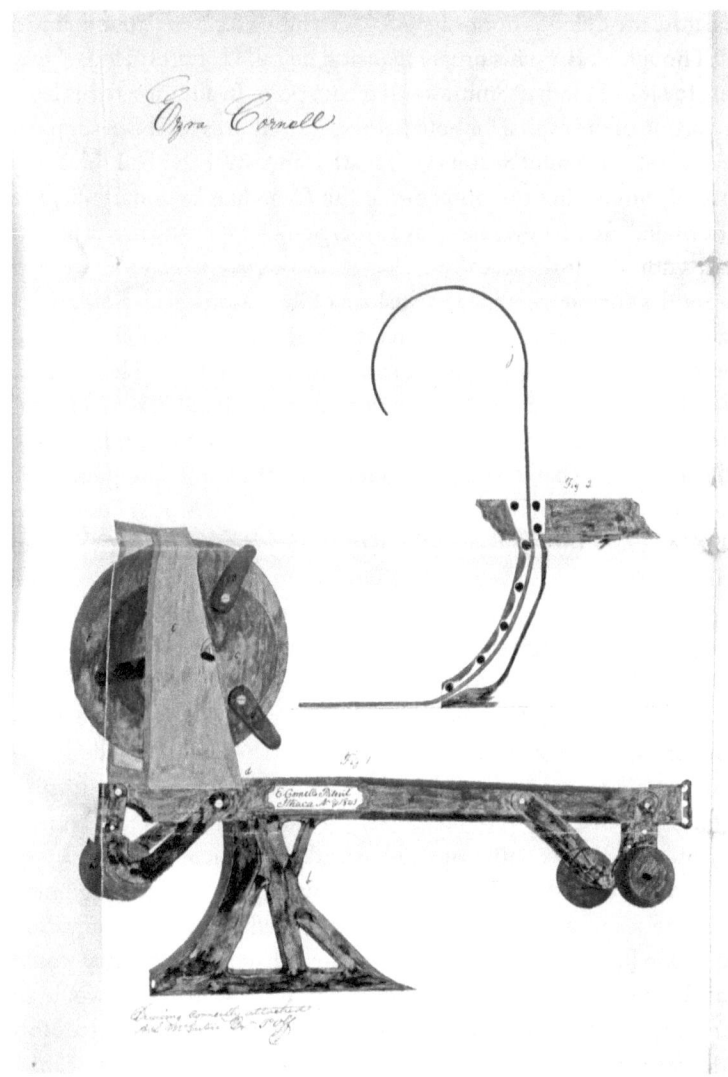

Patent drawing of cable-layer, by Ezra Cornell, (circa) 1843 (courtesy Division of Rare and Manuscript Collections, Cornell University Library).

frugality impacted him so strongly that he felt impelled to dispel the taint of extravagance by resorting to what would become one of his favorite obsessive tics: counting and tabulating—in this case, his travel expenses—to the penny for the entire preceding week.[13]

If anyone was identified with waste and fraud, it was ex-congressman F.O.J. Smith. Before leaving his seat in Congress, Smith had been awarded by his fellow legislators the dubious honorific L.S.C. (standing for Liar Scoundrel and Coward).[14] Though work on his ornate mansion he called "Forest House" was flagging due to lack of funds,[15] Smith was far from poor. In addition to his legal practice, a portion of the capital for both his present ventures had been supplied by raiding a secret slush fund Secretary of State Daniel Webster had set up for the purpose of influencing the outcome of the Canadian boundary dispute with Great Britain.[16] Smith now saw a way to merge his least unsuccessful investment to date with his most successful.

With Smith's encouragement and the loan of Robinson's blacksmith shop, Cornell had set about modifying the Barnaby Mooer plow to combine the two tasks of trench-digger and cable-layer into one smooth operation. The contract with Bartlett specified a sum per mile for trenching *and an additional and separate sum* for laying in the cable. The redesign involved mounting a roll of cable between the handles so that it would run out behind the blade simultaneously as it cut the furrow and then bury it. Since the Barnaby and Mooer design was a double-bladed plow, this conveniently permitted a channel in between the blades through which the cable could be guided into the ground. The cable could be laid and covered at a rate of what Cornell estimated to be three miles a day. If this estimate proved accurate, it would not just bring Smith's costs in line with what he had promised Morse, but leave a hefty profit besides.[17]

The plan now exuded both economy and invention, the two qualities Morse prized most, and Morse quickly assured Smith he could provide a continuous roll of cable for the new machine.[18] The firm he contracted to furnish this was James E. Serrell of New York.[19] Serrell had recently patented a new means to create just such a continuous coil of lead pipe which would work perfectly with Cornell's machine.[20] The whole trenching operation, from beginning to end, would be seamless. Through serendipity, the efficiencies of innovation indeed seemed to be self-multiplying, just as Morse had predicted they would be. Most of these economies, with the exception of the cable-layer, would in the end, however, prove false, and the trenching contract would constitute a thorn in Morse's side for years after.[21]

By this time word had gotten to Col. Beebe in Ithaca that Cornell had gotten waylaid by Smith on his way to Augusta (Mooer apparently had seen him there). Beebe was not happy to be left in the dark: "I wrote you some days ago at Augusta and so did Mooer.... I did not know you had stopped at Portland til I was told today, *after enquiry*, that he saw you there. He says that you are contriving something to lay lead pipe for F.O.J. Smith and he thinks you may succeed. I hope you will succeed in making some money."[22] (Beebe did not know just how prescient was this last remark.)

If we are to believe the account that Cornell provides, which has since become more or less folklore,[23] Morse had a coil of Serrell's cable shipped up to Portland. He himself arrived a few days later on August 19.[24] The three of them rode out to a nearby farm where Cornell's shiny newly finished implement stood hitched to a team of oxen, the brand-new roll of coil nicely mounted between the two handles. Preliminary tests had supported Cornell's estimate that the machine, with a man and eight mules, on flat ground, would be capable of laying up to three miles of telegraph cable per day,[25] but now Morse would see it with his own eyes. In Cornell's version, perhaps the sight of the strange men in their suits spooked the animals as they took off at a run, dragging the machine and the poor, cursing Irish corn farmer behind them. When the team finally came to a panting halt, the cable had all just magically disappeared. When Smith handed him a shovel, the farmer dug out the filled-in trench, revealing to an astounded and gratified Morse the entire length of cable, neatly deposited underground at a precise depth of eighteen inches.[26]

21

Sins of the Father

F.O.J. Smith had returned to Maine not only to take the reins of the newspaper but to resume his profession as an attorney as well. That summer he had tried a rather high-profile murder case, acting as the defense in what was apparently a rather lurid love triangle.[1] While the brutal Jamestown murder had not been exactly a lovers' triangle, it certainly was lurid and notorious enough to capture the public's attention. The murder itself, the trial and the subsequent commutation, the last thanks to the timely intervention of T. Romeyn Beck, all had been splashed across the papers from Syracuse to Charlotte. It was the stark contrast—the eruption of such a violent act of passion in the context of the normally staid Quaker community—that had no doubt made it a favorite topic of drawing room conversations throughout the Northeast, just as had Governor Bouck's last-minute act of mercy.

Likely none of this was news to Ezra Cornell. Even though he had been in the backwoods low country; he had been staying with relatives, so the news blackout in February by MaryAnn probably had been ineffective. With notices carried in all the New York, Baltimore and Philadelphia papers at the time, he could not have been unaware. By the time we find Ezra Cornell furiously hammering away in a Portland blacksmith shop, his cousin Alvin would be rotting away in the jail cell in Mayville, New York. And while there is no direct evidence that news of the gruesome murder in Jamestown preceded him there to Portland, it seems certain that Smith would by then have become aware of it as well. As a criminal defense attorney, the Alvin Cornell case would have held a great deal of interest for him too, especially because Alvin's lawyer had raised what was at the time a new and unique defense: innocence by reason of temporary insanity.[2]

Alvin was found guilty and condemned to hang, but as to the question of insanity, T. Romeyn Beck, at Governor Bouck's request, had begun an investigation of his own. Despite his initial dubiousness, Beck had eventually uncovered mitigating testimony[3] that in the end would convert the death sentence to incarceration. Alvin, in a final outburst, would reject the act of mercy, claiming that it placed an ineradicable stain on his honor and that of his family. Preferring death, he insisted on his original sentence and raged against his sister, who had invoked Beck's help in allowing him to escape the hangman's noose, "treating the imputation of insanity as an unendurable stigma upon his character."[4] Following Alvin's resentencing, the commutation was almost bigger news than the murder itself, and despite MaryAnn's news blackout, it is almost certain that Ezra had found out about it by the time he had rejoined Smith in Portland. But as to how Ezra dealt with these events psychologically and to what degree they affected him is a question for speculation as there is not a jot in his hand. So while we may construe these events as entirely unrelated, on another level they appear to us as seismic tremors stirring the geomorphic strata of Ezra Cornell's psyche, constituting a shift that would further strengthen the diaphanous threads tying Cornell to Romeyn Beck and thus, by extension, to Morse and Joseph Henry.

Given the Cornell family's history with jealousy-inspired murder, there would have been a natural tendency to fixate again on the recurring specter of hereditary madness; despite the lack of mention, the murder and the likely retributive consequence must have thrown a pall over Ezra's psyche. Perhaps MaryAnn's reticence to discuss matters convinced him by now of the prudence of silence, as there is no mention of it in his letters either. So, just as he is about to embark on what will be a major career change, we see this specter of familial madness rise and hover silently over him for a moment, only to be dispelled (in part) by the hand of Joseph Henry's mentor and teacher T. Romeyn Beck.[5] For the present it seems Ezra will be content to bury the thoughts of his cousin's actions, like his new friend Morse's cable, out of sight, and for a while at least, out of mind as well.

For Beck, the commutation represented primarily the triumph of a humane and rational juridical system. For Cornell, it was the first appearance of a black-sailed ship on his horizon, one that would make several further appearances throughout his life. This pattern would repeat itself: it seems whenever Ezra Cornell was about to mount a wider public stage there was some corresponding disturbing irruption of insanity in his family that would somehow manage to invoke Beck's influence. Just when a vaster horizon seems to be beckoning, this morbid soul-drenching curtain falls, a pattern of last-minute discouragement that will repeat itself almost identically when he embarks on his next greatest achievement, the creation of Cornell University. This time it would not involve the interference of Beck himself, but that of two of Beck's disciples, Sylvester David Willard and Edward van Deussen.[6]

We have already noted several striking similarities between Cornell and Morse's characters; the obsessive enumerative tics, the valuing of acquired skills rather than pedigree, the ruthlessness in pursuit of a given aim—as well as the contrasts: Ezra Cornell's vigorous physicality and lack of formal education, his devotion to agriculture throughout his life, which was partly due to social and familial pressures and partly his own belief in the power of hard physical labor (perhaps, as we noted, to dispel the encroaching overtures of madness and despondency). As for physical labor, Morse was about to provide Ezra Cornell ample opportunity.

Section III

The Madman and the Telegrapher

22

A Federal No-Show Job

While we have speculated on the state of Ezra Cornell's psyche at the time of his introduction to the telegraph, if there were two events in the 19th century that would shake the foundations of the American psyche in general, the first would have to be the advent of the telegraph, and the second, the firing on Fort Sumter. And while it may seem a stretch to compare the two, separated as they are by a gulf of almost two decades, and being of apparently widely diverging character and purpose, yet somehow they both embody the explosive annihilation of a former set of certainties; in the case of the telegraph, the very concept of the fixity of space and time; in the second, the indelible unity of the nation itself.

Whatever else he might have been accused of, Morse could never have been slighted for lacking foresight or an underappreciation of the possible extent of the effects of his invention. It was evident, even at these early stages of his project, that Morse had set his sights well beyond the Test Telegraph, on a nationwide network and perhaps even a global one. For the latter, as Morse was well aware, this would require crossing substantial bodies of water, and Colt's demonstrations had convinced him this was possible.[1] The next time Morse wrote Henry on the subject was in 1844 when Morse had appealed to him to engage in some discussion with the War Department in support of Colt's submarine battery.[2] The best that can be said about Morse's missive on Colt's behalf is that it was studiedly neutral on the subject of Colt's achievement and effusive about his own.

Though the original bill had authorized the Postmaster General to oversee the operations of the Test Telegraph,[3] Postmaster General Charles Wickliffe had taken such little notice of this that he neglected even to include it in his annual report of 1843 or 1844.[4] Morse apparently had taken this as a mandate to create a semi-autonomous government agency with himself at its head, awarding himself the title of "Superintendent of the Telegraph." With the

Original Morse telegraph apparatus, held by Cornell University (courtesy Cornell University Libraries).

exception of Smith, he had staffed his project not with men of any particular political or commercial savvy, but instead displayed a spirit of meritocracy, surrounding himself with those individuals whose intellect he admired—so long as that intellect was suffused with personal loyalty to him.[5] Among his first appointments was longtime supporter Alfred Vail, and shortly following this, his NYU colleagues, Professors Fisher and Gale, as technical consultants.[6] Smith had been awarded neither position nor salary, but (as noted) had already set his sights on the lucrative trenching contract. The qualities Morse logically should have been looking for, instead of personal loyalty and scientific erudition, were those of a passable civil engineer, but Smith had seen to it that this did not occur.

Shortly following the successful demonstration of the cable-layer in Portland, Maine, Morse and Smith, along with Levi Bartlett, had set out for New York, where among the first stops would be James Serrell's pipe factory. Cornell had remained behind in Portland, hoping that now, with Morse's endorsement, he perhaps could sell a few more Barnaby Mooer plows in the neighborhood. Quickly disabused of this notion, he had boarded the steamer *Kennebec* bound for Augusta, where he closed up his business and headed down to New York

City to join Morse and Bartlett, arriving in the city on October 12, 1843.[7]

Cornell had met up with Morse and Fisher at Serrell's factory on the west side of Manhattan on that Friday. Serrell had already promised to have the initial ten miles ready by the time Smith and Morse reached New York,[8] however, when they entered his smoke-filled factory, they were treated to a rude shock. They found it in a disorganized state with pipes and wires scattered everywhere. Serrell was already three weeks behind in the manufacturing process.[9] In what was apparently another cost-cutting move, Morse and Fisher's specifications, instead of using Colt's recipe of beeswax and asphaltum, required four conductors and a ground wire—all covered with only a layer of shellac-saturated cotton and twine.[10] With what amounted to the bare minimum of internal insulation, it was absolutely critical that the pipe itself be watertight, and this evidently was not the case. Vail had been doing some testing on the lead conduit, and the joints in the pipe seemed to leak. Air bubbles were evident when they inspected the lead casings.

Photograph portrait of Samuel Morse by Mathew Brady (courtesy Library of Congress, Prints & Photographs Division, Reproduction Number LC-USZ62-110084).

Even having built considerable leeway into his schedule, Morse was beginning to panic, making contingent plans of which he had already apprised Gale.[11] He generously awarded Serrell a new deadline of September 26, by which time Serrell agreed to have the first ten miles ready, with the balance no later than November 14. If he did not deliver the first ten miles by the 26th, the entire deal would be off.[12]

Alfred Vail for the present had remained at the factory on the west side of lower Manhattan, testing each length of pipe for leaks as it came off the rolling machine.[13] He had already discovered that the air bubbles in the lead that they had observed tended to burst under the annealing process. Introducing the wire incurred further problems, causing some of the latent bubbles in the lead to burst as well. To solve the latter problem, Morse had developed (and patented)[14] an innovative mandril for efficiently introducing the electrical conductors into the lead casing while it was being rolled.

In the meantime, fearing he might lose his trenching crews, on September 8, Morse advanced Bartlett $500. It was harvest time, and with the pipe production lagging, Bartlett had already decided to return north temporarily to manage his farm. The payment was in part surety of his return. In his absence, Bartlett had assigned Smith as attorney-in-fact for the company with the right to sign contracts on his behalf.[15] Smith had quickly ordered Morse not to divulge this arrangement to the Secretary of the Treasury, saying, "It is to you immaterial who makes the receipts if you have the requisite evidence they are authorized."[16]

Finding himself also largely redundant for the time being, Cornell had decided to absent himself to attend another event that happened to be going on at the time.[17] That year's American Institute Fair was sponsoring a plowing contest, and the event was being held (somewhat ironically) across the river in Paterson, New Jersey.[18] The Barnaby and Mooer side-hill plow had taken the gold and silver for this in 1840 and '41. With Cornell still holding the sales rights for the plow in Maine and Georgia, he was naturally eager to see how this year's new entrants stacked up against the model he had invested so heavily in time and money promoting. The opening address had been delivered by respected Massachusetts agronomist, J.E. Teschemacher, who had begun by cautioning the crowd, "Our agricultural youth are apt to forsake their paternal fields and flock in clusters to populous cities ... imagining that these are a high road to fortune and prominence...."[19] These words must have stung Cornell's ears as he stood on the precipice of his own decidedly non-agricultural-looking future; yet, forsaking the plow for the telegraph would turn out to be the best business decision he could possibly have made, and Teschemacher, in his case at least, could not have been more wrong.

The following day Cornell had continued on to Baltimore, where he was met by a newly energized Smith, who promptly informed him the cable-layer had not yet arrived as it had gone the long route by sea.[20] Though Serrell had already missed the first deadline, Morse remained reluctant to fire him. He had been painting a far rosier picture for Secretary Spencer than was actually the case, and abruptly firing Serrell would have raised several uncomfortable questions.[21] Since they were still awaiting the delivery of the cable-layer, the effect of Serrell's breach was more contractual than actual. The initial two-and-a-half mile run[22] had finally shown up on the Philadelphia docks on October 19 and

22. A Federal No-Show Job

was shipped down to Baltimore the following day, where an expectant Ezra Cornell had been eagerly awaiting it.[23]

The plan, which roughly followed the suggestions made by Morse's old pal Ellsworth, was to run the cable southwest from the Mt. Clare Rail Station at Pratt Street and Poppleton Streets in time for the sitting of the new Congress. Morse had already obtained the necessary rights of way from Louis McLane,[24] who was then in charge of the Baltimore-Washington branch of the Baltimore Ohio rail line.[25] Thanks in large part to Serrell and the decision to ship the cable layer by sea, they were already three weeks behind schedule, so the "can-do" Cornell had decided, in the interim, they would start digging the trench by hand, which he proceeded to do at 8 o'clock Saturday morning, to very little fanfare.[26] Morse and Smith had both been on hand for the groundbreaking, and neither should have been disappointed when the papers took no note of the event, nor had they organized any hooplah. By that evening Cornell had cleared the railway station, and the four heavy boxes of Serrell cable stood at the depot next to the tracks, ready to be mounted on the still delinquent cable-layer. Work would start in earnest when the rest of the cable and the cable-layer machine finally arrived, which was supposed to any day.[27]

Cornell had traded in his team of horses for mules, finding that the horses had been spooked by the proximity of squealing trains traveling just yards away.[28] Smith, still intent on "making hay," or in this case "making hay disappear," promptly had the fine horses shipped to his mansion in Portland for use with his elegant new carriage. Despite the fact that operations were proceeding at a snail's pace in Baltimore, due to confidence in Cornell's handling of matters, spirits remained high elsewhere. Both Morse and Smith had retreated to New York. Cornell had written MaryAnn, "Professor Morse is at New York and Smith has gone to Portland. They will both be here again in a few days.... We have commenced laying the pipe and laid 2,000 feet by hand that carries us out [of] the Baltimore Station, so we can commence laying with the machine which has not yet arrived." Evidently he had been receiving some pressure to return home to Ithaca as well, as he adds, "Mr. Smith says ... I must stay ... and he will fix me with a salary of $1,000.... Professor Morse ... approves." This would come to pass, but not until several months later, by which time Cornell had made himself indispensable.

In his spare time, Cornell pursued his first of many stabs at becoming an inventor himself. Smith had been brimming with suggestions about the cable layer, from the diameter of the drum holding the wire, to the method of securing it to the plow shaft, none of which Cornell was taking kindly, given Smith's continued insistence on referring to it as "our machine."[29] Despite the setbacks, Smith's personal stock had never been higher. He had told Cornell he was going back home to Portland, but instead Morse had shanghaied him, bringing him to New York to find a new supplier for the pipe. Virtually salivating with newly invigorated greed, he was not only confident he could line his pockets anew should the Serrell deal fall through,[30] but was now also claiming to have himself invented not only Cornell's pipe-layer but Morse's new pipe mandril as well![31]

The section Cornell had laid was already proving defective. Still not convinced that he needed to resort to Colt's formula, Morse had ordered the next run of Serrell's wire double-covered with shellac and cotton in hopes of alleviating the failures.[32] The rest of the cable never arrived. On October 28, a week after shipping the first run of pipe, Serrell had signed a release forfeiting the remainder of the contract.[33] Morse had already ordered an extra ten miles of wound cable from another manufacturer.[34] Three days later Smith had written Cornell from Boston informing him of the new contract with Tatham Bros. of Philadelphia.[35] Tatham not only had a reputation of being more reliable, but they had two manufacturing facilities, one in New York and one in Philadelphia.[36] There was, however, the question of whether the new Tatham cable would work with the cable-layer. Though they were unable to provide an exact size for the new reels,[37] Smith had rushed into the deal only two days after severing ties with Serrell. The cable-layer, once it arrived, apparently needed some additional modifications. Instead of specifying more exacting requirements to Tatham, Smith had directed Cornell to remain flexible, advising him to "omit all alterations in the drum part of the machine" on the model[38] and "if you can, so alter *our* machine"[39] to fit the new cable.

23

Cable Problems

It is difficult to do justice to the depth of mutual ill-will that Alfred Vail and Ezra Cornell harbored for each other during the course of their long association with the telegraph. Though expelled from the Quaker meeting because of his marriage to a non-Quaker, Cornell came from a background where modesty and plainness in speech and dress were paramount indicators of good breeding and manners. Vail was the pampered scion of a semi patrician dynasty of Northern Jersey iron makers. He fondly embraced all the trappings of gentility that his family connections afforded. Safe to say, men of such disparate background and values, thrust together into close contact, in the roiling stew of frantic activity surrounding the birth of the Test Telegraph, most likely would have come to despise one another anyway. But even in that age of hyper-polite communications, the hostility would eventually emerge in the starkest of terms, even in the written correspondence culminating in Vail's repeated calls to Morse by 1846 to get rid of Cornell, whom he characterized as both meddlesome and disloyal.[1]

The initial bad blood had no doubt sprung from an incident at the Serrell factory when Cornell had criticized Vail's method of testing the Serrell cable, but over the next few weeks in Baltimore, the sniping had become almost habitual. Cornell had, as was his habit when confronted with overt hostility, "clammed up," which only confirmed Vail's opinion of him as an uncommunicative oaf.

While the mood out on the line was tense, the overall sense was one of exultation at finally getting started on the project. Relying on Smith's somewhat optimistic estimate of three miles per day, they were still well within the window of good weather, and Smith had negotiated what appeared to be a favorable new contract with Tatham Bros. that would provide them with the new cable in a mere week and a half. Tatham claimed to have invented a new process that promised to be a great time-saver. Rather than using Morse's mandril for inserting the wires, Tatham intended to seal in the wires at the same time they annealed the flat metal into tubes, in one continuous "hot" process. In the meantime, Cornell still had ten miles of the Serrell cable to lay.

Having openly criticized Vail's method of testing the Serrell cable back in the shop, Cornell remained skeptical, now that they were in the field, about Vail's methods and acumen. To test the cable thus far laid in the ground, Vail was relying on a simple continuity test that he had devised. He had his assistant, Thomas Avery, hook a battery to each red, black, green and yellow color-coded wire at the beginning of the exposed pipe run and attach a galvanometer to the same colored wire at the other end.[2] Deflection of the needle indicated a current was passing. It was evident that this method would show a break in a wire but not necessarily any "cross-cuts" (short circuits). Eager to show his newly acquired electrical acumen, Cornell proposed a different method; but reluctant to offer up his observations so freely this time, having experienced Vail's withering scorn already, he tried keeping his opinion to himself. His patience, however, was running thin as he was forced to dig up run after run of buried cable for visual inspection when they got inconsistent test results. Rather than confronting Vail directly, he had buttonholed Vail's assistant, Thomas Avery, to help him conduct a covert, midnight experiment.[3] Instead of testing the continuity on each wire separately, they would instead test for current passing *to the opposite color wire*.[4]

Just as Cornell suspected, the midnight experiment proved that it was "cross-cuts," and not breaks in the wire, that were the major problem causing the dead line. This was naturally the kind of delicate situation for which Cornell had very little patience. Vail had already intimated to Avery that he should "confine himself to his own business."[5] Avery agreed to pass this information on to Morse, but cowed by Vail's demonstrated intolerance for criticism, he too had chosen to keep mum.[6] True to form, Vail, even after learning the results of the experiment, stubbornly continued to test the wires using the simple continuity test. As the crews moved slowly south along the rail line, problems with short circuits were sprouting like dragons' teeth in a rainy Thebes. Vail seemed entirely at a loss for an explanation.

The resulting finger-pointing was beginning to damage morale as they encountered delay after delay, laying each new section of pipe only to dig it up a few days later when Vail's test signals again failed to pass. Vail, having personally approved the pipe himself back in New York, instead of recognizing the cause of the problem, instead blamed it on poorly made joints, requesting that Morse order up five hundred new ones from Serrell, and then was perplexed, when the new couplings arrived, and the problems persisted.[7] Sometimes Vail's test indicated the wires already buried were sound, and then when

23. Cable Problems

they were retested, no current would pass. This was not only putting them behind schedule, but labor costs were mounting. Morse had already exhausted much of the $500 advance he had given Bartlett in September in New York with very little to show for it except for half a mile of intermittently working cable that had already been buried and dug up several times, no doubt to the growing amusement of the laborers, who were being paid by the hour.

No doubt discouraged by what he saw as Vail's incompetence and willful disregard for efficiency and common sense, Cornell was growing increasingly frustrated. He had already pinned part of his own hopes for financial reward not on the success of the telegraph project itself, but rather on his own invention, the cable-layer. While waiting for the cable-layer machine to show up in late October, he had written excitedly to MaryAnn, "I am making a moddel [sic] of the machine [the cable-layer] to deposit in the Patent Office."[8] Once his working "machine" finally arrived, Cornell began laying an average of a half-mile to a mile each day, and while this was far less than half the three miles he and Smith had promised Morse back in Portland, it was still respectable enough under the present circumstances. If they could keep up the pace, they still might finish before the January freeze, in time for the sitting of the new Congress. This had been Morse's express goal, but with Vail demanding each day that they dig up another section of the already laid run, the progress was halting; and as the length of the line grew, it became even more unpredictable and difficult to pin down the actual locations of the irritating failures.[9]

All this discord and drama had been papered over, partly by the unflagging optimism Morse exuded in the press. By mid–November less than seven of the proposed forty-miles had been laid. Money was getting tight. Like Cornell, Smith had more or less given up on making a killing on the project itself, but, not satisfied with what he hoped to skim from the trenching and the team of horses he had "borrowed," sensing in all this upheaval an opportunity to turn another quick dollar for himself, the grasping Smith had already set other plans in motion.[10]

It appeared Tatham's new "hot" process was proving less a time-saver than expected. The contract specified delivery of ten miles by November 10. When the promised cable still not had arrived by the end of November, Cornell had only another three miles of the (known to be defective) Serrell wire remaining. So, by December 1, having reached the Relay House in Maryland, just ten miles outside Baltimore, he was forced to stop, having entirely run out of cable.[11] The freeze was coming and within weeks they would be forced to discontinue operations anyway. It was becoming clear Morse's goal of finishing by the seating of the new Congress was not to be, and with funds dwindling, success of the entire project seemed in doubt.

Following Vail's departure for Baltimore back in October, Morse had left oversight of manufacturing the cable in the hands of Professor James Fisher, his trusted scientific advisor.[12] When Tatham's cable had not shown up at David Burbank's warehouse in Baltimore, Morse had queried Fisher about the holdup. Fisher informed him that the cable had been loaded on board ship in New York harbor as per the contract, but that he was still in the process of proving it.[13] Morse became furious hearing this news, saying that given the season and increasing urgency of the situation, Fisher should have been conducting his testing at the factory instead of waiting for delivery to the ship.[14]

Back in Maryland, an early winter snowstorm had covered the ground. Once the new snow melted, Vail once again found water inside several of the cables, rendering them nonfunctional. Once again Cornell had been forced to hand dig up section after section to make a visual inspection. Vail was by now firmly convinced that continuing with the Serrell pipe had been a mistake, but instead of taking full responsibility, finding an excuse, he now encouraged Morse to lash out at Fisher, and then also (somewhat uncharacteristically) at his and Fisher's trusted NYU colleague, Leonard Gale. When Morse did lash out, it was in somewhat uncharitable terms, saying (with regard to the new Tatham pipe), "The delays in New York and the illness of Dr. Gale are embarrassing me very much."[15] He then virtually ordered Fisher to approve the Tatham cable immediately: "I want matters closed up at New York without delay ... the Serrell pipe is laid to Relay House ... from the Relay house. We will commence with Tathams."[16] This was literally, at this point in time, a pipe dream.

Professor Leonard D. Gale of New York University. Lithograph by Enberg (courtesy Library of Congress Prints and Photographs Division, Reproduction Number LC-USZ6-110).

23. Cable Problems

The long-awaited Tatham cable finally arrived on December 6, three weeks behind schedule, and it was soon clear the results were no better than had been the case with Serrell's—in fact, worse.[17] Coil after coil proved defective even before it was laid in the ground. When the lead casing was opened up, the reason became immediately apparent. The wires and insulation had been scorched at regular intervals starting about ten inches from the joints, probably as a result of the new "hot" annealing process—yet another false and failed efficiency. Unlike the dithering exhibited with regard to Serrell, Morse had acted promptly, ordering Fisher to stop Tatham's production that same day.[18] Vail was still busy hunting down short circuits in the Serrell cable that was already in the ground. Having had nothing to do with overseeing or "proving" the Tatham cable, he this time at least had the luxury of laying the entire fiasco squarely at Fisher's doorstep—which he promptly did,[19] piling insult on injury by whining about now having to do Fisher's and Gale's share of the work in addition to his own.[20]

Despite another early winter snowstorm's having blanketed the tracks, Cornell had forged ahead, laying a half mile of the new run, but not before suggesting an alternate plan of action to Morse that would enable them to save face in what was appearing a more and more doomed effort.[21] What followed is possibly the best known of all the accounts of the Test Telegraph, which is also probably largely apocryphal (as are many of Cornell's recollections regarding his early career with Morse). As Cornell tells it, it was at Morse's behest—that Morse took him aside and ordered him to wreck the cable-layer which he then obligingly did, yelling the picturesque phrase for the benefit of the gawkers and newspaper reporters, "Hurrah boys! Whip up the Mules." Another, far more likely scenario follows: having until recently spent his time touting the Barnaby Mooer plow throughout the South, Cornell no doubt, in the course of those travels, had to listen to the laments of various plantation owners regarding the "lazy niggers"[22] who had wrecked a good plow on purpose to get out of working the field. Thus it probably had occurred to him to wreck the cable-layer and to do so intentionally. Either way, it had no doubt delighted the beleaguered Morse, who had no doubt been the one to arrange to have the papers there to cover it.

In the realm of reality, with Gale ailing and Fisher suddenly unresponsive,[23] Morse had little or nothing to do but fume privately at real or imagined betrayals, while desperately trying to conceal from the press and the government the fact that the project was mired in contention and difficulties. Vail had done nothing further useful to this point but whine. To be fair, he must have been at this point overworked and frustrated as well, his most recent tests having confirmed that virtually none of what they had laid in the ground was able to convey more than an intermittent signal.[24] Gale, probably already sensing the maelstrom of despair and vituperation that was brewing in Maryland, wanted no part of it. He wrote Morse at first that his health was improving and he hoped to be well enough to travel to Relay by Monday, December 11, the original date Morse had set for the meeting at Relay, and Fisher should be there by then as well.[25] But Fisher just as quickly disabused Morse of that notion: Fisher would, of course, never arrive, nor would Gale.

24

Big Confab at Little Relay

The Relay House, situated about ten miles from downtown Baltimore on the Baltimore Washington Rail Line, was a three-story, 32-room frame structure that served as hotel, restaurant and tavern, as well as the transfer point for all of the Baltimore and Ohio's western lines. This would also be the termination point for all Morse's efforts with the underground cable in 1843. With winter fast approaching and twenty of the original thirty-thousand dollar appropriation already expended, Morse's plan seemed doomed. They had been working frantically since March and there was precious little to show for it except some exhausted mules and ten miles of intermittently functioning telegraph line.[1] In a desperate bid to conserve his remaining funds, he had closed up the shop in New York where Chase, Taylor and Smith had been laboring away since September insulating the cable, and on November 25, he had written Gale, "If you have not paid Tatham and Serrell's, you will retain it in your hands.... I have already paid for more than has been delivered. This must be looked in to...."[2]

Ordering further payments to Serrell suspended and closing the shop clearly was too little too late.[3] It was clear that the line would not be in the ground before the freeze, and worse, it looked increasingly like, barring another substantial congressional appropriation, they might not even be able to resume once the thaw came. With work stopped, Morse had decided to summon his entire staff to Relay, Maryland, for this meeting.

The choice of Relay House had not been coincidental. Two of Morse's salient characteristics—his strict experimental empiricism and his conviction that he was God's instrument on earth—had come into play. Morse, the empiricist, had seen a ten-mile run of cable work without the benefit of relays when his former collaborator, Samuel Colt, had informed Secretary of State Abel Upshur back in March that he had successfully sent a signal ten miles, initiating an explosion, and had done so without the necessity of any intermediate relays.[4]

If the bothersome in-line relays were in fact necessary, it would be at distances of greater than ten miles.[5] With Relay House being placed conveniently at exactly ten miles from Baltimore, this was a natural stopping point. The very name "Relay" conjured up the problem he might face in extending the line further. It also held a special place in his private iconography/geography, and this was possibly why he had chosen it: it represented to Morse not just the terminus of his present endeavors, but a fixed point of certainty in a now swirling universe.

Morse knew he had to drastically cut expenses if he was to resume work in the spring, but he was also acutely aware that above all he must continue to project a positive face in the press. With newspapers carrying the thus far falsely glowing accounts of their progress, hopes of any additional monies were pinned to the public perception that the decision to halt had been voluntary. If this pretext wore thin, clearly, he would need a scapegoat, and Morse had already decided where to lay the blame.

Smith was not himself a salaried employee, and still holding the bona fide federal contract in hand that could consume the remaining monies, Morse could ill afford to antagonize him. Blaming the loyal, if incompetent, Vail was out of the question, and Gale had proffered the excuse of ill health, blaming it on his exposure to the fumes in Serrell's factory. In a rare show of pique, echoing Vail's whiney tone, Gale obliquely blamed Morse, saying, "I have never seen a well day since I superintended the work in that damp cellar of that [Serrell's] place."[6] Clearly the blame had to fall somewhere. So, with the project stopped, he had summoned Fisher, Vail, Gale and Smith all to Relay House in Maryland, ostensibly to make plans for the spring, but where he no doubt intended to explain that he could not afford to carry all their salaries through the winter.

Though he was loath to dispense with the services of his "scientificks," Gale and Fisher's combined salaries together amounted to almost double his own.[7] With less than one-third of the appropriation remaining, Morse was well aware he could not continue paying them through the winter. For similar reasons of economy he hoped, if at all possible, to find a way of delaying any further payments to Tatham Bros., but he remained adamant that the remainder of the pipe had to be delivered before the hard freeze. For this he needed Fisher's continued cooperation, and also some show of making good on the contract with Tatham once the cable arrived. When Morse had requested Fisher and Gale join them in Relay for a big "Pow Wow" on November 29, Fisher had written back temporizing, "Dr. Gale is slowly recovering.... I hope that we will soon be through here so that we can come on and be at work with you there."[8] Gale, also probably sensing the real reason behind the summons to Relay, had written the following week begging off, stating that Fisher could probably depart for Baltimore in the morning, but that as for himself, "God has been exceedingly merciful in raising me from the gates of the grave."[9] Morse had written his brother Sidney a few days before the Relay House meeting regarding Fisher that he would "give him the opportunity excuse himself if he ever gets here [Relay],"[10] finding out only later that day (or the following) that Fisher had no intentions of showing up.[11]

Morse's deftness at managing complex and hostile circumstances to his own ends, and also the personal ruthlessness that he could display when faced with failure, would soon become evident.[12] With the project inextricably mired in ineptitude, aware that work had come to a complete halt, Morse sought to divert attention from his problems by requesting an additional authorization from the Secretary of the Treasury to put on several "solderers" for the upcoming sections.[13] Clearly he knew he would have no need of them before the freeze and probably had no intention of using them, but this late request to Spencer was an attempt to put a positive face on a dismal situation. He justified this by saying that Cornell's cable-layer had so far outstripped Vail and his soldering crews that this had become necessary (neglecting to mention the recurring problems with the existing pipe and the lack of new cable that would soon force them to stop completely).

All the sustained high dudgeon and public tap-dancing had set the scene for the dramatic meeting at Relay House, which, despite the absence of Fisher and Gale, proceeded as planned on December 17[14] with only Morse, Vail, and Smith in attendance[15] and with winter fast closing in, the "temporary" decision to stop work was made permanent and official. The following day Morse sent a note to Secretary Spencer stating they were suspending operations in that it was "impossible to superintend the work in the winter."[16] A week later he would send a self-exculpatory press release to the *National Intelligencer* saying, "The lateness of the season embarrasses my further operations till spring."[17] Always ready to excuse himself or Morse from blame, Vail had thrown the "scientificks" under the bus, summing up what must have been an embarrassing and wrenching circumstance thusly: "Through the negligence of Dr. Fisher and perhaps Dr. Gale, our lead pipe is defective and wires also. Mr. Smith and Prof[essor] Morse and myself stop together at the Relay House."[18]

The fact that Morse had made up his mind to halt work well before the meeting at Relay was evidenced not only by Cornell's engineered cable-layer "accident," but also the fact that on December 6, Morse had ordered Fisher to ship the balance of the Tatham cable directly to the Patent Office in Washington, D.C., instead of to Burbank's warehouse in Baltimore.[19] At Relay Smith had evidently insisted that the underground approach had failed totally and that when and if they resumed in the spring, it should be with an overhead solution.[20] One after another, the chickens were coming home to roost. Even Morse's offhand snub of railroad magnate Benjamin LaTrobe, Jr., with regard to hiring the civil engineer, Niernsee, had come back to haunt him. Four days after the confab at Relay, Louis McLane had written, saying he was having second thoughts about granting Morse the railroad right-of-way free of charge: "I believe I mentioned ... an objection exists in the minds of many of members to granting gratuitously the use of the road."[21]

There was, as always, the question of the money. According to Cornell, Vail had apparently refused to retire from the project, or surrender his salary even temporarily,[22] but probably he had been convinced otherwise by Morse. He and Cornell would continue on through the winter, laboring in the chilly Patent Office rooms overlooking G Street NW,[23] and Cornell in the even chillier basement, the only two members of the project still technically "on salary" (but probably without actual pay, with Morse paying only their room and board).[24]

A week after the meeting at Relay House, Cornell found himself in charge of several miles of unused Tatham cable plus the dug-up defective Serrell cable, all of which had been unceremoniously dumped in the basement of the Patent Office, some on rolls, some haphazardly gathered into bulky twine-tied bundles, deposited there apparently with the blessing of Commissioner Ellsworth. Morse had written to Vail and Gale, "Mr. Cornell will take entire management of the ... coils of pipe ... under the ... superintendence of Dr. Gale *with the assistance of Mr. Vail.*"[25] Cornell had been instructed by Morse to salvage as much of the copper wire as possible by stripping off the lead casing.[26] Though he had not been among those invited to Relay, Morse had essentially handed Cornell the entire task of laying out a plan for the new *overhead* cable.[27]

25

The Trouble with Fisher ...

The decision to fire Fisher had been wrenching, not only to Morse but to Gale as well. Fisher and Gale were not only colleagues but also the closest of friends. Neither would have lightly tolerated poor treatment of the other at Morse's hands (under ordinary circumstances).[1] Fisher, like Samuel Morse, the son of a pious man of the cloth in the Protestant reform tradition, had attended Yale University, graduating in 1826, having entered at the advanced age of fourteen. Unlike Morse he was an avid student and had done well at school, graduating from the New York College of Physicians and Surgeons in 1831.[2] In 1839, following a trip out west (to Michigan),[3] he had become full professor of mineralogy and chemistry and the University of the City of New York (NYU), succeeding Leonard Gale in that position.[4]

Though there were the vague outlines of a new plan in place for when the weather broke, Morse was fast running out of funds and options and the project was clearly on life support. Morse had been desperately trying to explain the technical reasons for the delays in the press to a mostly uncomprehending public with little success.[5] Though his accounting had been scrupulous, he had already expended more than two-thirds of his original $30,000 appropriation with little or nothing to show for it. The press were circling like vultures around what increasingly looked like a government boondoggle on a vast scale. Even with the project at a standstill, Morse continued hemorrhaging cash. On December 6 he had written promising Tatham their next installment of $920, instructing Fisher at the same time to tell Tatham to suspend manufacturing providing they had at least 25 of the contracted 30 miles of pipe ready.[6] His chances of obtaining an additional appropriation for the coming year were looking increasingly slim.

Blaming Fisher was an easy way out but it should have been clear by now that most of Morse's problems stemmed from water intrusion into the pipes, for which he really had no one to blame except himself. Had Morse proceeded with his plan to adopt Colt's treatment for the telegraph wire all these problems might have been avoided, but for some reason, he had not. Colt's method called for the wire to be wrapped in a layer of shellacked cotton,[7] and then drawn through the mixture of beeswax and asphaltum. Morse had deemed this second step unnecessary,[8] probably thanks to Joseph Henry,[9] who had suggested instead of "cement," that a coarse winding of twine be used to create an air space to separate them. Henry had advised Fisher, "Galvanic electricity has never been made to project itself through a stratum of air of the ordinary density so as to exhibit a spark although the experiment has been tried with a battery of several hundred plates.... *I should therefore conclude that it would be of little importance to fill the space between the wires with cement*[10] provided the metal can be as well secured from contact by less expensive means."[11] This had been sufficient to convince Morse, especially since air was certainly a much cheaper insulator than asphalt and resin.[12] Morse still regarded Henry's pronouncements as gospel,[13] and while perhaps Henry may only have been trying to be helpful, his advice was about to almost cost Morse the entire project.[14]

The problem was that, unless the outer lead casing could be made entirely waterproof (which events would prove it could not), those air spaces that had been such a good insulator when dry, when it rained, would fill up with water; and unlike air, water is an excellent conductor of electricity. Any engineer worth his salt would have figured that out immediately. Henry's air space/twine solution was attractive not only because it was inexpensive, but because it also avoided the possibility of a future patent clash with Colt over the cable design. The potential cost savings over forty miles of wire had blinded the usually scrupulous and careful Morse, and Fisher was about to bear the blame.

Colt and Morse, along with Professor John Draper of NYU, had jointly conducted some of their earlier experiments at the Bloomingdale ropewalk of Ebenezer Chase on the upper west side of Manhattan. Chase was originally a Boston rope-maker[15] who had relocated to this rustic section of Manhattan (around present-day 55th Street).[16] There he had built a ropewalk, which was essentially a long straight path for laying out and twisting strands of rope for nautical use, but it was also ideal for testing the long wires with which Colt and Morse were working. When Morse again needed someone to cover the long wires, he had naturally turned back to Chase.[17] Three weeks after the passage of the congressional appropriation (and several weeks before Morse would hire on his staff), he had asked Chase to submit a bid for winding the wires, and Chase's response indicates he understood that the process would be the same they had followed for Colt wire: "The following are the terms for which I will contract to cover your wires; I will put on the first covering for $2.75 per mile.

25. The Trouble with Fisher ...

You furnishing the materials, and or, I will cover and find all the materials for $5.35 per mile, of the best quality cotton twine, of the large size *saturated in India rubber and tar*."[18]

Chase's initial bid to supply materials had been withdrawn,[19] and a new bid for the cotton "twine" to cover the wires had been approved on May 4 by the Secretary of the Treasury, as had the purchase of the wire itself two days earlier.[20] Morse had concluded with Chase for the wrapping and also for his services, and had ordered that the wire be delivered to Chase's ropewalk in Bloomingdale,[21] sending Fisher up to oversee the operation.[22] Morse also supplied Chase with a laborer (Fisher's brother-in-law) to help with covering the wire.[23]

Whatever guilt Morse may have felt over the decision to fire Fisher, he had disguised it well, approaching it as one would a military campaign. Hostilities had commenced on December 2 when the initial salvo was fired, centering on a pair of bank drafts Morse had sent Fisher for delivery to Tatham. Under cover of a letter of that date,[24] Morse had included two bank drafts (monies Fisher had requested earlier), totaling $673.20. In the same letter, ordering Fisher to suspend operations, he had demanded, "Inform me also if you have *faithfully* tried the wires with the Battery in each coil."[25] When Fisher did not acknowledge receiving the drafts in his reply,[26] Morse had again queried about them and Fisher had written back, neglecting again to make any mention of the missing drafts. By December 12, still with no response from Fisher regarding the fate of the bank drafts, Morse had written indignantly, "I am much surprised ... your letter of the 9th to find no mention of the drafts sent you on the 2nd Dec ... search for the package without delay." This was somewhat brusque, but adding even more brusquely, "Inform me of the result."[27]

Perhaps vaguely conscious by now that he was being set up, Fisher's reply on the 15th still does not acknowledge receiving the drafts. He instead attempts to turn the tables on Morse, accusing him of neglecting to send them: "I have written two letters at different times ... requesting that drafts might be sent on to the Messr. Tatham. I was very much mortified in not hearing from you in reference to them."[28]

Fisher, unlike Smith and Cornell, had continued to invest all his future hopes in the success of the telegraph project, and he knew Morse valued loyalty above all else. As far as he was concerned, this was a simple misunderstanding, so Morse's sudden display of rudeness seems not to have not registered with the same impact it might have otherwise.

By this time David Burbank's bustling lumberyard, located next to the Baltimore Rail Station, had in effect become the *de facto* operational headquarters for the telegraph. To add to all the confusion, Fisher's replies to Morse were now being routed through Burbank's office, and Burbank had undertaken at Morse's behest to perform his own set of tests on the Tatham cable, yielding different and far less encouraging results than Fisher's. Citing the discrepancies, Morse would fire Fisher on the 20th.[29] The move evidently shocked Gale, who took leave from the project, citing health reasons incurred in the course of his duties, thus relieving Morse of the necessity of firing him and solving some of his payroll problems.[30] Levi Bartlett would be furloughed,[31] leading to a vicious sparring match with Smith over the trenching contract that would go on for years thereafter.

On December 22, two days after receiving Morse's letter firing him, the question of the missing bank drafts was cleared up when Fisher discovered them stuck behind a bookcase at the Observer's offices in New York.[32] Morse's brother Sidney had apparently misplaced them or just forgotten to notify Fisher they had arrived.[33] Fisher had hurriedly compiled the receipts and Tatham's bills of lading and posted them to Morse that same day—within hours. When he returned home he found his wife in premature labor and Morse's most recent letter dismissing him from his post. Despite the pressing domestic issues, Fisher nevertheless took time to then draft a second letter to Morse (that same afternoon) expressing profound shock and surprise over the cause of his firing, which seemed to him clearly based on a misunderstanding (concerning the misplacement of the bank drafts) and one not in the least his fault.[34]

Morse only sent a curt note saying, "I have no further need of your services as my assistant."[35] The notice was peremptory and lacking in reasons, as Morse would later insist, deliberately so as to spare his friend's feelings.[36] This, however, had led to Fisher's mistakenly believing he was being fired for the confusion stemming from the missing bank drafts—not his mishandling of the Tatham cable. But Morse had already set himself the painstaking task of compiling a point-by-point indictment of Fisher's actions, labeling it "Evidence of Fisher's neglect,"[37] but he had held it in reserve, supposedly to avoid embarrassing Fisher. On the 29th, he had let loose with both barrels.[38]

The reason Morse had held back most likely had little to do with Morse's concerns for Fisher's feelings and more to do with the fact that Morse still had not yet decided who to blame for the Tatham fiasco. Smith had been pressing him for full payment on the trenching contract on the clearly disingenuous pretext that the government was in default due to their inability to supply the cable. The major part of Morse's complaint directed at Fisher concerned his handling of the cable integrity tests, but if all or part of that blame could be shifted back to Smith and Tatham, then that might go a long way to getting Smith off his back. Most likely he had withheld any explicit criticism of Fisher for the time being until he knew exactly how much of the blame might reasonably be shifted to Smith.

25. The Trouble with Fisher ...

According to Morse, it was only when the cable arrived at Burbank's in Baltimore that the problem with the Tatham cable had been uncovered. As for the integrity tests, the bad faith, if any, certainly was not on Fisher's part alone. The tests done by Burbank had been under Morse's instructions. Morse had insisted vehemently to Fisher that Burbank used the same strength battery in his tests as had Fisher. Ignoring his own part in the matter, Morse shot back (referring to the discrepancy in test results), *"It was your unfaithfulness to your trust,* (since you oblige me to divulge it) is the principal cause of your removal."[39] Fisher had replied, obviously stung by the injustice, "I was furnished with a battery consisting of two pairs of plates [and] a galvanometer ... the wire was proved ... in so far as it was possible to judge ... the insulation was perfect, but it appears ... they were tested with *a more powerful battery* at Baltimore."[40] In his indictment of Fisher, Morse continued to insist that the test conditions had been equivalent![41]

Fisher had been instructed to employ a two-plate battery on a one-mile-long cable, whereas Burbank's tests at Relay House had been with a four-cell battery and supposedly on the same length cable, but Fisher had unknowingly handed Morse the perfect rationalization for the difference in wire length. Henry's letter back in April, of which Fisher had reminded him, implied that doubled wires in proximity acted essentially as a single conductor with regard to voltaic potential. So in Morse's mind at least, Burbank's results may have pertained to a line with an effective distance of one mile as well.[42]

Clearly a knock-down drag-out with Fisher now conducted in the newspapers was not in Morse's best interests,[43] but privately he had been complaining to Gale that Fisher's actions were due to either criminal negligence or inexcusable ignorance.[44] Fisher's lengthy and defiant reply to Morse's letter of the 29th had amounted to a virtual indictment of Morse's bad judgment: "I am sorry your memory is so short that you cannot remember what took place in regard to these very charred places between you I and the Messrs. Tatham ... they proposed to cut them all out *you objected* on the ground of the great loss of wire."[45] Fisher went on to describe how Morse himself—after seeing the burning to the insulation, when he visited New York—pronounced the Tatham cable serviceable. So not only was Morse responsible for the problem with the inadequate insulation of the Serrell cable, if we are to believe Fisher, he shared at least part of the responsibility for the defective Tatham cable as well. Word would eventually leak out but this intricate slur engineered by Morse, ending in Fisher's dismissal, would cast a shadow over Fisher's reputation from which he would never really recover.[46] Clearly the quaint notion of a meritocracy, in which Morse had clothed his endeavor, was now dead.

26

On the Third Floor of the Patent Office

Vail had been told in Morse's closing-up-shop letter of the 18th[1] to "render him [Cornell] *assistance[!]*"[2] As Morse well knew, the two of them (Vail and Cornell) would be cooped up in the Patent Office for the coming winter months together, and by using this particular turn of phrase, Morse had virtually guaranteed that what had been until now a simmering tiff would turn into a bitter blood-feud. It was not only the phrasing of Morse's instructions that had rankled Vail (who had always considered himself above Cornell's station, superior in both intellect and breeding), but the fact that Morse had addressed them in this case to both Vail *and* Gale was a further embarrassment.[3] In the meantime, Vail chose to divert himself during the seasonally shortened days by provoking Cornell in the manner that a child might provoke a dangerous animal in a burrow with a stick: sniping behind his back, spying on his mail, interfering with efforts to patent the cable-layer by diverting his mail to Ithaca.[4] There were other equally sophomoric jokes he played: when Cornell turned to Patent Examiner Charles Grafton Page for help with his electrical studies, Page had prepared a list of books for him on the subject.[5] After he was notified the list of books and periodicals he had ordered from the Patent Office Library were ready to be picked up, Cornell had found them suddenly unavailable. They were all checked out by Vail that same morning.[6] Cornell instead obtained them from John Meehan at the Library of Congress.

There was perhaps yet another reason for Morse's apparently temporary and uncharacteristic obliviousness to protocol. He had taken up residence for the winter in Commissioner Ellsworth's comfortable Town House in Washington, D.C. With the patriarch, Ellsworth, distracted by his own set of looming battles at the Patent Office, Morse would often find himself at home in the company of Ellsworth's attractive 17-year-old daughter, Annie, whom he had first

View of the Patent Office circa 1846, photograph by John Plumbe (courtesy Library of Congress, Prints & Photographs Division, Reproduction Number LC-USZC4–3596).

encountered in the course of her duties as a copyist at the Patent Office. So, perhaps distracted by the blow-up with Fisher or the battle shaping up with Smith over the trenching contract, or the tantalizing presence of Ellsworth's nubile daughter, Morse's sudden elevation of Cornell to Vail's equal, or implied superior, had not been handled with any great degree of tact.

On December 27, partly as an olive branch to Smith, with whom he was now engaged in a battle over the trenching, Morse made good on the promises Smith had been making to Cornell regarding plans to elevate him (Cornell) to the position of mechanical assistant to the telegraph. This would make him equivalent in title and salary to Vail. Morse, writing to Secretary Spencer, had said, "My present labors require the services of an efficient mechanical assistant whom I believe I have found in Mr. Ezra Cornell,"[7] adding as an apostrophe something that must have further scalded Vail's pride: "He will take entire charge of procuring and applying the materials ... subject only to the scientific superintendence of Dr. Gale or myself."[8] Smith no doubt had been the one encouraging Morse to fire Fisher, and Spencer no doubt would have questioned the extra expense otherwise, so Morse had started the letter, "I have the honor to report that I have dismissed Professor Fisher" to allay his concerns in this respect.⁹

By the New Year, Cornell and Vail were (at least theoretically) at equivalent salaries, and this had only added fuel to the already blazing fire.¹⁰ Vail had grudgingly tolerated Cornell's "two-cents" at the Serrell factory regarding the forced air pump test, and then the illicit midnight expedition with his assistant Avery to test the Serrell cable in the ground, the latter only because Cornell had gotten so far ahead of him in the trenching operations that it was embarrassing him. Cornell's "two-cents" had now turned into a (promised) thousand-dollar salary.

Halftone portrait of Henry Leavitt Ellsworth (courtesy Library of Congress, Prints & Photographs Division, Reproduction Number LC-USZ62–11615).

The only real consolation Vail had left was that, while he had been consigned to the airy and spacious third floor, Cornell was banished to the dank basement, stripping the insulation from the reclaimed wires. But even thus banished, Cornell had found a way to show up Vail. Amongst the various items stored away in the dusty, dank basement, Cornell had discovered what turned out to be Benjamin Franklin's original printing press from London.¹¹ The press would eventually be installed in the National Institute as an exhibit and prove to be one of the most popular. In the meantime, while he stewed over recent events and found new ways to provoke Cornell, Vail seemed otherwise marginally content to spend his time perusing the bounty of the Wilkes Expedition that had been deposited there and studying in the Patent Office library, working on his notes. He allowed himself only the occasional early morning excursion along the Potomac to augment his own personal collection of less than exotic petrified hickory.¹²

The intellectual resources were far from meager. In addition to the library, Vail and Cornell had access to the eclectic and exotic artifacts from the three-year-long American Exploring Expedition, some still in their crates, others scattered about the vast marble third floor—tantalizing, unassembled vignettes of some far-away civilization.¹³ The fabled Ex-Ex, under the direction of the

26. On the Third Floor of the Patent Office   141

equally fabled (and by many, hated) martinet Charles Wilkes, after three years roaming the oceans of the world, had returned home to America with this remarkable cache of artifacts and booty. They had touched the skirt of Antarctica and darkened the waters of the Pacific with bloody effusions. Among those who had not make it back had been Wilkes's own nephew Wilkes Henry, who was clubbed to death in the surf of a Fijian atoll while on an ill-advised fruit-foraging expedition. Aside from the critical nautical charts the mission had produced,[14] there were also hundreds of carcasses, dried or preserved, flora and fauna, seeds, 500 living species of plants, war clubs, shields, Polynesian masks and jewelry—indeed all the confounding and wonder-inducing spoils of the mysterious South Pacific, along with the copious notes from the expedition's prestigious, loquacious and energetic corps of naturalists. The specimens, along with the accompanying linguistic, botanical and zoological journals, had been carefully arranged and catalogued and packed in color-coded crates before traveling east overland, most arriving in Washington well ahead of Wilkes himself, who had taken the long way via Tierra del Fuego.

If one is tempted to sympathize at all with Vail's patrician frustration regarding Cornell's "good ole country boy" manners and persona, one need only refer to the amused tone of Cornell's letters to his wife regarding the intellectual spoils which now lay spread before him. Ensconced between the seat of power and never-before-seen wonders of the natural world, in typically sanguine (if possibly tongue-in-cheek) fashion to MaryAnn, he had written, "The objects are but few but still would be interesting to you; the Patent Office with the Gallery of the National Institute, the White House with the President, and the Capitol with the pictures in the rotunda and the Congress, are the only objects worthy of much notice here."[15] As a form of *noblesse oblige* and perhaps to disguise some of his more nefarious hijinks, Vail had offered to tutor Cornell in drafting. With the spoils of the expedition lying scattered about, there was certainly plenty of subject matter. Cornell, in his idle hours, using the skills Vail had taught him, began making sketches of the Polynesian armor, crafts and weaponry, that he sent home to Ithaca.[16]

Having bonded over their mutual antagonism to Vail during their midnight experiment, the two New Yorkers, Cornell and Avery, had partnered up with for occasional hikes as far south as Mount Vernon, sometimes in the company of their other roommate at Fletcher's, Charles Munroe.[17] On lignite hunting expeditions north of the city, commenting about the charred-looking rock, tongue in cheek, and obviously feeling lonely, Cornell remarked "This may be a place where Noah threw some fire brands over from the ark during his solitary voyage."[18]

It was no wonder that the subject of ocean voyages preoccupied Cornell.

The entire catalog of the American Ex-Ex had been given under Henry Ellsworth's care, and Ellsworth evidently was not at all happy about that,[19] a fact he had made plain to Joel Poinsett as well as to Wilkes himself.[20] The "objects" (Cornell had so loquaciously described to MaryAnn) now occupied a significant portion of Patent Office real estate, but remained under the authority of the National Institute (not that of the Patent Office). Ellsworth lacked both official custody and the color-coded catalogue lists that would have properly identified them, so uncrating them promised to be a frustrating and complex task; and while Ellsworth had *de facto* authority to catalog them, thus far he had mostly declined to do so.[21] Those that had been uncrated lay scattered about the expansive marble floor, stashed in alcoves, stacked by columns. It was a mess: skeletons half reconstructed without benefit of the documentation which had not arrived; Titian Peale's careful drawings strewn about the cavernous chamber in a disheveled jumble; some of the animal specimens reconstructed using the bones of other animals.[22]

In 1844 Wilkes had been summoned back to Washington to face a court-martial for his overzealous punishment of his men during the expedition, which included the whipping of his cabin boy in Morecambe Bay, Australia, for stealing soda crackers. As co-founder of the National Institute, to which Congress had consigned the task of organizing the collection, Joel Poinsett had agreed to take up some of Ellsworth's slack, but the man Poinsett picked to curate it had made such a mess of things that they finally had to call in Charles Pickering, the expedition's naturalist, to rectify matters. On seeing what had transpired, and lacking the crucial color-coded inventory that had mysteriously disappeared in transit, after a few months Pickering too had thrown up his hands in disgust and left for a tour of Egypt and India. By the time Vail and Cornell arrived, fresh from the telegraph fiasco, it was all under the care of Wilkes himself. Wilkes had begun construction of two huge greenhouses behind the Patent Office for the botanical specimens, with the second completed just after Vail and Cornell appeared on the scene.[23]

Among the other items that occupied the vast chamber overlooking F Street NW, in addition to the American Ex-Ex artifacts, was the library of prior work, physical models submitted in support of patent applications. These occupied the entire other half of the upper floor.[24] Though clearly distracted by the ongoing tiff with the National Institute over them, Ellsworth evidently liked Cornell well enough that he took pains to make both the collection and the Patent Office Library freely available to him.

It was a unique opportunity for an autodidact like Cornell. Aside from the miniature models, the Patent Office library bulged with books and periodicals on the very subjects that formed the crux of his present interest and wherein lay, as he saw it, at least part of his future fortune. Starting with works on cements and mortar,[25] he had gradually moved on to more abstruse topics like electrical circuits. Having learned from Page that Gale thought the phantom signals that plagued the underground cable installation might have been caused by some new force called "induction" instead of simple "conduction," Cornell decided to read up on Joseph Henry and Michael Faraday's treatises on that subject as well.[26] He seemed to be girding himself for some upcoming intellectual combat, saying, "Realizing the importance of more definite information on electrical science, I decided to utilize the long winter evenings in study."[27] As the field of electromagnetism was still in its infancy, though lacking a formal education, Cornell was beginning to feel he was not at an insurmountable disadvantage when it came to that particular subject. While his cable-layer may have posed no threat to Morse, his newfound interest in electromagnetism would eventually lead to a titanic clash with his erstwhile benefactor.[28]

Morse, though himself a corresponding member of the National Institute since 1841, had evinced little interest in the Wilkes artifacts. There were tasks that needed to be done for the Test Telegraph if they were to resume work in the spring. First, there was all the valuable copper wire in Serrell and Tatham's coils that could be reclaimed. Despite Cornell's recalcitrance due to his growing interest in more erudite subjects, Morse asked him to press ahead with the stripping and re-insulating chore.[29] We can assume that while studying early experiments in electromagnetism, Cornell had come across mention in Henry's article in Silliman's describing Lewis Beck's more economical method of wrapping the conductors for electromagnets. Since he had little desire to remain locked up in the basement with the unreclaimed pipe, this spurred him to make some experiments[30] of his own in a similar vein. Cornell soon convinced Morse he could devise a machine to rewrap the insulation more efficiently, writing MaryAnn, "I am engaged this week … in building a machine to rewind our wires with cotton yarn. I intend to rewind 4 wires at a time."[31] Having by now learned to put some stock in whatever Cornell said, Morse wrote brother Sidney asking him to forward immediately any stores of cotton material he had in left New York from the previous fall for Cornell's use.[32]

The wire he was extracting from the pipe was #16 copper, wrapped with cotton.[33] Cornell had recently learned from one of the patent examiners who had befriended him, Gale's old friend, Professor Charles Grafton Page, that if Morse proceeded as now planned in the spring with the overhead system, bare wire could possibly be used, making his present task seem even more pointless. Despite Morse's continued urging and his own awe of "the Professor,"[34] Cornell, in this case, knowing Morse's plan, decided to spare himself the needless drudgery of rewinding insulation onto 30 miles of wire, focusing instead on his own projects. Intent on finding other sources for the wire,[35] he had gotten an estimate for brand-new copper at a very reasonable 35 cents a pound[36] and found a way to economically dispose of most of the un-reclaimed lead pipe and wire by selling it back to David Burbank.[37] This was the more logical course. Burbank could process the wire at his Baltimore warehouse far more efficiently than Cornell could in the basement of the Patent Office by simply melting down and reselling the lead covering, and burning off the cotton insulation from the copper wire inside.[38]

27

The Burden of Big Science

Government-sponsored scientific endeavor certainly had come a long way from the days of SEUM and SPUA. The Morse telegraph had been labeled an "experiment," and it was an experiment not just from the scientific standpoint, but from the standpoint of the government's participation in what would come to be called "Big Science." The Test Telegraph and the Wilkes Expedition had both been examples of this new and expanded role the government was willing to assume in scientific affairs, and it was thus somewhat ironic they had both ended up as orphans, sharing space at the Patent Office.

The National Institute itself, though lacking physical facilities of its own, had been formed largely for the purpose of deciding how to spend the massive Smithson bequest.[1] The unprecedented expense of the American Ex-Ex had been an object of intense political controversy and the choice of Wilkes to head it was extremely unpopular in Navy circles. Somewhat ironically, it had been Morse's old friend Joel Poinsett from South Carolina, now secretary of war, who had been instrumental in selecting most of the officers for the expedition, and he had hand-picked Wilkes, not yet even officially of captain's rank, over several more experienced Naval officers. On their return, it had been once again Poinsett who had formed the plan for using the spoils of the Ex-Ex to leverage Congress into creating a national museum with funds from the Smithson bequest.

The Wilkes Expedition had been under the auspices of the Navy, but oversight of the telegraph, at least technically, had been assigned by Congress to the Post Office Department. Partly due to Morse's reputation on the hill as a notable crank, the Assistant Post Master General, John A. Bryan, had looked on the whole project as "an abominable plan to ruin me."[2] Finding the Post Office *abdico consulatum*, Morse had been mostly running things himself. The Department of the Treasury, retained approval over all contracts and expenditures and in the case of the telegraph, the "crank," Morse, had been scrupulous, detailing his expenditures down to the penny. Every contract had been submitted for approval beforehand and each and every expenditure approved or disapproved in advance. Colt's big grant from the War Department had, in a sense, been Big Science as well, but the economic model was different and the oversight had been far more lax. This was due to the fact that his support had not initially come from a direct appropriation but from the budget of the War Department in the form of stipends, explosives and ships to blow up.[3] If the higher-ups disliked the way Colt was spending their money, their option was to just yank his funds entire (which they had in fact done from time to time). For Morse's appropriation, all the money had to be expended and accounted for. Despite Morse's scrupulous accounting, the government was, thanks to F.O.J. Smith, about to get another taste of what it was like to be on the hook for the success or failure of a major project over which they had little or no direct control.

Morse once had written his good friend Levi Woodbury, former Secretary of the Treasury, "It is indeed gratifying to me to find that the greatest intellects, the highest scientific minds of the world ... are precisely those who have manifested the most enthusiastic interest in the Electro-Magnetic Telegraph."[4] No one was more aware than Fog Smith that, despite his evinced enthusiasm for the project, he himself did not stand amongst Morse's "greatest intellects" or "highest scientific minds." Amongst the corps of fledgling telegraphers that Morse had assembled, Smith stood out like a sore thumb. He had only his roommate Cornell, the ditch digger and plow salesman, for company, and now it looked like even the stodgy, down-to-earth Cornell was about to defect to the ranks of the "scientificks."

Though he whimpered at the prospect, Smith clearly had no real objections to being the fall guy, so long as he got a sufficient payday, for which aim he now appeared ready to sacrifice any hope of resuming work in the spring. The remaining $4,000 of Bartlett's contract was at stake, and Smith was about to go all in, demanding he get paid even if it meant the end of the project. Cornell had been rooming with Smith at Fletcher's Boarding House near Union Station, and so Smith no doubt had been apprised of the ongoing tensions at the Patent Office. Despite Smith's constantly egging him on to undermine Vail to Morse behind his back,[5] Cornell had refused to take the low road, and thus far that strategy obviously had worked for him. By the beginning of the New Year, however, it seemed Smith's constant harping had begun to work on him as well.

As far as a assigning a scapegoat was concerned, Smith had been the more natural target than Fisher. Indeed, it seems evident that his strategy from the beginning had been, through sniping and back-stabbing, to intentionally exploit

27. The Burden of Big Science

any cracks in Morse's little intellectual meritocracy whenever and wherever he could. It is not out of reach to say that Smith may have in part engineered Fisher's disgrace himself. This kind of maneuvering had been allowed for a while unopposed, but only while Morse was preoccupied with the "cross-cut" issue. Where the "scientificks" like Gale and Fisher had been the highest paid of Morse's assistants, Smith had been consigned to skimming what he could off the trenching contract. That and his patent stake had been his saving grace till now. With both Fisher and Gale now out of the picture and the trenching operations shut down for the winter, Morse had plenty of time on his hands to reflect on whether or not he too had fallen prey to Smith's machinations.

Gale had taken Fisher's firing harder than anyone, but still retaining some influence over Morse, he was determined to keep the management of the project, what remained of it, in scientific hands. Obviously unaware that Cornell had been appointed to the position, he had proposed that Morse replace Fisher with Charles Grafton Page, chief examiner at the Patent Office. Page, like Fisher, was both an accomplished scientist and Gale's good friend and colleague, and perhaps more significantly to Morse, he was morally pliable.[6] Page as well as Ellsworth had befriended Cornell during his winter confinement at the Patent Office, and even offered him guidance in his newfound avocation, the study of electricity.

Though Gale had contritely agreed to forego his salary for the time being,[7] he vigorously denied any part in the cable foul-ups. Once recovered from his illness, he suggested a series of experiments to definitively determine whether the "cross-cuts" were in fact the fault of the singed conductors, by proposing that, over the winter, Morse run two parallel lines, using the Tatham cable, from the Patent Office to the Capitol[8]—one underground and one above ground—so they might see exactly where the failures were occurring and why. This had not sat well with Smith, who just wanted to get paid for the trenching contract and be done with it.

With little or nothing to show for the past three months of labor, Gale's suggestions were clearly in part a "Hail Mary" designed, like Cornell's fake cable-layer accident, to allow Morse to save face in the press (and also generate a little revenue, as they would charge for the experience of sending a message).[9] Morse was by now desperate for a new appropriation. Demonstrating the telegraph's workability, even on a reduced scale, to the Congress and the public, might make it appear that a new appropriation was not throwing good money after bad.[10] Morse was sensing the imminent collapse of the entire enterprise, running out of funds and finding himself under attack from every quarter. The problem was that he had already informed Bartlett that he could go home to New England for the winter.[11]

The idea seemed too good to pass up in several respects.[12] Not only would it put Bartlett's idle trenching crews back to work, it would provide some use for all the copper wire Cornell so assiduously reclaimed from the Tatham and Serrell cables and without endangering any new efforts in the spring by using possibly faulty wire. While not the 40-mile demonstration Morse had wanted, in addition to possibly settling the "cross-cut" issue, a line inside the city would at least be something he could point to as a solid achievement.

The main problem for the "in-town" demo was the ongoing uncertainty over the Bartlett contract. For Morse to implement Gale's parallel configuration, he needed the trenching crews, but he had already tried once to send Bartlett home. Evidently hoping to see the matter closed, Morse had proposed a compromise: paying Bartlett an additional $1,000, making the total paid out $1,500, or one-quarter of the full contract price on the project (roughly equivalent to the work already done on the test line), with the stipulation that Bartlett would resume work on April 1 (*except for "a certain distance between the Patent Office and the Capitol"*).[13] With ten miles completed, this would only cover his actual work to date but with no compensation for the stoppage, and further obligating him to do the trenching for Gale's experiment. Apparently Bartlett was initially agreeable to this new arrangement, as Secretary Spencer sent him a new bond attesting to it.[14]

The main hypotheses that Gale wished to test was not that the "cross-cut" resulted, as Henry had previously suggested, from increased voltaic potentials, but rather that it might be the result of induction (not conduction)![15] But in encouraging Morse to pursue this course, Gale was putting himself in a position to suffer financially. Gale had bought up tons of raw lead with his personal funds "on spec" to be resold to the government for Morse's underground cable. If his experiment proved the superiority of the overhead approach, Gale stood to lose all or most of the $2,500 investment.[16] Showing himself firstly a man of science, Gale had nevertheless pursued this idea as a means to determine if the cause of the cross-cut was due to *in*duction or *con*duction, noting that if it was in fact induction, then "there is no remedy but to remove the wire from the pipe and suspend them in the air on supports of wood."[17]

Morse had set the crews to work on the double line between the Capitol and the Patent Office, but they did not get far before work was stopped. It was mostly smoke and mirrors anyway. Morse had already decided on the overhead approach, and the real show was what he had been planning with Cornell—to get twelve miles of overhead line up towards Baltimore before the thaw.[18] By mid–March the test wires were still not completed inside the city, but the point was moot. Gale's answer would have to wait.

28

Bartlett's Contract

The meetings at Relay House had continued until the 20th of December. The decision to stop operations had been made almost immediately. On the 18th Morse had presented Smith with a proposal about what to do about the rest of the trenching. Bartlett was to run Gale's test line inside the city and then go home until April.[1] Smith had refused. It was clear to Smith it had been Morse's aim since first learning of the pipe defects to set up either Fisher or him as the fall guy. On the 20th Morse had played his one card, firing Fisher. Smith had seized on this immediately, saying it was clear that the present situation was due to the government's default and demanding full payment on the balance of the Bartlett contract. While he now had his fall guy, if Smith prevailed in his claim, Morse might have to cut a $4,000 check to Bartlett out of the remaining $7,500 appropriations money, thus effectively killing the project. Smith, realizing he was in the clear, had gone to Washington, where he drafted a letter for Morse to sign excusing Bartlett of any liability and admitting full responsibility for the default.[2] Morse had refused to sign it.

Probably, Levi Bartlett was already fed up with the whole project and certainly wanted nothing to do with this new inspiration of Morse's for winter trenching inside the city for a measly $500; however, unlike Smith, he was not at all keen on back-stabbing Morse, and this is where he parted company with Smith. But for Smith to intervene more forcefully, the "man behind the curtain" had to finally reveal himself entirely, and this of course represented one more thing (besides the skimming on the Tatham contract) about which Morse thus far had neglected to inform the Secretary of the Treasury.

Smith was well acquainted with the ins and outs of government funding from his experience on the Commerce Committee. He was also personally ignorant, deceitful, crass and crude and took few pains to conceal these faults, but he knew how to play a poker hand. He was aware all of Morse's most recent problems regarding the cable could possibly be traced back to him and the

Tatham contract, and that Fisher might not have been the only one in Morse's gun-sights—maybe there was another shoe to drop.[3]

Morse had apparently gone directly to Bartlett as on the 23rd, as he informed Spencer that he had reached an accord with Bartlett over the trenching.[4] The Secretary had then sent back a bond for Bartlett to sign.[5] Everything seemed fine. Morse, in fact, felt confident enough to appoint Cornell as mechanical assistant at a salary of $1,000 per year.[6]

Any default leaving Bartlett's crews idle had stemmed from the fact that Tatham's cable had proven unusable. Since Smith himself was the one who had promoted and negotiated the contract with Tatham in the first place, for him to charge default could only have been construed as a remarkable exercise in self-rationalization. Morse, thinking he had the upper hand, had then taken a misstep; he demanded compensation from Smith for the defective Tatham pipe, proposing he make the government whole by seeking restitution from Tatham.[7] Smith could only have viewed this, with the project seemingly on the ropes, as Morse's trying to pick his pocket. Nevertheless, Smith had dutifully whined to Tatham, "Under the circumstances it is proper that some mutual understanding be devised with regard to the loss…. I do not feel it proper that I should be exposed to bear it alone."[8] Smith, however, was not going to stand for this situation for long. On the 29th, Morse provided Smith the ammunition he needed when he wrote to Fisher laying out the charges against him in full, saying, "*Your unfaithfulness to your trust* (since you oblige me to divulge it) is the principal cause … the insulation was unsound … it would not have been left to me to discover the disastrous defect at the moment of putting the pipe into the ground … compelling me to suspend the trenching operations, *making the Government liable*."[9]

Morse had initially gone along with Smith's plan to disguise his role in the Bartlett contract from the Secretary of the Treasury. With no further qualms about embarrassing Morse, Smith now stepped forward as "attorney in fact," and claiming that since the government had defaulted, he was due payment in full of the balance of $4,000 on the contract.[10] Morse and Smith both knew this was a reach, but explaining why to Spencer would have meant embarrassment for both. Morse had reached out to his friend Henry L. Ellsworth, who had counseled him not to accept Smith's role as attorney-in-fact, realizing the position in which this might put him, but Morse had not heeded his advice.[11]

Morse could have simply said "no" to Smith, but there was, for him, a second and far larger fly in the ointment. When Smith had signed the contract with Tatham Bros. back in late October, he had been able to convince a distracted and desperate Morse that in so doing, he had saved the project over $1,000 and demanded that they jointly pocket half of that sum each. Morse

28. Bartlett's Contract

had caved and awarded Smith his half of the purported savings immediately,[12] but prudently declined to pocket his own half, crediting it back to the government instead, having this latter transaction witnessed by Henry Ellsworth.[13] Nevertheless, he had failed to report Smith's malfeasance to Secretary Spencer at the time, and thus had put himself now in a position to be blackmailed.[14] With Smith in a position to reveal all this, it seemed Morse had little choice but to cave in again and pay Bartlett in full for the unperformed trenching.[15]

Once Smith had stepped forward, naturally all hell had broken loose. Despite the feeble attempt at conciliation, and while Morse was still fuming to anyone who would listen regarding Fisher's malfeasance,[16] it was Smith, however, who now became the epitome of evil and the devil incarnate. Morse declared to brother Sidney at the Observer offices of Smith, "Where I expected to find a friend, I find a FIEND!"[17] Smith remained adamant about getting his entire cut for the trenching, and when Morse had balked, he had threatened to go over his head (directly to the Secretary of the Treasury or the president). Vail, for his part, was just glad to be out of the line of fire and had written his wife, "There is more going on here than you can possibly dream of. But you cannot know; only that I am not in the scrape."[18]

On December 29, a furious Spencer had written back, "Since it now appears he [Bartlett] has assigned his contract, I do not see why the government should be limited in its dealings with the person whom he may choose to designate."[19] When Spencer had summoned Bartlett to his office, Smith had shown up instead. The negotiations with Smith evidently did not go well as Secretary Spencer, having initially agreed to negotiate the default,[20] now flat-out refused to sign any new contract.[21] On January 3, Morse had finally come clean about Smith's skimming on the Tatham contract and laid out the circumstances in full regarding the pipe.[22] Obviously fed up and feeling deceived, Spencer had put Morse, as the "agent of government" in the whole matter, on the hook for damages, instructing him not to pay Bartlett any damages since any indemnification had to be authorized directly by the Congress.[23] When he learned of Spencer's decision, Smith had made good on his threats to go directly to the Congress[24] and the president,[25] but it was to no avail in either case.[26] As everyone but Smith seemed to be aware, the circus was over, at least for the year, and the clowns were being sent home. Ten days later Morse sent a follow-up note saying that further repairs to the Tatham pipe would cost about $500, clearly suggesting out of whose pocket the payments should come (Smith's).[27] Five days after that, Smith was instructed to discreetly pay Tatham off in full, ostensibly closing matters with them as well.[28]

While Smith's more dubious financial stratagems had failed by now, his relentless psychological and legal attacks had taken their toll. Morse had essentially laid himself at Spencer's mercy. Increasingly isolated and despondent, Morse had sought refuge with Patent Commissioner Henry L. Ellsworth. Even in the dismal NYU years, he had never been at a lower point, having written to brother Sidney just after Smith exposed his role in the Bartlett matter, "I have no heart to give you the details of the troubles *that almost crush me*."[29] With the payments to Bartlett and Tatham to close out their contracts, he had expended all but $7,000 of the appropriation. There was little hope for a new one thanks to Smith's shenanigans. Morse had little to show for all his years of effort except nine miles of intermittently functioning telegraph line, plus thirty miles of singed, unusable wire, most of which now sat in rolls or bales in the basement of the Patent Office.

He had summed it up in a letter to brother Sidney resignedly: "Fisher I have discharged for unfaithfulness, Dr. Gale has resigned from ill-health, Smith has become a malignant enemy and Vail only remains true at this post"[30] (neglecting to make any mention whatsoever of Cornell). Relieved that Smith's larceny had finally been fully exposed, he was still beset with doubts regarding the firing of Fisher. It was all too much for Morse, who was now sick in both spirit and body. He confided to Ellsworth in a moment of candor, "My temperament, naturally sensitive, has lately been made more so by the combination of attacks by deceitful associates without and bodily illness within."[31] Faced with what appeared to be insurmountable circumstances (for which he partly had himself to blame), he would convalesce at Ellsworth's house, where through the dark winter he would console himself with the charming company of Ellsworth's daughter, Annie, and visions of a national telegraph network constructed on the thin fabric of Cornell's optimistic estimates.[32] He did nothing except emit the occasional press release, sitting in on several sessions of Congress, hoping to see some way through to a new appropriation, but clearly the case was hopeless. It was basically a choreographed ballet with the dark angel, with Morse playing the "back-stabbed" dying swan through the waning days of a dismal January.

29

Cross-Cut!

For Samuel Morse, engineering hurdles, purely practical decisions of a technical nature, always seemed somehow to become intertwined with great moral questions and issues. With time now to reflect, it must have become clearer that it was entirely possible his former dear friend and colleague, Professor Fisher, had not in fact deliberately betrayed him or lied about his test results. But despite Fisher's having pointed out the reasons in some detail, Morse had remained seemingly genuinely perplexed, unable to fully comprehend why Fisher had gotten different results with the Tatham cable from Burbank on the integrity tests.[1] It was entirely illogical, but it nagged at him like a scab he could not help picking.

Gale, meanwhile, had raised two further troubling possibilities, ones that went toward Fisher's defense: that a damp atmosphere might have caused different results, and also that the "cross-cuts" might not have resulted from "*conduction*" at all, but to a force called "*induction*."[2] Induction was a force that passed through any insulation, regardless of the soundness, quite free and unaffected. All the back-and-forth sniping with Fisher, coupled with Gale's new speculations, thus had congealed into a discomfiting prospect: perhaps he had been wrong to blame Fisher.

A phrase Fisher had quoted in defense of himself still rang in Morse's head: "In a long line, no mode of insulation but the air will prove effectual."[3] In his rush to condemn Fisher, Morse may have forgotten all about that previous conversation. Fisher had indignantly raised it, saying, "I wrote to Prof. Henry,[4] *whose letter in reply you saw*" (referring to Henry's letter back in April[5]); that voltaic potentials ("tension") might inevitably increase in tandem with line length (common surface area), thus making short-circuits virtually inevitable in a long underground cable—no matter how good was the insulation.[6] It was as if Morse had blocked out the unwelcome news and, as his subsequent actions testify, it truly seemed at the time to be entirely new information to Morse when he confronted Fisher.

There was no question Fisher had approved the initial run of Tatham pipe on November 10, but this had been, according to Fisher, also at Morse's instructions![7] Though the practical question regarding whether to go to an overhead line, for all intents and purposes, was already settled,[8] getting a definitive answer regarding Fisher's supposed perfidy had become an obsession for Morse; why would differences in the strength of the current create a short circuit in one instance and not in another? To answer this, Morse had first turned not back to Henry,[9] who had raised the issues regarding the "cross-cut" a full eight months prior, but to the other scientific minds still at his disposal: Leonard Gale and Charles Grafton Page at the Patent Office.[10]

Morse's obsession with finding the explanation for the short circuits had by now devolved into a personal preoccupation, in part prompting his plan to run the parallel lines in the city, but also resulting in a rather bizarre experiment with Charles Grafton Page that took place on the roof of the Patent Office Building. In late January, Morse had attempted to put at least the first of Gale's hypotheses to the test: to see if a large enough surface area of a conductor could induce a cross-cut even through the medium of the air. The second part of the test called for testing this also under humid conditions, and for that part, Morse had choose his timing of the test carefully.

The roof of the Patent Building in Washington was covered in 22,000 square feet of copper sheathing, seemingly ready-made for the purpose of testing Henry's theory about a large surface area of conductor. Morse and Page had set up their equipment on the roof. Using a Grove battery, Page sent a current through the vast copper roof. He then suspended a zinc plate a few inches above the flat roof, connected to a galvanometer. At first nothing registered on the galvanometer. Then it began to drizzle. The zinc plate and copper both became wet and were now separated by a humid rather than a dry interval of air. Suddenly, the galvanometer jumped. Morse had his answer. Apparently all of Henry's assurances about air being virtually an impassable insulator were only true in the absence of moisture.[11]

Possibly hearing of Page and Morse's experiment, Joseph Henry had proffered his own theory regarding the "cross-cut" phenomenon, spontaneously writing to Morse:

> During ... my investigations in electricity ... among other results I arrived at one which I think may have an important bearing on the success of the Telegraph ... while a current of electricity is passing through a wire, one part of the conductor is constantly plus to any other part which succeeds it, the difference in the degree of the electrical state constantly increasing as the distance of the two points is greater ... the insulation which would be sufficient to stop a current when applied to separate two consecutive portions of the same wire which had been divided would be entirely insufficient to prevent the cutting across....[12]

29. Cross-Cut!

Morse knew it was unwise to ever underestimate Henry's advice.[13] The fact that Henry had chosen of his own volition to write him about this matter must have only reinforced Morse's worry over the likelihood of "cross-cuts" in a long wire due to increasing voltaic potentials.[14] Henry's analysis in this case, however, would prove incorrect; the possibility of a short circuit occurring is equally great at the beginning of a long wire than at the end (and possibly more so).

Aside from what was contained in Ellsworth's annual report to Congress in February, Charles Page had also likely informed Morse by now that if he went to an overhead solution, the parallel (fifth) ground wire would become redundant. Grounding at the pole had already been employed successfully in Europe by Wheatstone, Steinheil and Cooke for several years.[15] If Page was right, all Morse needed were two transmission wires—one for sending and one for receiving—and a plate buried in the ground at intervals, serving as a ground.

With both Page and Henry's advice in hand,[16] by February 4, Morse had made up his mind irrevocably.[17] Come spring he would go to an overhead solution.[18] Based on this decision he now prepared a whole new set of interrogatories for Henry that clearly framed the issues that he would face in converting a buried cable to an overhead line. They also seemed designed in part to impeach Henry's former advice given to Fisher: "Could a stronger current penetrate even sound insulation?" "Was any insulation going to suffice over longer distances?"[19] Rather than having these sensitive questions answered in a letter, Morse had made the decision to take advantage of an unseasonable warm spell to visit to Henry in person. He wrote Henry at Princeton, apologizing for the delay in responding, announcing he was coming to visit him in a week,[20] placing an ad in the Washington papers for bids on seven hundred chestnut poles the day before.[21] Mohammed may have been going to the mountain, but Morse wanted it clear to everyone—he had made up his mind *before* consulting Henry, not *after*.

30

A Fight Over Pole Insulators

By the first week of February of 1844, Morse was firmly committed to the overhead pole solution, the same approach that Wheatstone and Cooke had employed successfully in England, and the same one that Morse had rejected only a year earlier.[1] This shift brought with it several fresh challenges, the main one being how to keep the transmission wires from grounding out or "cross-cutting" where they contacted the pole. Wheatstone had utilized a custom-made porcelain conduit affixed to the cross-pieces, but this was rather expensive and would require time to manufacture. Time and money were luxuries Morse no longer had, so he had instead turned to his two assistants, Vail and Cornell, to provide a better idea.[2] The relations between Vail and Cornell were already at an all-time low, and now Morse had pitted them head to head to come up with the pole insulator solution.

For Cornell, this abrupt shift in strategy already had yielded several unfortunate side effects. It effectively eliminated both avenues of revenue on which he had based his optimistic assessments of future riches to his wife, MaryAnn: his cable-layer (that he was about ready to patent),[3] and the new wire-winder-insulator that he hoped to patent in the near future. Both would be rendered largely irrelevant and unprofitable, at least as far as the future of the telegraph was concerned, by Morse's decision. Aside from these two soon-to-be-dead horses, there was another invention, the exact nature of which Ezra seemed reluctant to commit to paper, saying only that it was obvious, "like an old woman looking for her specks all over the house when they are snugly perched on her nose."[4] It evidently had excited him far more than the cable-layer, saying only it was "an invention in embrio [sic] that ... will open the eyes of the world. It will be far in advance of anything of the day and it astounds me that it should have been overlooked so long.... I am unsuspected of having anything of the kind in contemplation."[5] Obviously he wished to keep it secret. Perfecting it had formed the crux of his request for the loan of technical books from the

30. A Fight Over Pole Insulators

Patent Office Library—and also his reason for seeking out the advice of Charles Grafton Page.

Vail's solution was unimaginative; it required the expensive manufacture of an asphalted corset that would gather all the wires into a group.[6] Cornell, with his innate talent for improvisation, had come up with one that was not only innovative but inexpensive: using the glass knobs that were commonly found on the bureau drawers at Fletcher's rooming house (which were both cheap and readily available) as a spacer between the pole and the wire (just scoring them to seat the wire and then cementing them with asphalt). Unlike Vail's, Cornell's solution did not require gathering the wires close together. With Morse's concerns about "cross-cuts" running at full throttle, it would have taken very little to sway him toward an approach which not only kept the wires at a maximum of physical separation but was also cheap and effective—in other words, Cornell's approach was the more logical, but Morse was aware that preferring it would be taken by Vail as yet another slap in the face.

Having made up his mind, but alarmed by Vail's vehemence on the subject, he agreed to let Joseph Henry decide. So in addition to his questions on the cross-cut, in mid–February, Morse brought along with him samples of both versions of the pole insulator. When Morse returned to Washington, he let Vail know the bad news: Henry had clearly preferred Cornell's design over his.[7] This was the last straw for Vail. First Cornell had shown him up over the leakage test for the lead pipes at Serrell's plant; next he had criticized his method of testing the buried wires; then he had been the one to save Morse from public embarrassment by wrecking the cable-layer; and this, the final insult—Henry and Morse's combined imprimatur on Cornell's skills as an inventor. The former sniping and schoolboy pranks would soon take a decidedly vicious turn, and Vail would henceforth do whatever was in his power to undermine Cornell (either overtly or covertly) and not just Cornell, but the man whom he now saw as Cornell's ally, Joseph Henry. In both cases, Morse would eventually be held himself responsible and answer for Vail's antipathy, and the cost would not be slight.

31

Out of the Frying Pan

"Mr. Morse wishes to see Mr. Cornell before he goes to his tea."[1]

Cornell's frequent and affectionate missives home to MaryAnn, prior to and just after his arriving in Baltimore in October of 1843, while not containing any specific news of the telegraph, had been informative and perennially chatty and upbeat, brimming with the news of his other doings and questions about the status of matters at home (a constant topic: his weight) and the garden.[2] Then, just as had occurred immediately following the brutal events in Jamestown with the unfortunate Alvin Cornell affair, there was an uncharacteristic hiatus; the next two months passed without a jot from the home front except for one flimsy letter from his brother Edward concerning his creditors.[3]

Despite Ezra's undeniably brightening prospects, there was likely a financial reason behind this lacuna as well. With his debts at home for the abortive plow venture mounting, his creditors were clamoring for repayment. Sales of the plow had not covered the loans he had taken for the sales trip. He had, to date, received two dollars and a quarter from Morse (plus expenses), none of which obviously had made it home to MaryAnn. Smith till now, true to form, had also paid him entirely in promises and declined to buy any interest in the cable-layer.[4] To make matters worse, in his last letter to MaryAnn, Cornell had proposed that she relocate the entire family to Washington, a suggestion which evidently had not gone over well, and perhaps also partially explains the subsequent extended hiatus.[5]

Finally, on New Year's Day, Ezra broke his uncharacteristic silence with a typically lengthy, somewhat contrite but also slightly humorous missive making clear that he had (in case MaryAnn had not already gotten the point by his use of the Barnaby and Mooer flyers for stationery), been conducting himself in as frugal a manner as possible, going so far as to deny himself the normal diversions of the season—"while others have been making calls on the president

31. Out of the Frying Pan 159

and other lions in the national cages I ... have been confined in the ... Patent Office attending to my business."[6] Despite the renewed chattiness, it is clear he was bursting with some important news. He delivered it only after first acknowledging his over-inclination toward verbosity ("I wish I had a pair of boots that wear as long as my notes"[7]), finally letting slip in closing the fact their financial prospects were brightening: "The Secretary[8] has approved of my appointment as Mechanical Assistant at a salary of $1,000 per annum. So that is all right."[9] Morse was paying his room and board, but his first paycheck would not be due until mid–March, and the way things were going, doubtful if then. The creditors would not be happy.

As if to underline the fact that, just as he claimed, "every moment day and night is occupied either with my business or my inventions,"[10] his letters home had all been datelined the Patent Office instead of the rented residence at Fletcher's where he was rooming with F.O.J. Smith and others then on the payroll, like Thomas Avery and Charles Monroe nearer the General Post Office. Though the lengthy letters to MaryAnn show how guilty he felt about abandoning the family for such a long period of time, it is also clear he did not much miss the long, dreary upstate New York winters (though later he would grudgingly admit that while treading the carriage-churned, muddy streets of Washington, he did long for the twinkling purity of a freshly snow-covered pristine upstate New York landscape).[11] If he was looking for purity, however, as MaryAnn likely well knew, Washington was perhaps not the best place to find it.

Despite the fact that he had finally abandoned the practice of using the leftover Barnaby and Mooer handouts for stationery, it is clear Ezra Cornell still found himself unimpeachably impoverished and probably embarrassed about his situation. When, thanks to Vail's malicious meddling, MaryAnn mistakenly received the reply from the Patent Office regarding the pending plow patent, he had asked her to forward them back to him at the Patent Office address, instructing her to use the same envelope and cross out his address to save the $2.00 postage.[12]

Regardless of his financial straits, Cornell seemed determined to find Washington not entirely bereft of charm and entertainments. But since he had repeatedly assured her he had no life at all outside of work, any apparent departure from this policy would naturally need to be handled diplomatically. Thus, it had not been until a full month after his appointment as mechanical assistant that he finally allowed himself to acknowledge it with any celebratory excursions. Clearly in a buoyant mood, having decided to take in the Capitol and the sights of the surrounding countryside, he had written to MaryAnn giving her a detailed description of his visit to the Rotunda and all the artwork there (which he appraised conservatively, with a taxpayer's jaundiced eye, at "more than $10,000"): the Pilgrims, the signing of the Declaration of Independence, Pocahontas pleading for the life of John Smith, the surrender of Burgoyne to Gates at Saratoga, Andrew Jackson—all were described in scrupulous detail for her benefit and that of the children, Alonzo, Elizabeth, Charles and little Oliver Hazard Perry.[13]

While he was still rooming with Fog Smith it seems that some of Smith's unfortunate social attitudes had rubbed off on him: "While I am writing Smith is complaining of the 'nigerfide' [sic] condition of things in this region. He says you can't get a white ribbon without a black thread running through it. There is a black streak through everything here among the niggers and I think he is about half right."[14] Despite his casual adoption of the "n" slur, obviously Cornell's Quaker upbringing meant he could not long tolerate Smith's blatant and repulsive racism. While the business alliance would persist thereafter, the housekeeping arrangement with Smith would not have to last much longer. Smith, enmeshed in the blowup with Morse over the cable, would soon return to (presumably lily-white) Portland.[15]

Cornell was about to become the author of Morse's March Miracle, and despite MaryAnn's pleas, he had decided to remain in Washington to get ready for the big push. On February 16, Cornell had written MaryAnn: "I shall embrace the first opportunity that presents to leave and pay you a visit but it is quite out of the question for Prof[essor] Morse to get along with his telegraph as he is now situated without me ... he is anxious to get 12 miles in operation by the first of April."[16] Despite all his drawings and attempts at lively and vivid description of Washington life, things had not improved much. Having been absent from Ithaca for over a year, Cornell had begun to feel himself losing his grip on his family. In Ezra's letters to MaryAnn he had even asked her to hold up someone besides himself as a role model to the children: "If on the whole you are disposed to think my nature devoid of the milk of human kindness, I hope you will inculcate a better system of morals in our dear children."[17] He offered up Benjamin Franklin as a role model for them—surely a noble example but a poor substitute for a father. "Big Science," as Cornell was learning, had its costs.

Smith was no doubt filling his ear with tales of Morse's perfidy, most of which Cornell, for the present, would choose to ignore, even though Morse had adopted Cornell's "bureau knob" insulator design without giving him so much as a nod of credit.[18] With his prospects for raising quick cash soon to be shut off thanks to Morse's decision to go to an overhead pole solution, Cornell found himself more dependent than ever on his new benefactor, Henry. Not all of Smith's bad influence, however, had been wasted. Cornell had adopted Smith's facility for skimming, as he would pocket $100 for the new pole contract with

31. Out of the Frying Pan

David Burbank.[19] He also was learning to make use of his new acquaintances and had asked Ellsworth to make mention of the cable-layer in his next report to Congress, writing to MaryAnn rather over-optimistically, "Mr. Ellsworth will notice my machine in his report to Congress.... I had rather realize two or three thousand dollars for it at once than to be dreaming of anticipated millions."[20]

Ellsworth proved as good as his word, devoting an entire page to the cable-layer, along with an engraved woodcut of the machine (probably provided by Morse).[21] Also cognizant of the fact that Morse's recent decision to move to overhead lines might severely affect the value of Cornell's patent, Ellsworth had graciously added a blurb saying the cable-layer was "suggested originally for the purpose of laying pipe for Professor Morse's telegraph, but is no less adapted to the purpose of laying lead pipe for conducting water."[22]

Another interesting article appearing right above the notice of the cable-layer in Ellsworth's report was a reference to an article in that year's *Civil Engineers and Architects Journal* describing Cooke's experiments in England with the use of overhead wires using a ground return. Though both Morse and Vail would later claim to have invented it while constructing the overhead line, it is obvious that Morse first read about it here.[23] As mentioned, it is likely Gale had brought this to his attention.

In noting another of his recently expanded extracurricular activities (unrelated to either the telegraph or the cable-layer), in the same letter to MaryAnn,[24] Ezra described his encounter with "The Godlike [Daniel] Webster" that had taken place the previous week. Webster had been making an argument before the Supreme Court regarding Stephen Girard's bequest to establish a boys' college in Philadelphia.[25] It was a case that drew immense public interest, and one that would incidentally come to have great bearing on Cornell's future endeavors. Ezra had evidently dedicated an entire morning and much of an afternoon listening to the great orator's booming, sonorous baritone as he addressed that solemn body.

The substance of the case was this: Girard had determined to establish an institution for "indigent white males." This in itself was nothing new (the Lancastrian schools had long adopted this ethos), but what *was* new was that the school was to be entirely secular. The will stipulated explicitly that no clergy were to be hired, nor was it to teach or promote the Christian religion. While technically legal, it violated the sensibilities and social norms of the day; and on these grounds, Webster, representing other potential heirs to the estate, had challenged the validity of Girard's will, arguing that the government should not grant the college charitable status if they persisted in "excluding Christianity."[26]

While Ezra Cornell fidgeted in the packed courtroom in Washington, D.C., enjoying the "intellectual feast"[27] of an exquisitely honed intellect, in far Western New York, a poor figure, his distracted and only occasionally coherent cousin Alvin Cornell, sat alone in the Mayville Prison, facing the executioner's rope. The trial of Alvin had started on January 25, 1844. The Schenectady and Philadelphia papers had published the juiciest tidbits of testimony[28] and the dailies in every major city, including New York and Washington, brought breathless accounts of each new development. The appeal for clemency based on insanity was handed over to the scrutiny of Cornell's personal "Deus Ex Machina," Dr. Beck, and Amariah Brigham, who both, post sentencing, had been called in by the governor for a clearer estimation regarding Alvin's sanity.[29]

Though perhaps ignorant of the contemporaneity of these two events, they both would have profound and far-reaching effects on Ezra Cornell's psyche. Webster would lose his case, but Alvin would be spared execution mostly due to the influence of T. Romeyn Beck. After hearing the additional deponents, Beck had issued a medical opinion, differing from his own and Brigham's of just two months earlier, stating that, based on additional testimony he had solicited, Alvin Cornell was insane at the time of the murder.[30] Governor Bouck, as a result of this new report, had commuted Alvin's sentence. While the original story of the gruesome Jamestown murder had smoldered in the national consciousness, that of the commutation would consume newsprint like wildfire. Papers in Schenectady, New York, Brooklyn, Philadelphia and South Carolina all carried dramatic descriptions of Romeyn Beck's last-minute dramatic intervention and Governor Bouck's ensuing clemency as if it were a modern reprise of *The Merchant of Venice*.[31] Once again, just as had occurred following the murder, following the recommendations for clemency from Beck, there would be a resounding silence on these events from Ithaca's Cornell clan.[32]

Cornell had by now also moved out of his uncomfortable digs at Fletcher's and had been living and sleeping in a special rail car on the B & O line. Miraculously, thanks to soft ground, they had managed to erect the twelve miles of posts Morse wanted, reaching all the way to Beltsville, Maryland. On March 23 they had paused there to finish stringing the lines and perform various tests.[33] The line worked flawlessly—unless it rained, in which case, just as with the underground cable, it did not work at all (the upside being that it dried out much more quickly).[34] Undiscouraged and undeterred, Morse was determined to push the line through to Baltimore by May 1 in time for the Whig convention. With a few tweaks to the ground wires, the line to Beltsville was soon functioning almost flawlessly, and the goal of reaching Baltimore by then seemed eminently achievable—then everything ground to a halt.[35] The convention had slowed things down on the road to a crawl. All the rail cars were

31. Out of the Frying Pan

monopolized by the passenger traffic heading to the convention, so Cornell had none to move his poles. Cornell decided to take the day off to attend the convention himself in the congenial company of a chance visitor to Washington, the principal of the Ithaca Lancastrian school, William Pew.[36]

Still, even with the line just one-quarter complete, the news from Baltimore reached Washington an hour and a half ahead of the rail—an eternity for the politicians who hung on every strand of rumor or gossip. Each day members of the public and the Congress gathered breathlessly at the Capitol to witness Morse's tests, eager for any snippet regarding the Whig nominations relayed from Annapolis, where he had stationed Alfred Vail.[37] On May 1, Morse had written excitedly to Vail, "*Be very particular.* I shall have a great crowd today. I wish all things to go off well. Many M[embers of] C[ongress]s will be present. Perhaps Mr. Clay."[38] It is not known if Clay was on the Capitol steps to receive the news of his nomination in person, but a passenger on the afternoon train from Baltimore gave the following account of watching it being transmitted from Annapolis: "I noticed a young man seated on a rudely constructed platform, resting on a square pen made of railway ties.... He had a small machine before him and was engaged in manipulating it while reading from a manuscript which had been handed him by some one on the train.... I learned that it was the fact of Mr. Clay's nomination that this young man was sending to Washington. I have since, and very lately, learned that the operator was a Mr. Vail, of New Jersey."[39]

The convention had thrown off Morse's schedule by a full week. Cornell complained to MaryAnn, "If I had a gang of such men as I could pick up at Ithaca, I could have got to Balt[imore] a week ago."[40] By May 7, they were still only sending from Annapolis,[41] but when work resumed, things went at a faster clip. By Tuesday, May 13, they were already past Relay House, where Vail had again set up his test apparatus and his bulky Grove battery.[42] Cornell had written, "We have got the telegraph within eight miles of Baltimore ... it will take ten days yet to get ... all complete."[43] On May 20, Morse retested the line from Annapolis to Relay and to Washington.[44] All worked well. By May 22, both Morse and Cornell had arrived in Baltimore to help set up the receiving apparatus, and Morse began to pay off his crews.[45] Morse traveled back on Saturday's afternoon train to Washington to get ready for the grand event the following day. It was too late for the opening of the Whig convention, but politicking was still going on hot and heavy in Baltimore at the Democratic and Tyler conventions.

On May 24, the famous inaugural transmission, "What Hath God Wrought," scripted by little Annie Ellsworth,[46] traveled over the new line from the Supreme Court to the Mt. Clare Station near Pratt Street and then back to D.C. There was no denying the achievement, but it was clearly a triumph held together at the moment by spit, gum and varnish. Morse's in-line relays were still undependable, as well as easy targets for any boy armed with a slingshot who could bring down the entire system, as damage to any one of these delicate mercury switches resulted in the failure of the entire line. The wire had broken twice where the silver-solder had been poorly applied to join the segments. Cornell would describe the line as "frail" and "crude."[47] The telegraph worked, but it was far from "ready for prime time," and no one seemed more aware of that fact than Ezra Cornell.

By the 27th, the Washington papers were already carrying a small blurb titled "Telegraphic News."[48] Morse was putting out feelers for the Philadelphia and New York line, which, due to the distances involved, would magnify all these problems, but he still hoped to first offload them onto the government by selling them the patent and having the Post Office operate it. Ezra Cornell had focused instead on coming up with the necessary improvements that would transform the Test Telegraph from a propositional oddity to a workable commercial system, capable of extension to a national network. The centerpiece of his effort was his magnetic relay, a device he had been working on in secret since January that, as he had described it to MaryAnn, would "open the eyes of the world."[49] Cornell's prognostication would prove partly true, as it would certainly shortly open one person's eyes at least—that person being namely Samuel Morse himself.

The quotation, "what hath God wrought?" in a more unfortunate context, might equally well have applied to Ezra's cousin, Alvin, who, that same day, would be transferred to the psychiatric annex of Auburn Prison. Shortly thereafter he was moved to the Utica State Asylum, where, put under the care of Drs. Brigham and Beck, he would continue to deteriorate into an untreatable hysteric. Once consigned there, he would live out the rest of his wretched life in the "psych ward" annex (actually a despicable filthy hole). While it may seem odd to contrast what was certainly an occasion of unbounded joy for Morse, his friends and associates (those who remained) with such a solemn note, what for Alvin Cornell was "the beginning of a miserable end," a long descent in madness, was for his cousin Ezra only marking "the end of the beginning" of his ascent to fortune and fame.

Morse had promptly paid himself his $500 completion bonus.[50] The government contract over, on June 6, 1844, the Secretary of the Treasury conveyed Morse's letter reporting the successful completion of the electromagnetic telegraph between Washington and the city of Baltimore to the Congress.[51] The only thing still hovering over Morse's head was the Bartlett affair, as Smith had named him as party to the litigation after Secretary Spencer had hung Smith out to dry. But even the tenacious and feral Smith, with the successes of the project

31. Out of the Frying Pan

more evident every day, had begun to see more lucrative and less time-consuming prospects for filling his pockets with the public's money than a legal wrangle with the U.S. government. Smith was about to be proven right. Within a week, people were lining up on the street by the hundreds outside the B&O rail station in Baltimore, clamoring just to get a peek at the marvelous instrument. Alfred Vail had set "ladies hours" so that every "Tom, Dick and Harry" would not crowd them out. On the Capitol steps, politicians of every stripe and the public crowded around to witness the clatter of the miraculous machine delivering news of events in faraway Baltimore—the communication age had begun.

SECTION IV

Relay Race

32

The Magnetic Telegraph Company

Fisher had once accused Morse of base ingratitude, charging, "You were willing to avail yourself of my assistance ... and then whistle me off as of no further use."[1] Morse indeed had "whistled off," at this point, almost everyone. When Morse formally informed the new Secretary of the Treasury, George M. Bibb, that he would be returning the unexpended balance of $1,079.49 (minus $500 or $600 in outstanding bills), he also had asked for an extension of Vail's employment for the time being,[2] and that of Vail's brother-in-law, Henry Rogers, to set up them up respectively as the station masters for the Baltimore and Washington telegraph offices, asking to pay them with the unexpended funds. Not to leave himself on the hook if Bibb said no, that same day Morse had fired all the remaining employees of the telegraph (which consisted of Rogers and Vail).[3]

Though the government thus far had declined to buy the patent outright, a new bill for another appropriation stood before the Congress, one that would extend the line from Baltimore to New York.[4] The unexpended funds from the first appropriation had been quickly consumed, and despite being officially fired, Vail and his brother-in-law,[5] Rogers, both continued collecting their salary for the time being, and both had agreed to continue working at least until March of 1845 or longer with no pay, should that prove necessary to keep the line in operation.[6] It was not. On March 3, the Congress approved an appropriation of $8,000 specifically for the upkeep of the Baltimore-Washington line.

Morse had hoped that, once he proved its viability, the Post Office would step up to the plate and take over operating the telegraph. He had been confident in that outcome ever since the success of the project seemed inevitable back in late April, but they had shown no such inclination.[7] With the Post

Office still reluctant to take it over, the appropriation for the New York-Philadelphia extension had been tabled for the time being. Morse was sick of dishonest associates, sniping and back-stabbing, and he had accomplished his aims, so now his fondest hope was to go off to Europe and leave the entire project in the government's hands. To this end, Morse at first appeared ready to dispose of the entire patent to the government at the bargain price of $50,000.[8] Morse was so eager to divest himself of the project, he had gone so far as to swallow his pride and ask the odious Smith to frame a bill proposing the sale.[9] A sale to the government would also put to rest the recent concerns expressed in the papers about a dangerous private monopoly. He even had obtained a power of attorney from Smith, Vail and Gale to negotiate such a deal on all their behalf, but Vail had objected to the figure, and by time it got to the Congress, the asking price was $100,000.[10] It had been met with a resounding silence. The Congress was uninterested at any price, and the Post Office, which had displayed a prodigious lack of interest in managing the Test Telegraph thus far, remained skeptical of any new project. This attitude had changed little even with the arrival of a new postmaster.

The new postmaster was Cave Johnson who, as a senator, had found Morse's telegraph appropriation such a source of amusement that he had suggested an amendment devoting half the appropriation to the investigation of the science of mesmerism.[11] Though Morse could now rightfully claim to have mesmerized the entire country and could bask in their unblinking adulation, for all the hooplah, the accomplishment itself remained still somewhat frail and tenuous. Though it had been universally hailed in the newspapers as perhaps the scientific achievement of the century, no one knew how fragile the entire enterprise was better than Morse and his recent successes had done little to calm his nerves. Plus, there was still Smith's litigation hanging over the entire project like a dark cloud.

Morse may not have gotten any interest from the Post Office, but he would get Andrew Jackson's former postmaster, Amos Kendall, to run the business end of things. Kendall, a Washington power broker, cut an exceedingly odd figure: his ghoulish demeanor, the spidery penmanship, the high-pitched voice, all combined to suggest an almost aggressive frailty—yet all this concealed what was an indomitable will. When by March of 1845 still no action had been taken by the Congress, Morse had engaged Kendall as sole "agent" for the patentees.[12] Living under this cloud of uncertainty while the Congress waffled could not have been easy, but Morse could finally make good on his plans to go off to Europe. Kendall's job would be to handle not just the proposed sale, but *all* business matters, including those relating to Smith and Bartlett.

Apart from engineering the licensing of the patent, Kendall had quickly

32. The Magnetic Telegraph Company

shown his genius in regard to these other matters as well. His rapport with Smith was unequaled, and the rancorous and seemingly intractable controversy over the trenching contract had been finessed away as if it all had been a bad dream. The next thing Kendall had done was to double the selling price for licensing the patent to the government to $200,000,[13] in the meantime working to put together a group of private investors to buy the patent outright for $100,000.[14] If Kendall[15] failed to interest his fellow slave-state Democrat, Cave Johnson, in the project, then clearly the idea of the telegraph as a publicly run adjunct to the Post Office was dead as a doornail. With the government clearly uninterested, Kendall had finessed his offer on behalf of the private investors, dangling a carrot in front of Morse with an offer to put $10,000 cash in his pocket before Morse left the country in August for his European jaunt.[16] It was a tempting offer that had set Morse salivating.[17] He could effectively wash his hands and go off to Europe, this time flush rather than threadbare as he always had been in the past, a man not only of vision but now of substance, confident in the success of the American telegraph and his part in making it a reality.

The first question raised by Kendall's potential investors naturally should have been, "What are we actually buying?" The Baltimore-Washington line was officially property of the U.S. government. In addition, there was still a bill before Congress to license the rights of the patentees that might yet pass. Not only were there fresh challenges to the patent percolating, the old charges of intellectual larceny from Charles Jackson still stubbornly dogged Morse. The "port-rule," the only physical part of the invention that was unquestionably Morse's idea, had recently become obsolete with the advent of the telegraph key. What was there really left to sell? Despite all Morse's assurances to the contrary, they might well issue $400,000 in public stock and then find that the answer to that question was, "Nothing."

Consequently, by mid–April, all Kendall's efforts to raise a public stock company had failed. By then, clearly the congressional bill was a dead horse as well, so the Mexican standoff had resulted in no suitors for the patent left standing.[18] Kendall was discouraged, but had remained undeterred. He had pledged himself to Morse's interests, and that meant he would find a way forward even if he had to invent an entirely new kind of entity to fund the enterprise.[19] Morse had no doubt described to Kendall at some point the means by which artists generally raised capital: the subscription plan (something which Morse himself had used in the past and which had proven a great source of embarrassment when he did not deliver).

In any case, his first attempt at raising a stock company having failed, within a month Kendall had managed to cobble together a new plan for raising $15,000 through a joint subscription offering, the kind that in the past that had been formed to commission some of Morse's paintings, this time with 25 subscribers,[20] including $500 each from Cornell[21] and Charles Grafton Page, chief patent examiner at the Patent Office (with Kendall as president, contributing an equal amount[22]). It was a novel approach. The first person to subscribe had been William W. Corcoran, a prominent banker and promoter of American artists.[23] No doubt to everyone's surprise, the ever-shifty Smith also plunked down the lion's share of $2,750.[24] Among the other signers of the articles were Charles Page, Cornell, Vail, Gale, Smith, John J. Greenough,[25] and Greenough's law partner, Charles Keller.[26] Henceforth the investors would simply be known as "the capitalists."

As stated in its charter, the main purpose of the new Magnetic Telegraph Company was to extend the government test line first to Philadelphia and then to New York.[27] This would put the political capital of the country in direct, instantaneous communication with its two economic powerhouses. What is more, Kendall had acquired rights to the test line through March of 1847[28] at no cost to the company. What he needed now was someone to ramrod the construction of the new portion of the line. Though Morse's name was nowhere to be found on the list of investors, he remained as active as ever in the affairs of the company. Cornell's reputation as the "go to" and "can do" guy had no doubt reached Kendall's ears by now, but Cornell, who had been working since June without salary, seemed presently at odds with Morse for reasons that Kendall could only dimly understand.

33

A Red Herring

Vail's salary as station-master at Baltimore had continued uninterrupted, while Ezra Cornell's salary had stopped entirely in June of 1844. To put salt on the wound, Morse had hired Vail's brother-in-law, Henry Rogers, as station master for the Washington office. Cornell did not complain at the time, but clearly he bore a grudge against Morse thereafter.[1] With the end of the appropriation, there was no more American electromagnetic telegraph, and also not much for Ezra Cornell to do. Morse had managed to keep him busy; taking a leaf from the book of another consummate showman of the age, P.T. Barnum,[2] he had set up sideshows[3] in New York and Boston for the public, availing himself of Cornell's services as carnival barker, paying him in vague promises of future employment and promising to look the other way at whatever Cornell could skim from the profits.

There were in fact no profits to skim from the Boston demonstrations. The attendance was only sufficient to offset the costs of the demonstrations themselves. With his Ithaca creditors hounding him, after five months of this *pro bono* work, MaryAnn put her foot down. In a gesture designed to quell her objections and ensure a modicum of domestic harmony, Cornell had asked Morse to put his brother-in-law Orrin Wood to "work," as a lobbyist. Morse agreed, dragging poor Wood all the way from Ithaca to Washington without offering so much as a penny in compensation towards either expenses or salary.[4] With few funds of his own and no means of support, after three months, Wood felt obliged to sponge ten dollars off Morse, but he had been made to feel so embarrassed about it that he swore he would not do it again.[5]

By the end of 1844, Smith's unrelenting backbiting, combined with Morse's unfulfilled promises and preferential treatment of Vail, were beginning to make inroads into Ezra Cornell's psyche. His esteem for "the Professor" was waning, and he had even gone so far as to question whether there was anything to Charles Jackson's stale charges that Morse had stolen the idea for the telegraph from him.⁶ Smith had invited Cornell to New York to continue the demonstrations of the telegraph, to which Cornell had reluctantly agreed, although the only additional remuneration was an overcoat that Smith promised to have charged to him by his tailor.⁷ Alternating between Number 10 Wall Street and 30 Vesey Street, Cornell otherwise seemed reasonably happy and in good health, though he chafed at being confined to an office.⁸ He could not pay off his debts with good feeling and pie-in-the-sky alone and his old benefactor in Ithaca, the patriarch of Fall Creek, had been dunning him for repayment of the seven hundred dollars he had loaned on the plow venture.⁹ He tried to put Beebe off, but Beebe was not buying it, and the demands had grown more strident.¹⁰ Cornell, still working with no salary and no official duties, had been reduced to sleeping on two chairs set up in the offices of the express company.¹¹ He had even appealed to Smith for some funds, but Smith, in characteristic fashion, passed the request on to one of his associates.¹²

Wood had written his brother-in-law from Washington, D.C., "I get along finally with friend Vail. I think much more of him than I did 3 or 4 weeks ago."¹³ This was not especially welcome news. Cornell had obviously begun dropping some hints about an invention he had come up with on his own, even applying Morse to lend him the services of Thomas Avery and some wire for the project. He made this request without divulging any of the details,¹⁴ which had naturally aroused Morse's curiosity. Morse had encouraged Burbank and Vail to ply Cornell and/or Wood to find out its exact nature. Both Cornell and Wood remained tight-lipped—Wood, mostly because he really didn't know anything, and Cornell because he no longer fully trusted Morse. Wood had written Cornell about Vail's persistent inquiries, "I have discovered something new in him [Vail]. He is not so cross about your patent right or invention. I should like to know something more of it … if it is necessary I can keep it a secret."¹⁵ Burbank had simply inquired coyly, "You did not say what your new invention is?"¹⁶ This no doubt had raised further alarm bells in Cornell's mind—as it would turn out, both literal and figurative.

Cornell had put Burbank off the scent by suggesting he was working on several kinds of machines for allowing the telegraph poles to accept the wire hoops that guided the wire.¹⁷ He had answered Wood's query a bit more honestly, saying he had come up with a fire-alarm bell that would be operated by telegraph.¹⁸ This had satisfied Burbank and Wood, but had only aroused Morse's curiosity further. Finding himself under pressure from several quarters, Cornell disclosed his alarm bell idea directly to Morse, asking whether it would be all right with him if Cornell patented it under his own name.¹⁹ Morse's reply had been a disappointment but not entirely unexpected—a resounding "no." What Morse provided instead was a recital of his own patent, adding somewhat

33. A Red Herring

superciliously, "Now, an alarm is *intelligence by sound*, and if this is accomplished by *the electro-magnet*, it certainly comes within my patent."[20]

On receiving this, no doubt having heard of the new congressional appropriation of March 3 for the upkeep of the Baltimore-Washington line,[21] Cornell had sent Morse a modest bill for materials purchased in the course of past services in Washington and New York.[22] Morse's reply (in a letter marked "private") had been to decline to pay it, referring Cornell instead to Amos Kendall in New York. At the same time he warned Cornell to cease confiding in Smith, softening the blow by offering him more "pie-in-the-sky" regarding the prospects of his telegraph bill and future employment.[23] Privately he had been incensed about Cornell's temerity in presenting the bill for $10.70. For weeks afterward he had fumed about it, complaining to Smith, who had immediately written Cornell, "The professor came to me today quite [unreadable—exercised?] about your $10 bill."[24]

When Cornell finally did meet with Kendall in New York, ostensibly to press him about the alarm device, Kendall had simply referred him back to Morse. It was typical bureaucratic "ring-around-the-rosy," something for which Cornell had little patience. Disgusted, finally, on April 17, Cornell had let go with both barrels, reciting his own and Wood's[25] long list of grievances to date and adding, "In regard to the personal abuse I rec'd from some of the proprietors, I will say nothing."[26] Morse's response was just to offer what was essentially another slap in the face. He summoned Cornell to Washington, ordering him to inspect the entire Baltimore-Washington line, agreeing to pay him $15 total for performing this task.[27] Having few options but a total break, Cornell had grudgingly accepted.[28]

What were Cornell's supposed crimes that would lead Morse to treat him thus? The ongoing cause was that Cornell had refused to renounce Smith, and in fact Smith had become his sounding board and main source of gossip for all the preceding five months. Kendall had the finesse to pull off appearing conciliatory with Smith, but Cornell could not remain friends with Smith without seeming disloyal to Morse. He had the temerity to suggest a device beneficial to society utilizing Morse's telegraph—a fire service, but doing so under his own name—that and the bill for $10 were the most recent source of friction. Further, he had offended Vail on numerous occasions simply by being smarter than him.

Cornell's previous inventions, the glass pole insulator, the wire-winder, all had been folded under the great Morse umbrella without so much as a nod toward him. Selling his cable-layer, as far as the telegraph was concerned, ever since the move to the overhead system, was a dead horse. It seemed increasingly clear Morse would never, in any case, extend credit for any innovation to a subordinate, but

Cornell wanted to be absolutely sure of that, and the alarm bell was his means to test this hypothesis. Cornell no doubt had confided his plan to him, as Cornell's good friend J.J. Speed from Ithaca had written him winkingly, "How do you succeed with your plan for detecting rogues?"[29]

Why was Cornell, after a year and a half, suddenly intent on ferreting out Morse's true character? The answer was that he had something in mind far more valuable than a cable-layer or a fire alarm. The alarm bell device, in fact, had been a red herring. Having availed himself of the Patent Office library in the dreary winter of 1843 to educate himself on the subject, Cornell had devised what he considered critical improvements regarding telegraphic circuits. By integrating Henry's Intensity/Quantity distinction in a manner that had not occurred to Morse, he believed himself to have found a way to make the telegraph far more reliable over large distances. While Morse was still casting about for ways to improve his equipment, Cornell's device could ease the transition to a nationwide network. These vague ideas had jelled into this actual physical device he now called the "magnetic relay." Best of all, it could be employed at a receiving station without requiring any modification at all to Morse's apparatus. The question all along had been whether or not to show it to Morse. So the query over the alarm bell had all been a smoke screen, and Cornell now had his answer.

The impact of the device, if it worked as Cornell proposed, would be major. With the substitution of one of Henry's "quantity" magnets at the receiving/sending end and wiring it to a peripheral circuit with the magnet connected to the partner station, Morse's irksome in-line relays could be done away with entirely. Secondly, the relay made the return transmission wire[30] essentially redundant. Lastly and perhaps most importantly, it would allow for the creation of substations running on the marginal circuits without requiring running an independent wire for each substation.[31] Clearly all this had implications for the future of the telegraph far beyond those of any mere fire-alarm bell, and Cornell's decision not to reveal it to Morse over the preceding year must have now appeared to him now a wise decision.

34

The Mule Kicks Back

Morse, as had been illustrated in the case of Fisher, had no compunction about kicking a man when he was down. While Cornell no longer trusted Morse, he did trust Smith, and Smith trusted Kendall. At the end of May (probably at Smith's urging) he had put all his cards on the table, spilling the beans to Kendall in New York, divulging the exact nature of the relay device that he had thus far successfully concealed from Morse. Kendall's response had been to offer Cornell a job with the new Magnetic Telegraph Company.

Cornell not only understood better than anyone how to bulldog a telegraph line, no one understood the technical challenges involved with extending the Test Telegraph better than he. Attempting to build a nationwide network on the same configuration used for the test line, he knew, was foolish and impractical. Problems of marginal importance in a line of forty miles would be magnified and become major headaches on a line of 200 or more miles. Cornell had objectively described the state of the test line as "frail" and "crude."[1] Even Morse was aware that the strength of the signal was an issue and without the bothersome in-line repeaters, on a line of the length now proposed, it was doubtful whether the signal would be strong enough by the time it arrived at its destination to operate a device to physically mark the paper.[2]

Morse, planning to leave for Europe, was about to dump all this, along with the management of the rest of his affairs, directly in Kendall's lap. The one thing of which Kendall was sure, with Morse leaving shortly, was that none of the technical problems were likely to be solved by him any time soon. However, if the relay actually worked as Cornell described, it would solve most of the problems Kendall would soon face. Drawing up an employment contract, with Smith looking on, Kendall proposed to hire Cornell at a rate of $1,500 per year, prefacing the agreement by saying, "On reviewing what we now understand to be your views in respect to improvements you believe you have made ... which may come in aid of or be associated with Morse's inventions of the same nature, we are willing to arrange for your employment and services in behalf of the Magnetic Telegraph Company."[3]

The version of the relay that Cornell had shown Kendall in the office at #10 Wall Street was a crude working model, but according to Cornell it not only would improve the reliability of the telegraph, it would allow for integration of sub-networks without requiring manual transcriptions. This was key to what Kendall contemplated: a seamless network—one that, without operator intervention, could reach across the entire country. It also could render the telegraph robust enough to allow for the contemplated extension of the lines westward and northward without incurring the myriad problems and constant glitches that had plagued the test line.

Cornell's employment status over the preceding two months had stemmed partly from lack of funds, but also from his own fear that if he was an employee, Morse might simply subsume his relay under his patent as he had done with all his other inventions to date. Cornell was no fool, and he had insisted up front that Kendall grant him the sole right to ownership over certain "improvements which you conceive ... and which you also have in contemplation in the use of wires, magnets, mechanisms and galvanic electricity ... which may come in aid of ... Morse's inventions ... whether the device shall have been or shall be patented by you or your assignees." But the shrewder Kendall had added a caveat to the contract saying that the Magnetic Telegraph Company "shall have the right to use, improve and employ [the relay] at their option."[4] To Cornell's untutored eye, the contract seemed ironclad and gave him all that he had asked for and more. He could go back to work, this time at a reasonable salary, and still retain all the rights to his invention. This is what the agreement seemed to say, but Cornell had underestimated both Kendall and Morse's cunning. By using the word "improve," Kendall had essentially granted Morse a license to invent around Cornell's device.

The way the contract was worded had basically been a sop to Cornell's ego. While Smith was in his corner, by granting Cornell rights to his device, certainly Kendall now might have Morse's objections to deal with. Tasked with holding together a fractious coalition of investors, Kendall had all along been fighting the perception that he was "Morse's man." This in reality was a way out for him on that score as well, and all Kendall really had needed to do was leave Morse sufficient wiggle room and his genius for "reinvention" would supply the rest. Morse's "wiggling" would be sufficient to supplant and obviate Cornell's achievement and, in the end, erase it almost entirely.[5]

When Morse learned about this, he had been stunned, but once Kendall explained matters to him more fully, Morse had been somewhat mollified. Cornell had been struggling for the past year and a half to realize the relay on his own, constructing the model that now stood before Kendall and Smith in secrecy. This was the reason he had needed Avery's help and the loan of wire from Morse. Even after a year and a half, though, it was still a work in progress. The concept had been ingenious and novel, but the execution was clumsy and amateurish. The first thing he had done after the new contract was signed was send the plans for the relay to a silversmith in Ithaca named Sylvester Munger, asking him to fabricate it.[6]

The way the contract had been worded, Cornell's device was now fair game for improvement. Though clumsy of realization, it was a complex and intricate conception, and more importantly to Morse, it was thus far unpatented. Morse had immediately set in motion a complex (and what can only be called devious) plan to reinvent and claim Cornell's relay as his own. He had realized that to accomplish this he would need the legal room to outmaneuver Cornell, American ingenuity and European craftsmanship. Thanks to Kendall's craftiness, Page's compliant attitude and his upcoming European trip, he would soon have all of these at his disposal. By the time he was done, the relay would almost be no longer recognizable as the same device Cornell had shown Kendall, and Morse would own the patent.

Cornell was never really known for a forgiving temperament, and he was not about to just shrug off the ill-treatment he had received at Morse's hands over the preceding year. He no doubt viewed Morse's sudden interest in his device with suspicion. Perhaps in an attempt to rid himself of a sullen, brooding presence in the New York office, Kendall, as a first order of business, had ordered Cornell to travel to Bangor, Maine, to see about purchasing 3,000 cedar posts needed for the Philadelphia to New York line.[7] This would not only buy Morse some time to hatch his plans with regard to the relay, it had the additional benefit of throwing a monkey wrench into the cozy relationship Cornell had developed recently with David Burbank in Baltimore. Burbank had supplied all the poles to date and stood to lose if Cornell contracted them out privately.

Kendall and Cornell probably both had known from the outset that the trip to Maine was essentially a wild goose chase. Morse and Kendall had already begun advertising for new poles in the papers. There were plenty of lumber brokers who had standing contracts with loggers in Maine, and they would no doubt get a far better deal from them than any Cornell could possibly hope to strike on his own. Cornell, for his part, at this point entertained about the same level of enthusiasm for Kendall's "instructions" as he had for stripping the wire from the Tatham cable back in the basement of the Patent Office. His approach then had been to "invent his way" out of a tedious chore. This time, he had simply ignored the make-work instructions and gone for a long-overdue visit to Ithaca.[8] By the time Cornell returned to New York, Kendall had thought matters through, and instead of banishing him again, assigned him the task of acquiring the rights-of-way for the Newark to New York City part of the line[9] and the delicate task of running the lines within Manhattan and the city of Newark. They also now had to start thinking about how and where to cross the North River.

Kendall, in the meantime, had been hard at work lining up the resources

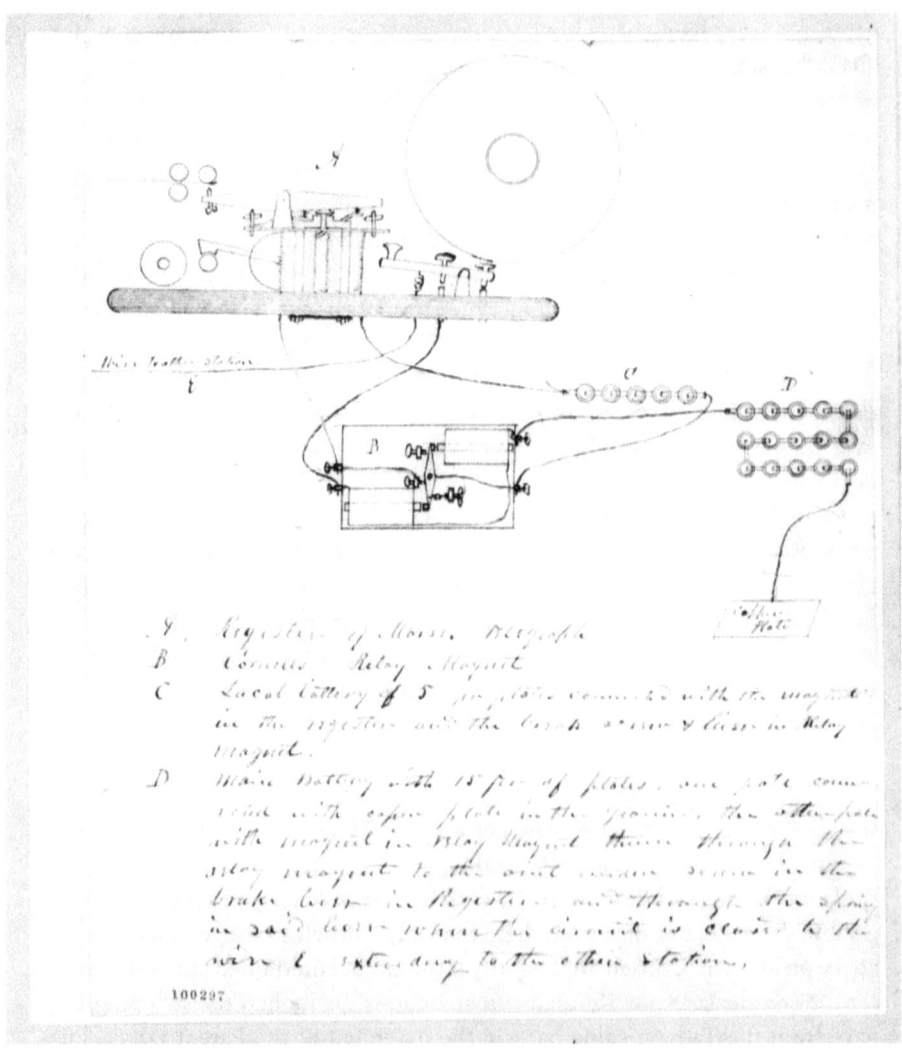

Diagram of the Cornell Magnetic Relay, sent to Henry O'Reilly by Ezra Cornell (courtesy Library of Congress Manuscript Division Morse Papers).

and materials for the proposed New York-Philadelphia line and engaging contractors for the other major lines he had in mind. By the time of Morse's planned departure for Europe in August, all the major contracts were in place, Pittsburgh, Buffalo and Boston; and there was even talk of a company that would take on the Washington to New Orleans route. By now Kendall was already referring to Cornell as superintendent of the magnetic telegraph[10] (Morse's former title), positioning him to deal not only with the New York-Philadelphia line, but with the other lines as well. Subcontractors were already pounding down the door, hoping to start stringing the lines across the country on "the Morse System." The first week of June, Henry O'Reilly, an old friend of Kendall's from the Jackson era, offered to contract to put up all the line west of Philadelphia clear to St. Louis.[11] By the 13th there was a contract in place for him to get to at least to Harrisburg by October.[12]

The contractors seemed to have had no difficulty in raising the subscriptions in a relatively short period of time (the cash was another matter).[13] Utica financier Theodore Saxton Faxton had partnered with John Butterfield and Hiram Greenman for the New York to Buffalo line; O'Reilly (with several other backers) had the Philadelphia to Pittsburgh run; F.O.J. Smith, along with Crawford Livingston and Henry Wells, in whose offices Cornell had camped out, had acquired the lucrative Boston to New York corridor[14]; David Burbank, the Baltimore lumber dealer, had taken on extending the Baltimore test line to Philadelphia.[15] The fifth, the New York-Offing line, involved an underwater cable to be laid by Morse's old collaborator, Samuel Colt. It would run from Fire Island near Long Island, ending at the Mercantile Stock Exchange in lower Manhattan, bringing commodity news obtained from ship arrivals, and it was to be the first purely dedicated commercial telegraph line. The only problem remaining was the robustness of the technology itself, and this was one that Cornell appeared ready to solve.

35

The State Fair

Though Page and Vail had teamed up to sell Page's version of the magnet in the field, both had soon become consumed with other business, so by the time the contractors were ready for them, they had decided to contract out the job to as-yet undetermined parties.[1] Cornell, on the other hand, was still bound and determined to get his version of the relay installed on the Buffalo and Boston lines. With his brother-in-law, Orrin Wood, bulldogging one effort and Smith in charge of the other, the chances for that were looking very good. Further ensuring this outcome was the fact that the energetic and ambitious Cornell had set himself up as the clearinghouse for any and all of Morse's equipment required by the contractors and employing his friends back in Ithaca[2] as the suppliers.

Rather than attempt to beat Cornell "head to head" in what was a controlled and limited marketplace, Vail had come up with yet another use for Page's device: to embarrass Cornell and paint him as the intellectual thief. Just as Cornell was preparing to depart for Utica, Vail had sent one of Page's relays to #10 Wall Street. By "happenstance," Leonard Gale stumbled on it; noticing Page's device sitting there on Cornell's workbench, he had remarked on it. Gale had promptly informed Page that Cornell was attempting to steal his device.

Cornell's personality was the exact opposite of Vail's. Unable to resist a "head to head" contest between his and Page's device, Cornell had taken the bait, hook line and sinker, asking Gale if there was any conceivable objection to his taking Page's version along with him when he left for Utica. Later, when Page had accused him (or Gale) of stealing "his" device, Cornell had defended himself, saying at the time, "[Gale] assured me that you was [sic] in no position to take out a patent for your improvement ... he said he saw no impropriety in letting me take it."[3]

In 1841, the New York State Legislature had appropriated $8,000 for the "promotion of agriculture and household manufacturers in the State," which had resulted, among other things, in the first New York State Fair. Held outside Albany, it had been a huge success. That year's fair was scheduled to take place in a farmer's field about one mile from downtown Utica and scheduled for the 16th through the 19th of September. In addition to the ninety sheep, twenty-five prize hogs,[4] and the exceptionally popular poultry surgery booth, there was to be a working telegraph line set up between the fairground and downtown Utica. Butterfield and Faxton had asked Wood to set up the demonstration for the fairgoers,[5] and Butterfield had asked Cornell to send Wood sufficient equipment for a line of one mile.[6] Not at all confident that he could accomplish this on his own, Wood had invited Ezra up to help with the setup, which involved running a line from downtown Utica to the fairgrounds, a distance of about a mile and a half. While a receiving magnet was not really needed for this short of a line, Cornell had already informed Wood's boss, Butterfield, that his relay was a critical component for a telegraphic apparatus of any length.[7] As a result, Cornell now found he had put himself on the spot much earlier than he had anticipated. He had turned to Munger in Ithaca to fabricate one in a hurry. Munger had complied, sending up the unfinished model to the fair with one of his assistants.[8]

On September 11, Cornell arrived in Utica to begin setting up the telegraph line and booth with the help of Wood.[9] By Tuesday morning, when the fair opened, their line was operating, but they could not have been prepared for the rush of over two thousand people who tried to cram themselves into the 30' by 40' pavilion that housed both the telegraph and the chicken surgery.[10] It was the former that had attracted the great bulk of the crowd.

Cornell had another reason to go to Utica besides family solidarity and boning up on poultry surgery. With Vail having refused to conduct any tests of his relay on the Philadelphia-New York line, Cornell needed to prove his device in the field. He also wanted to know how it stacked up against Page's, so when Page's device had "shown up" at #10 Wall Street, Cornell had taken it with him to Utica. Page's take on matters was that Cornell had brought the device to Utica specifically to show him up; to demonstrate to the contractors the superiority of his device. Playing the aggrieved party, Page had protested to Gale, saying it appeared Cornell had copied his device and was trying to sell it to the contractors under his own name. When Gale conveyed this to Cornell, he had asserted rather defensively, "Dr. Gale informed me that you[,] on sending your Relay Magnet ... expressed fears that I was disposed to invade or interfere with your interest in it, and wished me to write you and explain the circumstances of my having the magnet."[11] Cornell went on to explain why he had brought it to Utica, further asserting that he had not copied anything of Page's and had been working on his own version of the device for several years.

He then adduced Smith as a witness, saying they had discussed it back when they were both rooming at Fletcher's. "The Relay Magnet was a subject that had claimed my particular [attention] and my improvements were suggested to Smith over a year ago as he will probably recollect."[12] To clarify what seemed to be a hopeless position *vis-à-vis* Morse, Cornell wrote, "I have not however aimed at establishing an antagonistic interest to Prof[essor] Morse."[13]

Replying that he had entrusted his relay to Morse only on the basis of absolute secrecy, Page now issued a somewhat bald accusation. Without pointing the finger solely at Cornell, he had said, "There has evidently been a breach of trust somewhere," adding what was an unmistakable threat: "Mr. Kendall will undoubtedly see that justice is done."[14] This was the same kind of language that had resulted in Fisher's dismissal. Further, Page's letter also contained a list of his "claims," putting Cornell on notice that despite his position as examiner, he intended to patent them at some point in the future, which in effect, legal or ethical or not, constituted a caveat notice.[15] Clearly both men were girding for combat with Page as a stand-in champion for Morse and Vail. Page had privately promised Morse his "warmest advocacy," and this was his way of delivering. Cornell had closed his defense saying, "If my magnet is the best it will probably be used, if however yours should prove to be the best, you shall find no man more ready to acknowledge it and recommend its use than me."[16] Given Cornell's present position as clearinghouse for virtually all telegraphic equipment, no matter how well-intentioned his "may-the-better-man-win" policy might have been, it could not but have sounded disingenuous in Page's ears.

Page's letter, though, had shaken him to his core. If he was to retain any hope of patenting his own device, Cornell could not afford to make Page an enemy. In replying, Cornell had written, "If you had made the arrangement with Prof. Morse that you say you have and if Mr. Kendall has made the arrangement with you that he wrote me he had, then I cannot see the impropriety of my seeing your magnet. What is the difference if it is placed in my hands a few days sooner or later!"[17] It was a remarkable exercise in sophistry and quite uncharacteristic for the usually sincere and plainspoken Cornell.

The tiff with Page could not have come at a worse time for him. On October 30, Cornell, somewhat less confident now in his chances for approval, had sent in his application for the relay patent under the care of Charles Monroe in Washington to the Patent Office.[18] Page, as chief examiner, was the one reviewing it. Cornell had first met Monroe when they were rooming together at Fletcher's during the early days of the Test Telegraph. A week later Monroe had written back that, in addition to the patent claim and drawing, Page was going to require a working model.[19] This was bad news, especially since Wood and Smith had both been telling him that the workmanship on the Munger relays was poor and that there were still design kinks yet to be worked out. Maybe Cornell could get away with telling Kendall that the problem was want of experience in the operators, but clearly he was not going to get away with telling that to Page.[20]

Cornell was still holding firm to his "may the better man win" credo, and as far as he was concerned, the failure had not settled matters definitively. He needed a longer stretch than a mile and a half to determine whose device worked better. While accusatory letters flew back and forth between himself and Page, Cornell, true to his credo "may the better man win," had shipped Page's version to Buffalo for testing alongside his.[21] By October 30, it still had not arrived.[22] Two days later, Page's magnet finally arrived, and Wood wrote back the discouraging news, "I have had the relay magnets both in the long circuit this PM but did not succeed with either."[23] Thus far, in this relay race, everyone had been the loser.

Cornell had instructed Wood to continue testing both his and Page's versions of the device on the twenty-three miles of line strung between Utica and Oneida. The results had been discouraging, but this time more so to Cornell. Page's version performed consistently well, while his worked only intermittently. The problems with Cornell's relay persisted, and by the middle of November, Wood was at his wits' end as to how to get it to work. Apparently Munger, in fabricating them, had ignored Cornell's instructions regarding the core of the magnet and made them of hard iron. Wood had written despairingly "I could get not action upon my relay magnet ... there was so much fixed magnetism in my magnet."[24]

For Kendall, Utica had been a distraction from the far more important task of completing the Philadelphia-New York line. If they were to get it in operation before spring, the problem of crossing the Hudson had to be addressed and solved quickly. Kendall had since August been urging Cornell to come up with a plan, but despite his persistent queries, thus far Cornell had done virtually nothing.

Morse had departed Liverpool in late November of 1845 in the midst of a howling hurricane and had returned home to another storm, this one mostly of his own making. It did not take Cornell long after Morse's return to figure out that Morse had hatched some plan to supplant and replace his relay.

For Cornell, Utica had been a source of disappointment in several respects. Despite the success of the telegraph demonstrations at the fair, deaf to his ardent pleas, MaryAnn had failed to show up or bring the children—nor had Cornell time to make a trip to Ithaca. The strain on their marriage due to his extended absence had, as a result, grown deeper, and the tone of the letters to MaryAnn changed in tone dramatically from chatty and upbeat to somber and almost despondent.[25]

Theodore Saxton Faxton, the Utica financier, had made his fortune in transportation, most recently running a packet service on the Erie Canal.[26] Faxton had, besides the connection to Morse and the telegraph, in the small world of upstate New York, another somewhat more obscure link with Cornell and the Cornell family. As a philanthropist and one of the city's most prominent citizens, he had been instrumental in funding and building the Utica Insane Asylum that had been completed just two years earlier.[27] This was the facility where Cornell's cousin Alvin had been confined following his conviction for murder in January of 1844. By virtue of this, Faxton was also a longtime acquaintance of none other than T. Romeyn Beck. Beck was a trustee of the asylum, and his assessment of Alvin Cornell had likely taken place here at Utica.[28] So, even aside from his friendship with Cornell's brother-in-law, Orrin Wood, whom he had hired on as superintendent for running the line to Buffalo, there can be no doubt that Faxton was well acquainted with the Cornell clan, and in particular, the facts surrounding the sad case of Alvin Cornell.[29]

36

Raising Cash

For Morse, the European trip had promised a much needed respite from the trials and tribulations that had attended the erection of the test line. Ironically, it was that success that had finally provided him some of the long-sought approbation that he craved from those whom he still regarded as his true colleagues—artists.[1] In May he was asked to sit for his official portrait by the National Academy of Design, the organization over which he had presided.[2] In Europe he had been awarded honorary membership in the Belgian Academy of Archaeology[3] and would eventually be showered with a far greater number of accolades from every quarter: artistic, scientific and political.[4]

Increasingly, however, the main rationale for the trip was the relay. Aside from its technological implications, Morse had a very sound economic reason. Over the preceding decade he had diluted his original telegraph patent by offering partial interests to Vail, Smith and Gale in return for their work in obtaining the government appropriation. The relay was something that could be patented separately, and which, it was increasingly obvious, was nearly as indispensable to the telegraph apparatus as the register itself. Still more or less "out at the heels," and not eager to share his thoughts on the subject with his fellow patentees, he needed some other excuse to justify having the patent partnership underwrite the expenses of the trip.

When Morse put forth the main reason for the trip as his trying to secure the rights of the partnership for a European telegraph system, the patentees had seen through it immediately. Smith was the least shy about expressing his skepticism, saying bluntly, "In no case shall [we incur] any charge or expense or debt for us, or in our name, serially or jointly."[5] Smith had no faith by now in any effort to establish a line in England in competition with Wheatstone's, which was by now firmly established.[6] A firm believer in a bird-in-the-hand, he offered to relinquish all rights to a European telegraph in exchange for a settlement of the Bartlett contract.[7] Smith should have understood better than anyone the futility of this effort, already having had bitter experience with French and English patent systems on their previous trip together. Morse had finally conceded the point, writing Smith regarding the prospects, "It is more than doubtful I shall make anything."[8]

Vail had been similarly unimpressed with Morse's stated goals. To counter Vail's pessimism, Morse offered to exchange his interest in Vail's upcoming book for his share of the future European patent rights, but Vail was not buying it either.[9] He had expressed his pessimism even more baldly than had Smith, writing in a letter that found Morse on his arrival in England, "I shall not be disappointed if you return as you left, without accomplishing your object."[10] When those efforts to get the patent holders to underwrite the trip failed, Morse had resorted to selling off the patent rights to the telegraph in Cuba to raise some quick cash.[11]

The business reasons proffered, an attempt to sell the patent to the crowned heads of Western Europe, were both grandiose and ridiculous.[12] An idea for licensing the rights in Russia that Morse had cooked up was even more ridiculous. The technical challenges he faced now in the U.S. were simply beyond Morse's limited technical capabilities. He was, for the most part, a visionary, not a tinkerer, and the transition to the national network required the talents of the latter more than the former.

Recent events, though, had finally provided him an unrelated solid rationale for the trip in the form of Cornell's relay. The device had also clearly electrified Morse, and this now infused him with a new sense of purpose. A challenge to his intellectual hegemony was something that excited all his faculties. Using the expertise of Charles Grafton Page, he had set about miniaturizing Cornell's rather cumbersome device so it could be transported with him aboard ship discreetly. To Morse's delight, Page had informed him that miniaturizing it did not compromise its utility in the least; that Cornell had been laboring all the while under a misapprehension about the size of the magnet required.

Page and Kendall had been virtually sworn to secrecy, and Morse would go to greater lengths to conceal his aims. What would eventually come to be known as the "Morse's relay" would be nothing but a more elegant version of Cornell's device that had been realized with the help of Page and later the highly skilled "mechanician," Louis Breguet in Paris.[13] This, however, was a project that in modern parlance would be called "Top Secret"; it required the utmost discretion, and we shall see the lengths Morse would go to insure this.

As primarily a secondary or peripheral "quantity" circuit, it could be argued, Cornell's improvements were in essence derivative of Henry's ideas as well, constituting only a more robust implementation of the old "Intensity/ Quantity" idea. Morse's efforts would make this question moot as he would

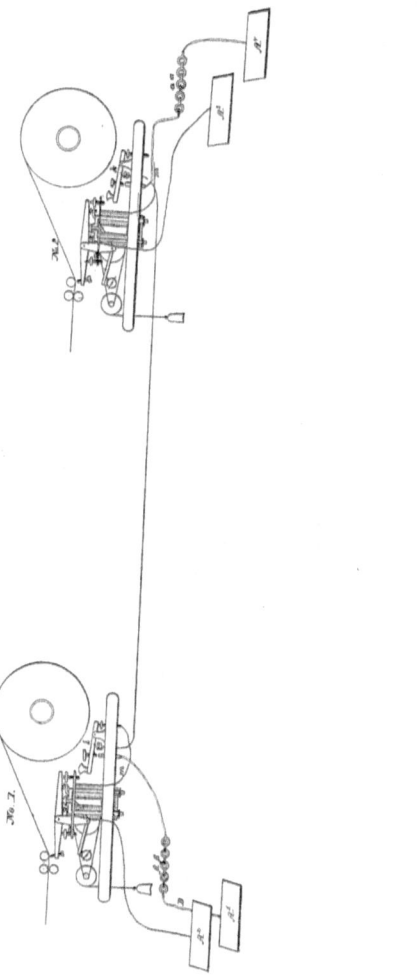

Patent #4,318, Diagram of Cornell's Improvement, December 20, 1845.

create a device that would eventually supplant and indeed virtually erase any claim of Cornell's. But asserting Morse's version was just a better adaptation of Henry's ideas or, as Charles Page would claim later, of Wheatstone's, Henry's or his own,[14] is to oversimplify matters and overlooks Cornell's contributions.[15]

The ingenuity embodied in Cornell's design was not just in adding a secondary circuit, but in the novel wiring arrangement wherein the line circuit operated an actuator for a secondary circuit (through the introduction of a solenoid operated on by an independent electromagnet) controlled by the sending station, no matter how far distant. In an innovation not previously conceived by Henry, the armature (later a solenoid) operated a switch that swapped the two functions of sending and receiving by automatically enabling or disabling the opposite function in its partner station, thus allowing for sending and receiving over a single line. Critically, unlike Morse's earlier configuration, the secondary circuit governing the sending and receiving functions operated entirely independently of the main trunk line.[16] "While the first invention of the special application called the relay is thus unhesitatingly ascribed to Professor Morse, *the practically much more important* arrangement of the terminal or local short circuit quantity magnet for reinforcing the power of the intensity magnet must as unhesitatingly be claimed for [Joseph] Henry and as an invention several years prior to that of Morse."[17] Henry had later written in the margin of his commemoration to Morse, "I am morally sure that Morse had no knowledge of a local magnet until he went to London in 1838."[18]

Morse's intent was to take the device with him to France, where he could employ skilled "mechanicians" to further refine and rework it so that it would be, apart from function, no longer even recognizable as the same device as Cornell had come up with. He intended, on his return, with the help of the "wiggle room" Kendall had built into the contract (and Page's promised "warm advocacy"), to then patent it under his own name as a totally new device.[19]

No doubt the "Intensity/Quantity" distinction had by now been adequately explained to Morse on several occasions, but he had only come to fully understand its implications when Kendall presented Cornell's version to him "in the flesh," so to speak, that June, in the Nassau Street offices of the Magnetic Telegraph Company. And just as Kendall had grasped immediately, Morse had also recognized it represented a quantum advance over all previous configurations of his telegraph. By November of 1846, Morse, with the new relay patented under his own name, would be referring to the telegraph as having two equally necessary components: "The Register" and "The Receiving Magnet," for both of which he claimed entire and sole credit (the latter being only his version of Cornell's relay device).[20]

37

When the Going Gets Tough ...

With the public success of the Test Telegraph assured, "the Professor" would inevitably find himself thrust into the scientific limelight, a role for which he was particularly ill suited. If anyone had reason to suffer from an impostor complex and all that it entailed, it was Samuel Morse. Even Annie Ellsworth's stirring quotidian supplied for the opening of the test line, "What hath God wrought?" may have concealed no small dose of sly adolescent sarcasm, an indication of how evident it was even to a seventeen-year-old girl that the emperor (or in this case the "lightning doctor") truly had no clothes. The scientific community at large, starting with Henry, whose indifference towards Morse's accomplishments had been expressed in thinly veiled sarcasm thus far, would be given full license for a less restrained and more public mockery.

To recapitulate the reasons Morse had for leaving the country at this juncture: when those four immortal syllables first clattered forth from that strange and birdlike, pecking apparatus, with the in-line relays[1] clearly obsolete and out of the equation, the only pieces of equipment to which Morse could legitimately lay intellectual claim was the transcribing apparatus and port rule. By 1845 the port rule had been replaced by the telegraph key (probably invented by Thomas Avery).[2] He had been, since 1837, fighting the doggedly recurring public charges from his fellow passenger aboard the *Paquebot Sully* in 1832, Charles T. Jackson, regarding the initial inspiration which Jackson claimed was, at best, a pretty piece of intellectual larceny. Despite his best efforts, these charges still clung to him like flypaper, and Cornell's treatment at Morse's hands had encouraged him to look at Jackson's charges in a new light."[3] Morse was also still fending off Smith's increasingly litigious claims regarding the Bartlett contract, and suddenly the whole Fisher mess re-erupted when Fisher, now destitute, had threatened to "show this correspondence to several leading scientific gentlemen," by this obviously meaning Joseph Henry.[4] Staying in the country just to battle Smith's and possibly Fisher's charges was an increasingly unappetizing, and in the end, unprofitable prospect. So perhaps to avoid this and rather than risk being exposed as a fraud, Morse, had decided to simply up and leave for the time being.

But while it might appear at first glance that, rather than confront an uncomfortable intellectual challenge, Morse had simply decided to turn tail and run, this is an oversimplification. There was a more compelling reason for Morse to make a quick and graceful exit. For the artist turned inventor, success could prove his worst enemy. While he had an undeniable flair for bravado, Morse certainly was possessed of a thoroughly anemic understanding of electrical circuits and their operation, and all the attention rendered it likely that this fact would be exposed sooner rather than later. His fear that he was about to become the butt of some observations of this kind, was well-founded. In February, shortly after his return, Henry would write Wheatstone, "Mr. Morse is a man of great ingenuity but of no scientific knowledge or habits of mind that could lead to the discovery of new principles."[5]

Given his previous reputation as a notable crank and a dabbler, the wonder is that more attacks of this kind had not publicly surfaced sooner, and if Henry had been biding his time, clearly he had missed his window. It was not until Morse decided to take a potshot back at Henry that the stern sentinels of the establishment would array themselves uniformly to take "the Professor artist" down a notch. As the Smithsonian Regents would somewhat belatedly but still acidly put it, "As an artist of repute, Mr. Morse had been appointed professor of the 'Arts of Design' in the newly established New York City University in the autumn of 1835; but with any literature of science he was remarkably unfamiliar."[6] This was not mere rhetoric. His scientific credentials were notably thin, bordering on transparent. His formal scientific training to that point consisted entirely in the chemistry courses at Yale taught by his friend Benjamin Silliman and attending the lectures by Professor James Freeman Dana on electromagnetism at the New York Athenaeum in 1827. He had contributed no articles of scholarly scientific worth except for the one article in Silliman's *Journal*[7] that had been ghostwritten for him by Professor John Draper of NYU.[8]

So, the Test Telegraph, when it first started in operation, was by no stretch of the imagination "state of the art." Not only had it omitted to fully integrate Henry's all-important "Intensity/Quantity" distinction,[9] but the whole specter of the "cross-cut" was one that had not been fully put to rest; copper wires were still being coated with cotton and shellac despite the fact that, for suspended wires, there was no demonstrable necessity for this.[10] Morse had grudgingly implemented Steinheil's ground return in April of 1844, when he was about half done with the test line, but only after Charles Page's repeated assurances that it would work.

37. When the Going Gets Tough ...

There is ample ground on which to excuse Morse's apparent ignorance and slowness in adopting improvements of this kind.[11] The principles on which the Test Telegraph was built were still poorly understood, even by the experts like Henry and Faraday, who still conceived electricity as some kind of a liquid that flowed in metal. So a conservative approach for someone who was (basically) a scientific layman, like Morse, was understandable. Despite the strides in practical applications (for which Morse himself was more than partially responsible), Henry's all-important "Intensity/Quantity" distinction was based mostly on observational data, and apart from Ohm's work, without a coherent foundation in theory, so we can excuse Morse's reluctance to implement it more robustly on these grounds as well.[12]

When, however, thanks to Kendall, Morse was presented not with a theory but an actual device, he had grasped the implications immediately. He had summoned Charles Grafton Page up to New York, to study it and to come up with improvements, particularly in relation to its size, as it was at present quite bulky and difficult to transport.[13] These were areas in which Page was an acknowledged expert and in which he had already published several well-regarded articles.[14] He had assured Morse it would be a simple matter for him to come up with a better design, a tenth the size and considerably more reliable than Cornell's, and the bonus—in time for his trip to Europe.

While his device had been *prima facie* innovative and conceptually brilliant, Cornell's working knowledge of electrical circuits, due to their haphazard manner of acquisition, was shot through with misconceptions, but unlike Morse, he seemed unaware of his own limitations. This had adversely affected the realization of his "gadget." Page's device would weigh only eight pounds. Since Cornell was laboring under the misapprehension that the relay magnet wire had to be the same gauge as the line wire, Cornell's relay required a magnet weighing 80 to 100 pounds.[15] He also evidently was having a hard time convincing his manufacturer in Ithaca, Munger, that the core of an electromagnet had to be of soft iron, not hard iron, to prevent hysteresis.[16] Munger's inability to grasp this simple fact would provide Cornell no end of difficulties in his dealings with the subcontractors.

After spending three entire days in New York[17] examining the device, Page had come up with his new design. He left this with Morse to fabricate, returning to Washington to tend to his Patent Office duties. Morse had evidently encouraged Page think it was he, not Cornell, who had devised it.[18] Even had Morse confided in him the true author, it would not have made any difference. Page's loyalties to Morse by this time surely extended to intellectual larceny. Page's role had evolved from simply providing technical advice and expertise, to helping with Morse's experiments with the "cross-cut," and more recently to interfering in a direct (and somewhat reprehensible, not to mention illegal) manner on Morse's behalf in the course of his duties as chief patent examiner. Having invested in Kendall's new company to the tune of $500, Page's "warmest advocacy"[19] soon extended to providing Morse confidential information on competitors like Royal House and others regarding their ongoing patent efforts.[20]

Since Page had already compromised his integrity by acting essentially as Morse's personal spy and advocate in the Patent Office, there was no reason to suspect he would not do so again in Cornell's case.[21] Much as Vail had done two years earlier, Page would throw up every variety of obstacle in Cornell's way to prevent him from patenting the relay before Morse's return.[22] In any case, his part in the relay fracas would come to haunt Page for the rest of his life, and he would come to view it as a mistake that he would later attempt to rectify (albeit in a somewhat bizarre and self-defeating fashion).[23]

Morse, prior to engaging Page, had obviously already discussed matters with his friend and Page's boss, Henry Leavitt Ellsworth. In the context of extending best wishes regarding the upcoming European venture, Ellsworth had written, "How does Page get on with his improvement—and how do you succeed in France and Russia? I sincerely hope you will be rewarded fully for your discoveries and great exertions and I hope there is no *mar plot* here to your plans."[24] Ellsworth evidently had adopted a largely laissez-faire attitude when it came to Page's extracurricular activities on behalf of Morse. This may have resulted solely from his affection for Morse or from a character defect of his own. In any case, Morse would be adduced at a later point as a character witness to answer charges against Ellsworth that he was an alcoholic.[25]

On the 21st, Morse wrote to Page[26] to show his progress thus far (and incidentally to inquire about Royal House, whom Morse knew to be already pursuing his own patent for a "printing telegraph"). In a letter marked "Private business," Page had written back, "Yours of the 21st duly received. House seems to be very secret about his telegraph [obviously for good reason]; he has not yet applied for a patent ... you must be particular in getting your coils made and have them at least as well made as those of the instrument. I should make the magnet bar of 7/8ths or 1help iron. I do not like your mode of mounting the lever. A bar of iron so small should not have a hole through it, and as you have it, it will not be so *steady* as if the shaft (which should be of brass) is a little extended on both sides."[27]

Clearly this was a delicate situation and one that Morse desired to conceal most of all from Cornell,[28] but even had he discovered Morse and Page's sudden interest in his device,[29] Morse could most likely have appeased him by pointing out that, as chief patent examiner, Page would most likely be excluded from

37. When the Going Gets Tough ... 195

lodging a competing claim. When push came to shove, Cornell would indeed throw this in Page's face.[30] (What Morse surely failed to inform Cornell was that it is not Page but he himself who would be lodging the competing claim later down the line.)

With Morse getting ready to leave for Europe, Page, probably not entirely trusting invention by correspondence, evidently had been busily working away on his own version.[31] Morse visited him in his rooms at the Patent Office the following week and returned to New York carrying a model relay on which Page had been working.[32] Evidently, while there, looking over Page's shoulder, Morse had done some "kibbitzing" which had not sat well with Page, as he could not retrain himself from taking a playful jab at "the Professor": "*Thank you* for the $20 received in yours of the 26th inst[ant]. I perceive you have a restless spirit with regard to improvements, but [I] advise you to 'let well [enough] alone.'"[33] Morse would have his magnet in time for his trip and apparently for a total price of $65 (with Page afterward issuing a disclaimer of credit for himself in the bargain)[34] and at a size that would fit comfortably into a piece of hand luggage.[35]

In Morse's absence, probably by then suspicious of Morse's motives, Cornell would set about making sure it was his version of the device that was integrated into all the new lines going up across the Northeast, hoping to present Morse thus on his return with a *fait accompli*. While undoubtedly it had evolved from the seed of Henry's ideas, the "embrio" of Cornell's invention, the magnetic relay, in Morse's absence, with the grudging help of the subcontractors, would thus come fully to its own parallel, somewhat strange, and stunted fruition.

Not to be left out of the chase at this early juncture, just following Morse's departure for Europe, Vail had teamed up with Page to sell Page's version of the relay magnet for $50 to all the subcontractors. There had been no takers at the time.[36] Unfortunately for Page and Vail, their timing was just a bit off. Nobody as yet had heard of a "magnetic relay"; they had no concept of what it was for and certainly no inclination to buy one for each and every telegraph office. Ezra Cornell was about to rectify that, mounting a campaign to assure the subcontractors his relay was not just desirable but indispensable to the operation of the telegraph. His approach had worked well at the New York State Fair in Utica, where in mid–September he had put on demonstrations for the benefit of the Albany-Buffalo line investors, but by late September, Kendall had informed Cornell they would be using Page's magnets exclusively on the Baltimore-Philadelphia line.[37] Vail meanwhile had flat-out refused to countenance the use of Cornell's magnet on any line over which he had authority.

38

... The Tough Go to Europe

Morse departed for Liverpool, sailing from New York on August 6, 1845, aboard the Packet Ship *Ashburton*.[1] He carried with him a model of his register apparatus and also a sealed box that was bundled up inside a diplomatic bag. Inside the mysterious sealed box was Page's new relay that now weighed only about eight pounds. It formed the crux of a convoluted stratagem Morse had devised for securing his future reputation and fortune.[2] The outlines of the plan offer a unique insight into his talent for arranging circumstances and leveraging friendships in such a way that sensitive and even somewhat (and sometimes grossly) unsavory business matters could be composed so as to expunge even the slightest taint of dishonesty.

As noted, Morse's plan hinged on absolute secrecy and he had involved only those individuals whom he trusted implicitly. He had taken all the necessary steps to ensure the secret of the relay remained safe, storing it in the custom box that was kept locked and for which only he and the American consul in Paris would have the key.[3] As a further measure, the box would travel in the diplomatic bag that his new status as diplomatic attaché entitled him to carry. The sanctity of the diplomatic satchel not only afforded protection from seizure or prying eyes; it carried with it the implication that the contents were to be considered U.S. property and subject to U.S. laws. In case that was not enough, Morse would carry a letter from Secretary of State Buchanan attesting to his diplomatic status.[4]

With all the "cloak and dagger," he may not even have shared with his usual confidant, brother Sidney, his ulterior motive, though he could not prevent himself from later hinting at it winkingly, referring to himself as "*porteur des dépêches,*" saying that his status as such had "cleared me from all difficulty in bringing my instruments."[5] Morse had also carried with him a working model telegraph apparatus but it had not been treated with the same secrecy as the relay and he had happily provided demonstrations of it to anyone that was interested.

38. ... The Tough Go to Europe

Cornell seemed either unaware or unconcerned regarding Morse's stratagems and he remained in a buoyant mood. His letters home were chatty and full of good-humored jokes and queries regarding the family and doings back in Ithaca. Writing to MaryAnn about the Fourth of July fireworks, he said, "I am glad to hear you had such a pleasant celibration [sic]. I don't see but you was both indipendent [sic] and patriotic." He at last chided her on her penchant for censoring unpleasant news: "But you do not alude [sic] to the dredful [sic] accident mentioned in the papers. The bursting of the cannon by which three men were killed."[6]

While still working under Kendall or Smith's direction in New York, he had become not only the bulldogger for the New York-Philadelphia line, but also the clearinghouse for all telegraphic equipment needed by the various subcontractors. Kendall's letters to him, at least when he was not writing directly under Smith's eye,[7] were addressed to "Superintendent of the Telegraph"[8] (Morse's old title that Cornell would somewhat reluctantly accede to in Morse's absence and which must have infuriated Vail). Cornell himself no doubt by now felt comfortable enough in adopting it. By August even Smith was addressing him as "Superintendent of the Telegraph." The Philadelphia line, under his supervision, seemed to be going up without a hitch: "We are getting nicely underway with our work and hope soon to have the lightning playing between New York and Philadelphia."[9] Still in a semi-joking mode, he had written to MaryAnn, "I am sorry to hear you are not in a good mood for writing as that will operate against the new postage law and I wish to see that law ... sustained."[10] Cornell's good humor would, however, shortly run dry, for a number of reasons.

39

Trompe l'œil

The weather for the crossing was good. The $100 cost of the passage was something Morse most likely had not had to pay out of his own pocket, making his enjoyment of the balmy ocean voyage complete.[1] His traveling companion was the son of his best friend, young Henry W. Ellsworth,[2] who had recently been appointed *chargé d'affaires* to the American Swedish diplomatic mission in Stockholm.[3] Having determined this time to travel in style, Morse had arranged his passage to coincide with that of Ellsworth's son, thereby gaining for himself status as an attaché to the legation, and the two of them had been duly feted onboard; for the first time Morse had a cabin above decks.[4] Hair now almost entirely gone gray, still wildly casting about his forehead like a surf of wayward laurel, real laurels daily being strewn at his feet but still clad in his wide-lapelled wool "Professor's coat," he no doubt cut a far more distinguished figure onboard than he had on the last trip with Smith back in 1838.

Morse and young Ellsworth had parted company temporarily shortly after arriving in England, and after a brief reunion aboard a steamer bound for Germany, they parted company for good. Morse had planned to follow Ellsworth to Stockholm and then proceed to St. Petersburg,[5] but he had abandoned that plan when he received Vail's letter containing the news that the Russian ambassador to America, Alexander de Bodisco, had told Vail after Morse's departure that the Emperor of Russia would refuse, under any circumstances, to adopt his system.[6] A hastily revised plan included a short stint for Morse in Hamburg, where he would stay at the Victoria Hotel, and a short detour to Belgium to accept in person the honor the Academy of Archaeology of Belgium had bestowed on him in June.

Returning to England on October 11, he set his sights now on France, arriving in Paris early the morning of Tuesday the 28th. Finding his favorite haunt, the Hotel Meurices, fully booked,[7] Morse took lodgings in a hotel frequented by English tourists of modest means, the Wagram, which was also a gathering place for up-and-coming American inventors pursuing a French patent for their products.[8] Located on the semi-fashionable Rue de Rivoli, it was walking distance from the Louvre Museum, but apparently Morse would not set foot in that or any other museum during his entire stay.

Though the focus of his visit was now even more clearly the relay, Morse still hoped to make some inroads for his system in Europe. Wheatstone and Davey had both successfully opposed the introduction of his telegraph in England.[9] Wheatstone's system was too firmly entrenched there, and thanks to recent efforts by François Arago, it was already gaining some traction in France as well.[10] His first official visit after arriving was to Arago, whom he had first encountered in 1839 on the trip abroad with Smith. Like Morse, Arago was an enthusiast of the Daguerrian method, and the two had got on well. Arago was respected and influential. Since the death of Joseph Fourier[11] in 1830, he had assumed the role of "perpetual secretary" for the division of mathematical sciences of the French Academy of Science. The efforts in England having come to naught, Morse was disappointed but not entirely discouraged, hoping that he might influence Arago to support his plan. He had written brother Sidney, "Wheatstone … may crow on his own dung hill, but the French cock crows here."

The configuration the French had adopted was really nothing like that of Morse's (except in employing electromagnetism). It did not involve his famous code at all. Most of the French operators it turned out were illiterate, so using "Morse's Code" was impractical. Based essentially on a miniaturized version of the old French railroad flag signals, the French system instead employed a series of ratcheted levers that, in combination, produced 128 semaphores that could be interpreted as railroad switching signals. Though Morse probably had realized even before setting out that the French would never grant him a patent, seeing all this must have confirmed to him that there was no way his version of the telegraph would be incorporated into the French system.

On Friday, Morse met with Alphonse Foy, chief administrator of the French telegraph, whose acquaintance he had also made on the earlier trip in 1838. Foy had invited Morse to inspect the French telegraph going up between Paris and Rouen in the company of Louis François Breguet, whom Morse denominated "a distinguished mechanician."[12] They had been joined by Breguet's wife in case Morse had not gotten the point that this was to be a pleasure, not a business excursion. The line had been completed as far as Lille, where they turned around.

Breguet had taken over his father's watchmaking business in Paris, but he was far more than just a watchmaker. As a member of the Bureau des Longitudes, the Société de Philotechnique de Paris and the Civil Engineers, he had become a prominent figure in government-sponsored technology.[13] It was his optical telegraph that had been adopted for the French rail telegraph, so his presence on this day trip to Lille had not boded well for Morse's efforts in this regard, but it did offer a fresh opportunity with regard to the relay. That evening on their return, Morse left the relay device with Breguet, asking him to reproduce it and if possible, improve it.

Morse spent the next few days, with the assistance of his son, James Edward, who had just arrived in Paris,[14] demonstrating the telegraph register in the lobby of the Wagram to the public. He visited dignitaries that included American poet William Cullen Bryant who, along with Morse, had been one of the founding members of the National Academy of Design.[15] James Edward had also helped with his father's demonstrations for the Chamber of Deputies and the prestigious Academy of Science.[16]

While he was forced to admit that the French apparatus was elegantly executed, Morse had challenged Foy to compare the speed of the two systems. Foy evidently had been favorably impressed,[17] and despite having been shut out of the running, Morse was able to do some crowing of his own in a letter to brother Sidney about the inarguable superiority of his system.[18]

At the time of his visit, Morse had learned that Arago was contemplating writing a complete history of the telegraph.[19] Arago's stature in France was roughly equivalent to that of Henry's in the U.S. He was not just an inventor in his own right but had an unimpeachable reputation for intellectual honesty and fair-mindedness. Arago had promoted Wheatstone's telegraph in France,[20] and Morse had visited him initially in the hopes of enlisting him in his efforts to establish his telegraph here; but that goal had been supplanted now with another; that of having his accomplishments being fairly and fully represented in Arago's upcoming history.[21] That same evening, under the new gaslights of the city, Morse had begun drafting a lengthy letter to Arago reciting his version of the entire history of the instrument to date.[22]

On November 6, the line from Rouen to Paris had been completed, effectively sealing the issue regarding his prospects in France, at least for the time being.[23] (What must have been particularly galling to Morse was the fact that it was an underground line.) On November 8 he received Breguet's new version of the receiving magnet along with Page's sample that he had loaned him.[24] He doposited two boxes containing the telegraph register and Page's relay with the American consul Robert Walsh, instructing him, "The sealed box please retain in private.... The other box containing the instrument which has been exhibited—to show it."[25]

Morse had never explained to Breguet the actual purpose of the device he had asked him to fabricate, claiming it had something to do only with the sending (not the receiving) apparatus. Walsh was an old friend of Morse's who had introduced him to Louis Daguerre on his previous trip and Morse felt he could be depended on if he perceived either his patent or his rights to the new relay were being infringed.[26] By asking that the relay box remain sealed in Walsh's possession, Morse had evidently anticipated that the perspicacious Breguet would eventually figure out the actual purpose of the relay and try to incorporate it in the French telegraph system (which is in fact what happened).[27]

Louis François Breguet, coming from a long line of watchmakers, was an unquestionable genius when it came to fabrication and miniaturization. His grandfather had been the famous Swiss watchmaker, Abraham Louis Breguet of Neuchâtel. In his Paris shop he possessed both the expertise and equipment to create devices of exquisite delicacy and precision. This had allowed him to further refine and reduce the size of the device Morse had shown him severalfold. When Morse returned from Europe in December he would have with him two scaled-down relays produced for him by Breguet at a cost of thirty-nine dollars that worked reliably and would fit comfortably into his jacket pocket.

Morse had taken one of Cunard's new side-wheel steamers for the return trip, landing back in Boston on December 4, having made new friends among the railroading crowd aboard the *Cambria*.[28] Barely able to contain his excitement, the first thing he planned to do on his return to Washington was to show the new relay to both Charles Grafton Page and Amos Kendall. Reaching Washington after a brief sojourn in New York, he brought Breguet's magnets (that weighed under two pounds), to show Page.[29] When he handed Page the Breguet relay, placing it on top of Cornell's model that had been submitted in conjunction with the patent application, Page exclaimed, "Here is a mouse on an elephant."[30]

Cornell had been trying without success since October to get his relay patented.[31] Unlike Morse, who had at his disposal seasoned Patent Office professionals like Page and attorneys like Greenough, Cornell had to resort to flyby-nights, like Charles Monroe, for advancing his application. The result had been somewhat predictable. Though his diagrams clearly show the independent circuit, he had failed to explicitly state in the claims, the necessity—one being for the receiving mechanism and the other for the line current—an omission Morse would exploit as part of the basis of his own patent.

Cornell had made the mistake of including a reference to Morse's apparatus in the diagrams, and so Page had cause to reject it. The patent as issued now was far from optimal; the illustrations were a far cry from the professionally drafted, complex diagrams that Cornell had furnished O'Reilly just two weeks earlier.

The new diagram still showed Morse's register but did not name it expressly. It had obviously been quickly thrown together, and aside from being amateurish, the accompanying descriptions had numerous technical deficiencies. As Morse well understood, Page, in passing it, would be demonstrating not an act of charity but an act of malice aforethought. Apparently Cornell had allowed himself to be convinced by Page to omit key elements of the receiving magnet,[32] possibly because the diagram suggested it impinged on Morse's patent. Page may have assured Morse that whatever Cornell now claimed as his invention, Breguet's new device would supersede it and also prove independently patentable. In what was either a remarkable coincidence, or the clearest evidence of collusion, the stripped-down version of Cornell's patent would be approved the following day after Morse's return.[33]

Apparently no letter was even sent to Cornell to inform him of this, and he had to wait until the 26th to hear the news in person from Kendall.[34] Morse's relay patent (Breguet's device) would succeed in April of 1846[35] despite the fact that, except for the size of the magnet, it emulated Cornell's design almost to a tee. When Page later had a change of heart about his part in this, he would call Morse's claim to authorship of the device "the most extraordinary fatuity" he had ever witnessed.[36]

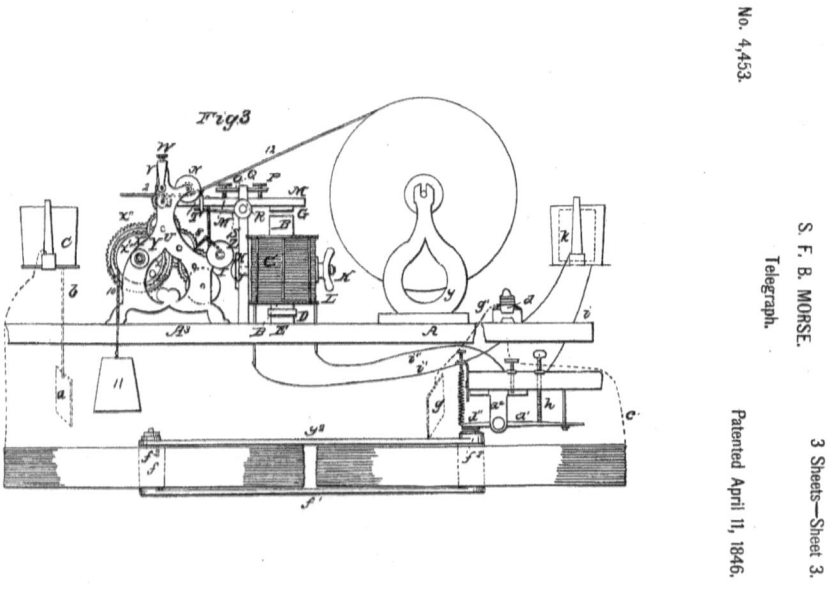

Patent #4,453, Diagram of Morse's Improvement, Sheet 3, April 11, 1846.

40

"The Telegraph for Dummies"

By mid–October of 1845, O'Reilly and Faxton both had been expressing confusion and anger regarding the slowness in obtaining equipment and the lack of instructions about how to configure it. They were increasingly at a loss about how to deploy it, especially when it came to the substations. The wizard Morse had distracted them from looking behind that particular curtain, but the passage of time and his absence had bred more incisive queries. Faxton had written Cornell, "I wish to ask you a few questions ... on a long line ... where there are several intermediate places at which it will be necessary to use it ... there are between New York and Buffalo say 10 stations ... what I wish to know is how all are to be accommodated?"[1] Faxton's subtext was, "would they need to run a separate line for each substation?" The answer, according to Cornell, was, "No, providing you use my relay." Vail had been working on a book designed to provide a history of the telegraph and to address this and other technical issues, but before leaving for Europe, Morse had put a gag on when it came to anything touching on the relay, claiming that any premature discussion would bias his chances of obtaining a European patent.[2] Vail had issued a slimmed-down version of his treatise, offering it to the contractors as an instructional manual but lacking a description of the relay, Faxton's queries could not be intelligently answered.

The contractors, having invested so much time and money in putting up the lines for the "Morse telegraph system," were by November clamoring for more specific information about what exactly was the "Morse telegraph system." At the moment, there did not seem to be a good answer. The only stable element seemed to be the register itself. Everything else, from the type of wire needed to the circuit configuration, seemed to be in flux.

Cornell, (still lacking a patent in November) though while claiming it was an essential component, had been stingy about providing any details about his device. Vail had been focusing on his book project, promising the contractors it would contain all the information they required. If the subcontractors wanted the information it contained, however, they were going to have to pay for it.³

When Vail had found himself ready to publish his book in July, Morse had been caught between a rock and a hard place. If a book appeared which purported to be a complete history of the telegraph containing a description of the relay, then all his careful subterfuge and planning would be for naught—if it did not mention it, then Cornell's claims to invention might appear reasonable, and Morse risked losing credit for what was certain to prove in the future an important component of the telegraph system. The propagandist Morse might welcome, in his upcoming absence, a more sympathetic narrative than any story Cornell might dispense regarding the relay, but the preferred form of that narrative was not clear yet, even to Morse himself.

When Morse finally got a chance to look at the proof of the book,⁴ it was in several respects more than he had bargained for back in February when he had agreed to be a silent partner in the project.⁵ What Vail had proposed then was a mere narrative of the Washington-Baltimore project (coupled with a "how to" guide for the subcontractors). What Vail showed him now was what purported instead to be a comprehensive history of the entire subject of electromagnetism going back to Wheatstone, Arago, Steinhall and Ohm, a two-hundred-page treatise replete with over eighty exquisitely executed woodcuts.⁶ There was one glaring omission: Vail had virtually excluded Joseph Henry's contributions from the account.

The omission of Henry was clearly intentional and vindictive. With Morse as a partner, even a silent one, this effectively gave his imprimatur to the finished product. There would undoubtedly be repercussions. Morse had immediately tried to back away, offering to swap his one-fourth interest in the book to augment Vail's rights to the telegraph in Europe. However, when the confrontation with Cornell over the relay had heated up, his whole calculus had changed again. Seemingly of two minds on the subject, Morse had tried first to suppress, then to encourage immediate publication.⁷ His waffling continued until the last moment before he stepped on the European-bound steamer, when Morse had appealed to Vail's brother George to have Vail to delay publication at least until his return in December.⁸ Vail, having spent close to $750 (half a year's salary) and six months worth of work on the project,⁹ was not about to just let the matter drop for the five months Morse was away.

Pressed for a more convincing reason for his objections, Morse had finally laid bare the main source of his hesitancy: it was not the treatment of Henry, it was Vail's description of the relay magnet. Without a proper description of it, the purported "history" would be woefully incomplete; but to describe the relay publicly would unravel all of Morse's careful subterfuges (to which Vail had obviously not been privy). A few weeks after Morse's departure, Vail had written a carefully crafted letter that potentially excused Morse regarding the slight of Henry and answered the stated objections, "I am glad to have it in my power to relieve your mind from the erroneous impression ... given in you last letter, namely 'has the *receiving* magnet and its necessity yet been published in any book or paper'? I can unhesitatingly answer this question by saying that, 'the *receiving* magnet and its necessity,' is no where described or alluded to in the *work*. I studiously avoided it."[10]

With the matter settled (to his own satisfaction at least), Vail felt there were no further rational grounds upon which Morse could oppose publication. Preparing Morse for the fact that he was about to defy him, harking him back to the abandoned values of the scientific meritocracy, Vail summoned the weight of opinion of the telegraph intelligentsia to his corner: "Hon. Amos Kendall and Dr. Page, the latter having read the proof sheets ... have stated to me that they cannot see what possible harm there is in its immediate publication.... Mr. K, in my conference with him, thought it might be an advantage to publish it."[11] To Kendall, Vail's work seemed to offer a uniform and coherent way to address the troubling questions the subcontractors were now raising with regard to the substation configuration and, sensitive to the charges of favoritism Cornell's actions had incurred, to offer it in a more even-handed manner. Page, for his part, having already disclaimed any credit for the relay, found nothing objectionable in it, relay or no relay. With Kendall and Page's support, Vail therefore felt justified in ignoring Morse's injunction—so long as he made no mention of the magnetic relay.

Still, hoping to enlist Morse on his return in his ongoing campaign against Cornell, Vail had hesitated, but having discerned by now the actual reasons behind Morse's objections,[12] he realized that either way, the villain in the case would be Cornell, not himself. Even thus self-assured and despite his "bad boy" bluster, Vail held back the main volume, publishing instead only a brief twelve-page extract that focused exclusively on the technical issues that the contractors were facing,[13] that he released just in time for Wood to take with him to sell at the New York State Fair in Utica.[14] The organization man, Kendall, had sanctioned the work but not Vail's separate enterprising. By the end of October the pamphlet was being distributed directly through Cornell along with all the rest of the equipment. Wood wrote Cornell, "Carter wants you to send three more pamphlets of Vail's and Davis' Manual.[15] I received mine last night."[16]

Vail, though, had still not given up on convincing Wood to act as sales agent for the expanded volume, and when it came out,[17] he had approached

Wood again, saying he could earn some "pocket change"[18] from the sales. Wood had written Cornell complaining that he had no real interest, but that Vail was irritatingly persistent: "I have rec'd another letter from Vail, Who is anxious to make me agent for the sale of his book."[19] When Vail persisted further, Wood had simply told him frankly what he could do with his "pocket change."[20]

Though Morse had succeeded in delaying any public mention of a relay until his return, Cornell's energetic self-promotion, coupled with the progress of the contractors in the interim, had made the omission somewhat glaring, and the issues remained pressing. With three different versions of the relay and two different descriptions of how to set up a receiving station now being offered, it seemed rather than being a well-organized and fully tested system to which the contractors had subscribed, like the biblical "Tower of Babel," the "Morse system" was starting to appear more like a jumbled anagram, a hodgepodge of different devices, suggestions and configurations, not entirely thought through, and one that, even with the help of Vail's pamphlet, still needed to be untangled in some as-yet undetermined manner to function properly.

In engineering terms, the equipment had been built but the design was not "hardened." Faxton and Butterfield, having observed the test telegraph at various stages of its completion,[21] had at least some understanding of how the various components were supposed to function. They had hired Cornell's brother-in-law Orrin Wood away from his job on the Washington-Baltimore line.[22] Having served there as Rogers's assistant, and being in close communication with Cornell, lent Wood more than a passing familiarity with the equipment and how it should be configured. O'Reilly, on the other hand, had no one really to rely on or consult except his young assistant, James D. Reid. Reid had been his assistant back at the Rochester post office and the bookkeeper for the *Rochester Democrat*, but besides having a quick grasp of things mechanical, that was the extent of his qualifications.[23]

Vail, having acceded to Morse's nervousness about giving out too much information, his pamphlet offered little new information that the contractors could not have ferreted out on their own.[24] The perception that the telegraph was still somewhat "a work in progress" had been compounded when Faxton discovered on his own, while running the Lockport-Buffalo line, that insulating the wires was probably entirely unnecessary.[25]

While the censoring of technical information was annoying, the implications of providing a skewed "history," even though Vail had been careful to provide Morse an "out," would in the end prove disastrous. Whether he knew it or not, Vail was dabbling in matters that were both extremely sensitive and precarious of construction. His "facts" amounted to nothing more than Morse's standard narrative. This was a virtual house of cards that had been carefully fabricated and arranged so that Morse would remain the heroic protagonist in any controversy that might arise in the future relating to the telegraph. Morse was well aware of the opinion within the scientific community that his part in "discovering" the electromagnetic telegraph had been purposefully exaggerated. The omission of Henry was about to bring about a furious confrontation the likes of which neither Morse nor Vail had ever seen.

Joseph Henry's advice had always been decisive for Morse and his influence almost mystical. Even his more bizarre pronouncements had carried enough weight to throw Morse into a panic. As late as January of 1846 we find Morse obsessed with a possibility Henry had raised about "clouds of electricity and magnetism" that might gather around and adhere to the telegraph wire like soup (a concern generated by one of Henry's offhand remarks).[26] In a larger sense Morse clearly envied Henry's semi-oracular status within the scientific community. He had left the country, in part, so that his own scientific reputation would remain intact and unblemished by overexposure or suffer by comparison.

Clearly, Morse hoped to somehow eclipse Henry in the future, even in scientific circles, but to do so without offending him was a delicate matter. This was difficult enough, but it had been rendered close to impossible by Vail's misplaced enthusiasm. With regard to Henry, Vail had intuitively grasped Morse's first point regarding the relay, but not the second: the book had insulted Henry, according him virtually no credit for anything whatsoever relating to the telegraph. His entire contribution was reduced to a single sentence noting one accomplishment for which (according to Vail) he had to share credit with the Dutchman, Moll: creating large electromagnets.[27]

In slighting Henry in this fashion, Vail had roused a sleeping giant within the scientific community of Washington, D.C. Shortly after the full book appeared in November, Henry had written his friend Alexander Dallas Bache, then superintendent of the U.S. Coast and Geodetic Survey.[28] In what was only the first salvo in what was soon to become a major battle, he asked, "Have you seen Vail's book? I should think he could scarcely be [called] an impartial historian."[29] Evidently Bache had, as he had written back disparagingly, "Vail's book is made up of all sorts of scraps ... foreign journals and ... *dits* (*sic* rumors)."[30]

The real question for Henry was how much Morse actually had influenced the contents of the book. There was perhaps not much love lost between the two by this time anyway. Henry had been "routinely intimating to his class"[31] about what he deemed to be Morse's outlandish and self-aggrandizing claims, but all these snide remarks had been in private. Once the book was published, Henry's sly jokes would turn to public scorn and thinly veiled threats. A long-overdue letter to Wheatstone,[32] which may have been in part occasioned by Henry's embarrassment over Morse's recent trip, that such a trifling intellect should have misrepresented itself as the pinnacle of American science, contained a thinly veiled threat against Morse, saying, "If he suffers any more such publications to be made ... he will array against him the scientists of this country and of the world."[33] Henry appeared willing to appear to give Morse the benefit of the doubt, saying, "However he may have had no knowledge of the preparation of the book,"[34] but it was clear he did not believe this himself.

Morse's response had been disingenuous at best. In a foreshadowing of his later disavowals of Henry's influence regarding his discovery, Morse asserted thinly that he could not possibly have had knowledge of the contents of Vail's book, since he was away in Europe at the time of publication. As David Hochfelder, the acknowledged expert on the subject of the Morse-Henry controversy, wrote, "While Morse may not have read Vail's book word for word, he was familiar with its contents."[35] Morse, however, had been at the time so terrified of incurring Henry's enmity that he had gone back and annotated Vail's entire correspondence with him on the subject to make it appear his interest in the book had been strictly that of a loan rather than an investment.[36] In a subsequent letter to Henry, probably written at Morse's prompting,[37] Vail had denied any purposeful vindictiveness, saying he had tried to get a student of Henry's to provide additional materials but that he had never supplied it.[38] Henry had rejected this half-hearted apology and refused to answer Vail. When Morse met with him later to try to patch things up in person, Henry had exploded at the mention of Vail's name, "Vail! I have nothing to do with Vail! What right does Mr. Vail have to write the history of electromagnetism?"[39] The fact that Vail's apology obviously had been penned by Morse had probably been apparent to Henry and this stratagem probably had backfired, convincing Henry finally that Morse had a hand in the book as well. Morse's apologies would fall on deaf ears.

So Vail's book had not only succeeded in thoroughly confusing the contractors, it had turned Henry from a lukewarm supporter into an implacable enemy. Henry certainly was no stranger to feuds over accreditation. His failure to acknowledge Lewis Beck in the original 1831 Silliman's article on electromagnets had led to a decade-long breach with both Beck brothers, one which eventually had healed, but which had left its scars on his psyche.[40] Henry's attitude, as a man of science, had been all along that his knowledge was there to benefit all mankind, which presumably no longer included Morse: "I have given him from time to time information on the subject of electricity but I think in the future I shall be more cautious of my communications."[41]

The intensity of Henry's reaction to what was obviously a second-rate

40. "The Telegraph for Dummies"

effort by a second or third-class mind was so obviously out of proportion it had surprised even him, and he had later found cause to attempt to explain its ferocity, if not to others, at least to himself.[42] Vail would excuse himself, suggesting a personal animus was operative, and that Henry had gone so far as to personally intervene to prevent a copy of Vail's book from being placed in the cornerstone of the Smithsonian.[43] The personal animus, if any, was assuredly on Vail's part. It most likely stemmed from February of 1844, when Henry's preference for Cornell's insulator solution over his had resulted in Cornell's acquiring the title that Vail obviously had long hoped to acquire for himself, "Superintendent of the ElectroMagnetic Telegraph."

In the proliferation of lawsuits following in the wake of the commercial network, Henry would find a more convenient venue for issuing payback. The plaintiff in these cases would be Henry O'Reilly, who now found himself the beneficiary of Henry's extremely incisive and guardedly partisan remarks. In a drama that eventually played itself out in front of the Supreme Court, Morse found himself the subject of a rather scathing (but mostly accurate) assessment in Henry's written deposition, relegating Morse's role to essentially *sous chef* of science, someone who had succeeded by "combining and applying the discoveries of others, in the invention of a particular instrument."[44]

The remarks contained in Henry's 1855 deposition were in essence no different from anything he had said before, albeit in slightly different words, but certainly the opinion now was offered in more unvarnished terms and no longer swathed in prefatory compliments. Henry had clearly consigned Morse to the same level of purgatory that Morse had once conferred Cornell: that of a mere mechanical assistant, a foil to those of loftier aspirations. He was no longer the "Lightning Doctor"; merely the "Copper Plumber."

41

O'Reilly

The subcontractors were, for the most part, neither scientists nor engineers. They were mostly (with the exceptions of O'Reilly and Colt) businessmen who had made their fortunes in the express business. They held little more acquaintance than the average person of their era with the workings of electricity or electrical circuits. They were, however, eminently practical men, and if there was a way to save money, they would find it.[1] Kendall, as an experienced power broker, (accurately) realized he might be creating new financial and political dynasties here and had taken care to involve men of not only of solid business sense, but sympathetic political leanings.

O'Reilly had first come to Kendall's attention as a prominent and outspoken critic of the anti–Masonic movement in Rochester. After serving as an editor for the *Rochester Democrat* rag, he had been appointed postmaster in Utica. The appointment was made by Martin van Buren, probably at Kendall's urging, as payback for O'Reilly's ardent and vocal support of Jackson and the Democratic Party and more recently his activities as recording secretary for the New York State Agricultural Society,[2] in which capacity he had been a vocal proponent of the Erie Canal.[3] His reward for his loyalty to the Democratic Party had been the Philadelphia to Pittsburgh contract.

Most of the rest of those involved in underwriting the new lines had cut their teeth in express or stagecoach businesses that had started with horses and canals and recently moved to railroads. Of the five, only Colt had any prior experience with the telegraph. Certainly none of them understood nor cared a whit about Ohm's Law, or waterproof cables, let alone Joseph Henry's more arcane distinctions regarding quantity and intensity.

Thus far they had been finding their way like blind men in a coal mine, by trial and error. Cornell had recently done the hard missionary work of convincing them of the necessity of a relay. They had been eagerly looking forward to Vail's pamphlet as a guide and counterbalance to Cornell's evident self-interest. Cornell had been freely disbursing Vail's pamphlet to the contractors, at the same time making sure they understood that, having left out the relay, the descriptions provided therein were partial and incomplete.[4] Because the pamphlet had been "sanitized" to remove the mention of any kind of relay or peripheral circuits, it therefore provided no good answers to the questions Faxton had been raising regarding the substations.[5] All this had only added to their already mounting frustrations.

For his part, Vail had not only forsworn any and all direct communication with his imperious counterpart in New York, he had adamantly refused to put any of Cornell's equipment in place on the Philadelphia-New York line. He had tried to get Page's relays substituted for Cornell's in other offices besides Philadelphia, but with Cornell controlling equipment disbursement, that had proven difficult. The relays and registers for the Philadelphia office all had been shipped to him in a locked box from Washington for which only Vail had the key.[6] Not just O'Reilly's but Goell's line from Somerville to Philadelphia had also suffered delays due to a shortage of glass insulators and other equipment thanks to Cornell's preferential treatment of Faxton, but the major obstacle to the larger effort had been the delay in crossing the Hudson, for which Cornell was solely responsible. This was in fact Cornell's ace in the hole, and he had insisted that his version of the relay be installed on the Fort Lee side of the Hudson. This demand must have set Vail fuming.

Stranded in the backwoods of Pennsylvania, by the end of November, a good portion of the Harrisburg portion of the line was complete; but even as the last few miles of cable were going up overhead, O'Reilly and his assistant, Reid, still lacked proper equipment to test it. Having, like Faxton, been led to believe that he had purchased a well-defined and complete working system, the "Morse System," O'Reilly found all this back and forth over relays to be mystifying. It was as though he had purchased a jigsaw puzzle with several pieces either missing, not cut to fit, or suddenly not available. The constant bickering between Vail and Cornell made it appear there was no consensus even amongst the purported "experts" as to what should be the final configuration of the circuits and receiving mechanism. Everything seemed to be in flux. For O'Reilly especially, it had been quite unsettling, and he was about to throw in the towel.

O'Reilly's contract with Kendall and the proprietors stated he must be through to Pittsburgh by December 13, 1845, six months following the signing of the contract, or he would be in default.[7] Without a better understanding of how the equipment worked and how it was to be set up, that goal was beginning to look all but impossible. Cornell's continued preferential treatment of Faxton also had been the major obstacle to his efforts. By mid–November, O'Reilly had applied to Cornell several times, without success, for receiving magnets and registers. By the end of November, Faxton already had a line set up and working between Buffalo and Lockport and was well on his way to connecting to Utica, while O'Reilly was still floundering, stuck somewhere between Lancaster and Harrisburg.

Cornell's blatant favoritism when it came to disbursing equipment had become a major issue. He had hit it off well with Faxton personally and had already been offered a job at $2,000 *per annum* on the Albany-Buffalo line (which fact Cornell did not neglect to immediately rub in his brother Edward's face.)[8] The cozy relationship between Cornell and Faxton had not escaped O'Reilly's notice. Not only was O'Reilly at a geographical disadvantage with respect to the suppliers, he did not understand how to handle Cornell's personality. His letters from Lancaster had progressed from supplications, to irate and then threatening, saying if Cornell would not supply what he needed, he would go behind his back and fetch it himself directly. Fed up with Cornell's dilly-dallying and imperious style, but still somewhat at his mercy, O'Reilly had resorted to begging, "For mercy's sake, don't delay sending me *at once* two sets of instruments to put the Lancaster-Harrisburg line in operation."[9] The next day O'Reilly though better of this and had instead reiterated his previous demand to know who and where the suppliers were: "Please write to me the name of the person you employed to make those instruments for me in New York."[10]

This had naturally incurred Cornell's instant distrust and dislike,[11] and with his focus still on the relay, instead of shipping the other requested equipment, Cornell's response had been to ask Kendall to frame an agreement for all the contractors to sign, acknowledging his (Cornell's) intellectual rights to the magnetic relay.[12] Given Cornell's deteriorating relationship with Kendall over the cross-Hudson line, the second letter from O'Reilly, under the circumstances, sounded more like a threat than a plea. Cornell had not responded the first time and no doubt found O'Reilly's renewed insistence on learning the names of his suppliers not only impertinent but downright sneaky. When the requested equipment had still not appeared a week later, O'Reilly demanded to at least see a sketch of this piece of equipment that Cornell now claimed was both essential and yet was entirely omitted from Vail's pamphlet, writing, "Please forward by return mail the *diagram you promised.*"[13]

The only diagram Cornell had at this point was his patent drawing that Monroe had submitted for him in November. Cornell had been (for good reason, evidently, given O'Reilly's sneakiness) reluctant to send this to him until the patent was passed. While Page was still giving him the run-around, Cornell had obviously been told by someone at the Patent Office that in the case of practical urgency, a signed agreement would offer him at least some protection:

that failing to obtain this might be construed by the Patent Office as relinquishing all his rights to the device.[14] Having Kendall's draft agreement in hand, Cornell relented, forwarding both the agreement and the schematic of the relay under the same cover to O'Reilly. It had crossed in the mail with O'Reilly's latest set of demands. Once he received it, O'Reilly immediately forwarded the diagram to Reid, who was presently shivering in Harrisburg, desperately awaiting the next shipment of equipment.

O'Reilly had been Kendall's hand-picked man, and with Kendall now in charge, this was not how O'Reilly expected things to go. The diagram Cornell provided O'Reilly was the same one he had been working on all November. It was well drawn, suitable for patenting, and clearly distinguished "The Register of Morse's Telegraph" (letter item 'A' in the drawing) from "The Cornell Relay Magnet" (marked as item "B"). Rather than placating him, this instead incited O'Reilly to begin questioning whose system it was he was actually buying. When O'Reilly heard back from Reid, his doubts were only multiplied. Reid said, "The diagram Mr. Cornell sent you could surely never have been designed for any use…. How am I to work a machine that is steadily charged as a permanent magnet?"[15]

Shortly after Cornell's letter reached Reid, the promised equipment had shown up as well. Based on Reid's negative assessment, O'Reilly had refused to sign Cornell's *ex tempore* agreement.[16] Still under the impression that he had purchased the rights to a complete working system, he began pressuring Cornell to provide him with only "original" Morse equipment. Apart from the confusion, he was furious now that clearly Cornell now intended to charge him separately for the relay, writing, "I am much embarrassed by the machinery, & diagrams, & that *other paper* you sent ... we are entitled to all *Morse's* apparatus."[17] With this, O'Reilly decided to remove himself and Reid to Philadelphia for the winter, abandoning work on the line until matters could be cleared up.

Alvin C. Goell, who had been contracted to put up the Somerville to Philadelphia section of the line, was an inventor in his own right, having devised and patented a new method of boring the casings of projectile explosive shells.[18] He had developed this while working at the Washington Armory. He was, however, when it came to Cornell, proving somewhat petty: sniping to Kendall behind Cornell's back, complaining the wires were strung too tight, causing them to break; and further casting Cornell as stingy with materials, specifically the glass insulator caps and wire.[19] Cornell had been shuttling back and forth from Manhattan to Somerville trying to keep the existing lines in good repair. Still under pressure from Kendall to get across the river, with little or no expertise in the area of submarine cables and the rest of his attention consumed by his attempts to patent his own magnetic relay, Cornell was indeed neglecting

Goell.[20] Beset with criticism from all sides and under constant pressure to complete the cross-Hudson piece, Cornell seemed unfocused and cranky. Goell, like O'Reilly, had resorted to begging him for glass caps and wire,[21] even while Cornell furnished them freely and in overabundance to Faxton in Albany.[22] Even Smith, in Boston, had been treated with some degree of disregard.

Morse had just returned from Europe and O'Reilly obviously hoped he would set matters straight. The same day he received Reid's disparaging assessment of Cornell's relay, O'Reilly had sent Morse a copy of the proposed agreement sent by Cornell along with the relay patent diagram, asking for further instructions: "You have doubtless learned from Mr. Kendall the difficulty, delay and disappointment to which myself and my associates have been subjected by relying on the *inventor*! Of the 'Relay Magnet' [Cornell]."[23] With this, O'Reilly had packed up and left for Philadelphia, putting his entire operation on hold. O'Reilly was already behind the eight ball. His contract was in default. He was still looking for equipment and information, but now more importantly, for assurances that he would not be sued. Barring that, he intended to remain in Philadelphia for the duration. He was already a week past his deadline and if the blame game was about to start, he was not going to be the one out of the loop. Vail and Smith had both been conciliatory rather than combative. Vail had informed O'Reilly on his arriving in Philadelphia that Morse had with him two magnets that he could ship to him, but they were presently in Washington (these were obviously the Breguet magnets Morse brought).[24]

When O'Reilly had not received a timely reply from Cornell, he followed up with an even more baldly accusatory letter: "The *instruments are defective* ... the relay magnets were *wrongly sent by you* with a claim for a patent, contrary to what good faith required from an agent of *Morse's* telegraph ... the relay magnets [are] of a character not only different from, but ... actually hostile to Morse's Telegraph System."[25] O'Reilly, in talking with Kendall, had no doubt already lain the full-blown fiasco entirely at Cornell's doorstep, even bragging now to Cornell that Vail had offered to provide him relay magnets of Morse's "original design."[26] What O'Reilly did not know was that Morse would not be able to actually supply them until February, and then would have only two.[27]

Reid had accompanied O'Reilly to Philadelphia, but his earlier assessment, aside from the relay, had contained a litany of complaints of how all the equipment received thus far had been of inferior workmanship.[28] A similar letter had been received by Cornell from Smith,[29] but only regarding the relay, since he had not relied on Cornell's Ithaca connections to supply anything else but that. Daniel Davis had been fabricating the rest of the apparatus for the Boston line. Davis, who had his shop at 428 Washington Street in Boston, was

41. O'Reilly

not only a superb mechanician, he was, like Morse, an avid Daguerreist and a scientist as well, sponsoring lively discussions on the subject of electromagnetism, he had even published a respected work on the subject before the advent of the Test Telegraph.[30]

With the Lockport-Buffalo line nearing completion, even Orrin Wood had been making polite but insistent demands for a "full set" of receiving equipment. Fed up with Munger's incompetence, Cornell's response had been to ask Wood to create his own relay device from a schematic Cornell provided. For a period of weeks there ensued what can only be called invention by correspondence, as Wood clumsily tried to iron out some of the kinks in Cornell's design.

With dissatisfaction growing in every quarter, Cornell had finally admitted (albeit in somewhat grudging terms) to Kendall that the main problems, at least on the Buffalo-Lockport line, had emanated from his relay, but blaming it on "a want of experience in their [Wood and Carter's] management of it."[31] When O'Reilly wrote next from Philadelphia (under his own name but clearly in Reid's hand), it was to announce that both he and Reid were in firm agreement on abandoning the project, blaming his decision on Cornell for illegally substituting his own magnet for Morse's on a plan that was evidently hostile to Morse and the patentees: "The instruments and relay magnets will be boxed up as soon as the clerk who understands the arrangement can be at Lancaster and Harrisburg ... the difficulty in using these articles ... led us (however reluctantly) to suspend the operation of our line.... I would not for double the cost ... have the delay and trouble occasioned to us by your not furnishing Morse's apparatus."[32]

With Morse back in the country for several weeks, by now Cornell had ferreted out his actual intent with regard to the relay. A sane man might have regarded this as an opportune moment to throw in the towel. Cornell's reaction instead was similar to the one he initially displayed when subjected to Page's charges of disloyalty and plagiarism in October: appearing both stunned and offended at first, and then not deigning to reply—and when pressed, lashing out like a caged beast, which is what he now did to O'Reilly.

Everyone except O'Reilly had been addressing Cornell as "Superintendent of the Telegraph." O'Reilly had continued to insist on addressing his letters to "Morse's Telegraph Agent," a phrase which now infuriated Cornell:

> I will now answer your pet[t]ifogging letter ... [I] *am not "Morse's Telegraph Agent" nor "The Agent of Morse's Telegraph System"* ... you say "the relay magnets [are] of a character not only different from, but as you claim a patent therefrom, actually hostile to Morse's Telegraph System." ... I will in the first place inform you that Professor Morse had already (when you engaged relay magnets of me) abandoned his own relay magnet and made

arrangements to use the invention of Dr. Page ... one of which he carried to Europe with him.... I will remind you of the fact that I mentioned this to you and tride [*sic*] to explain the advantages of Page's magnet over Morses's to you, also the advantages of my arrangement over either Morse's or Page's and that your reply to me was that you did not understand enough of the subject to appreciate the difference and desired me to furnish you such as I deemed best.[33]

The one remaining Page magnet Cornell had already shipped to Utica for testing on the Albany-Utica line, so Cornell's claim that he would have sent O'Reilly one of Page's design as well (just as he had done for Faxton), but for the fact that Page had been otherwise occupied ("I learned from Kendall ... that Page was to furnish a pair of his magnets for the NY Phila[delphia] line, but I have not yet learned of their completion"[34]) was a lie.

42

Saxton Faxton's Love-Hate Relationship

The upright "Hickory Quaker," Cornell, by 1845 found himself in a prickly moral position: accused of treachery by O'Reilly and of poaching by Page, despised by Vail and threatened with legal action by Kendall over his sales of the relay, he no doubt had himself begun to doubt where his actual loyalties lay or were owed. The icing on the cake came when Cornell again found himself being hard pressed by Beebe in Ithaca for repayment of all the monies he had advanced Cornell since 1843 on the failed plow venture.[1] Despite this, aware of his precarious position, shortly before the New Year he pledged an additional $400 to purchase shares in Kendall's company as an additional voucher of his loyalty.[2]

December had proven exceptionally cold and miserable, and though he had spent much of it in Audubon's comfortable basement, with everything coming to a head at once, his health had suffered. When MaryAnn accused him rightfully of not looking after himself, he half-jokingly answered she must have found out this information by "mesmeric hypnosis" because the telegraph was not yet operating to Ithaca.[3] The relay remained his only means to raise ready cash, though he was faced with complaints about its reliability. Finding himself again stock rich and cash poor, Cornell could not repay Beebe if he wanted to; he could not even afford to buy a new winter coat and was wearing the same one he had used on the plow trip.[4] He had also (somewhat on the sly), begun encouraging his friends back in Ithaca to take up a subscription to create a line between Auburn and Ithaca.[5] This may have been just an attempt to spread the misery and buy some time, but before he had time to catch his breath, Vail had leapt into the breach immediately, accusing him of conspiring with Smith and Livingston against Morse in a hostile takeover.[6]

By now Cornell's patent had been granted (defective though it was), but he still lacked the funds to manufacture the relay he was so desperately trying to sell to the subcontractors. In the meantime, to fight the charges of favoritism, Cornell had furnished Faxton with a similar agreement to the one he had sent O'Reilly, stipulating specifically to his ownership of the relay. He met with Faxton on the evening of January 8 in New York to discuss this and Faxton's offer of employment. Faxton left for Utica the following morning somewhat perplexed and angry. Despite what had otherwise been a cordial meeting, Cornell had remained adamant on the issue of payment in advance for the relay. Perhaps sensing he had somehow crossed a line, hoping to hammer out the situation just as he had with Kendall before signing the Magnetic Telegraph Company contract, that same evening Cornell forwarded Faxton his copy of his original employment contract with Smith and Kendall stipulating his rights to his inventions with respect to the various companies.[7]

Obviously having second thoughts about their relationship, Faxton, before leaving New York, had fired off a letter to Morse complaining about Cornell's behavior.[8] Morse wrote back confirming his reservations: "I would say that I am exceedingly mortified at the course Mr. Cornell is pursuing ... if continued it will compel me to take legal measures to stop him ... if you can wait a week or two.... I believe I can furnish you two [relays] at a cost not exceeding $15 each."[9] These two were of course the same two Breguet magnets Vail had already promised O'Reilly, but which Vail in the meantime had given to Clark in Philadelphia to duplicate. Faxton had then acquiesced to Cornell's "declaration of ownership,"[10] but under protest.

Morse now openly classed Cornell with his former friend and champion Fog Smith: a false friend, a grasper; the pair were two vines climbing up the great Morse pedestal.[11] However, despite both Vail and Kendall's (and now O'Reilly's) intense and open hostility toward him, Morse remained disinclined to have Kendall fire Cornell. Up until the flap over the relay business, Cornell had never represented any real threat to Morse or to his patent. Morse may have treated him badly but, like Kendall, he understood and respected Cornell's "can-do" mentality; and along with Smith, Cornell now controlled much of the manufacturing of the equipment need by the subcontractors. Furthermore, Morse still hoped he could solve the problem of crossing the river. Still, Cornell had clearly gone off the reservation, and Vail for one was growing increasingly impatient for some concrete punitive action. There was really not much to be done for now. It was, however, as Morse well knew, only a matter of time. The offer of employment from Faxton was already a matter of public discussion and the only question remaining was whether Cornell would stick around long enough to solve the cross-Hudson problem.

To Morse, Cornell must have represented something like an angel with two aspects: one face a destroyer, the other a creator. The relay was a menacing lightning bolt that Cornell held in one hand and in the other was the new vulcanized rubber-coated cable that might hold the solution to Morse's longstanding problems with crossing large bodies of water. The cable might even hold the solution to his long-held dream of uniting Europe and America via the telegraph, so Morse had cut Cornell enough slack to either hang himself or hoist up the flag of success. In the short term, Morse must have realized, Cornell's relay was no longer a real threat to him, and getting Cornell to accomplish finishing the New York-Philadelphia line expeditiously was critical (as was managing the growing dissatisfaction of O'Reilly and the other contractors). Kendall, by dealing Cornell a hand at the invention table, was the one who had inadvertently turned it into something of a cat-and-mouse game, and Morse was by now ready once again to step in and play the bigger cat.[12]

If Cornell was waiting for the axe to fall, he would not have to wait long. By the first of the year, with the cross-Hudson line appearing to be working, Kendall would abruptly fire him, asking that he deposit any remaining equipment in his possession at the telegraph office at #10 Wall Street.[13] Though technically no longer an employee of the Magnetic Telegraph Company, astoundingly, Cornell for the present just continued acting as if he was in charge. He remained at Audubon's, working on the submarine cable. Neither Morse nor Kendall seemed inclined to do anything about it. The day after Morse's unambiguous reply to Faxton repudiating Cornell, Vail had written Morse, also reviling Cornell in the strongest terms, demanding he be dispensed with once and for all,[14] to which Morse replied in a temporizing fashion: "Let us get along as quietly as we can with the plague till we can cut loose from him ... until we get the line through to Phil[adelphia], then [if necessary] we will have a reckoning with him should he be found on my ground."[15]

Morse may have been losing an employee but he was not going to really lose financially by Cornell's move. Shortly after Morse's departure to Europe, Faxton had incorporated under the title Springfield, Albany and Buffalo Telegraph Company. Morse had been a major investor. The receipts from Faxton's line outstripped everyone's expectations. Once it was in full operation, an ecstatic Morse had written Vail: *"One receipt on the Buffalo line for 23 days only* (not one month) *were $2,960.00 odd dollars!!"*[16] The Buffalo line, with Morse as a major stockholder, would become one of the most profitable lines in the United States and a major component of Western Union. When the company was folded under Western Union, this would make Ezra Cornell a very rich man as well. In any case, it was clear the telegraph was no longer an experiment but big business, and everyone, it seems, had recognized that it was a whole new ball game.

43

Organization Man

Amos Kendall was at first glance an unlikely trustee of a man's life's work. He was a frail, ghoulish-looking individual crowned with a wispy shock of white hair and looked like he had stepped straight out of a Charles Dickens novel. His spidery handwriting, sprinkled with expressions of malaise and ennui, gave one the impression he could barely hold a pen upright let alone run a company, and his high, reedy voice was not one that commanded instant respect. All these were misleading clues to a personality driven by an iron will and possessed of a flair for bringing contentious individuals and situations to heel with an ease that was unsurpassed in Washington and perhaps even the country. Morse had recognized these qualities immediately: they were precisely the ones he himself lacked. Before leaving for Europe, Morse had hired Kendall to manage his telegraph affairs and even considered appointing him as his Attorney in Fact, but just before boarding the ship for Liverpool he had changed his mind about the latter.[1] Despite this last-minute hesitancy, it was clear Morse trusted Kendall implicitly; that trust was not misplaced, as Kendall would remain loyal to Morse for the rest of his life and become one his few friends. He was in any sense his intellectual equal.

As a Southerner and a Jacksonian Democrat, Kendall had risen to prominence amidst the populist movement for debt relief following the Panic of 1819. Serving as Andrew Jackson's postmaster, he fell out of political favor with the advent of the Van Buren administration, but remained on in Washington as a power broker for the new party. According to his biographer, he was "the classic American self-made man."[2] Morse had called him "the most competent man in the country."[3] In a few short months he had been able to pull together all of the feuding parties involved in the telegraph enterprise under the umbrella of the new company and to do so without so much as a whimper on their part.

Following its formation in March of 1845, the Magnetic Telegraph Company had begun contracting out the vast network of lines that would march across the country on chestnut poles. However, Kendall had ensured one key link had been reserved, one segment carved out of this telegraph pie to remain directly under his thumb: the New York-Philadelphia line.[4] As the nexus connecting all the telegraph traffic between the major cities of the Northeast, it was sure to become in the future, by any standard, the spinal cord, the nerve center of the entire system and hopefully therefore the most lucrative. The plan was to have it built and superintended on the New York end at #10 Wall Street in Manhattan by Ezra Cornell, with Alfred Vail, his opposite, sitting at the telegraph office in #31 in the Merchant's Exchange building in downtown Philadelphia. Vail was responsible for the Philadelphia end of the line and the connection to the Baltimore Washington Test Telegraph that David Burbank was currently constructing.[5]

In Morse's absence, the relationship between Vail and Cornell had deteriorated to the point that neither was speaking to the other. Any communications between the two was being routed through Kendall.[6] Vail and Cornell stood at their respective postings like two potentates of old, feudal lords with only contempt and ill-will for their rival and counterpart. There had been other problems as well, not necessarily of either Vail or Cornell's making. The New Jersey railroad had refused to grant them a right of way, forcing them to run the line through Somerville,[7] New Jersey, along the old stage road from New Hope,[8] rather than along the main rail line connecting New York and Philadelphia. This had likely been retaliation for Kendall's unfavorable policies in relation to the railroad during his tenure as Postmaster General.[9]

Kendall obviously had no qualms about leveraging any ambient ill-will to his own advantage, but progress on the line, even with the even-tempered A.C. Goell interposed between them, had become mired in machinations, secrecy and backbiting. Since coming to head the new company, Kendall had made a determined effort appear to be above the fray and even-handed when it came to dealing both with employees and with the subcontractors, but Cornell's having acquired Morse's former title of "Superintendent of the Telegraph" no doubt had galled Vail no end and made him more determined than ever to somehow embarrass or destroy him. Given his long association with Morse, Vail might have prevailed sooner; but Kendall, having no dog in the fight, remained focused on getting the New York Philadelphia line finished as quickly and cheaply as possible. Increasingly, that was coming down to getting the line across the North River, a task for which Cornell, at the moment, was indispensable.

In addition to his other responsibilities, Cornell had become the clearinghouse for all equipment going to the subcontractors. Having taken little note of the benefits of Kendall's "even-handed" style, Cornell's insistence on parceling out knowledge and equipment in a preferential manner had rendered him the object of almost universal dislike among the subcontractors, incurring general dissatisfaction and accusations of strong-arming. Even the rough-and-tumble Faxton, who had been the main beneficiary thus far of Cornell's largesse, had taken umbrage at his tactics, saying, "I feel that Mr. Cornell has not pursued a very dignified or upright course with me."[10]

Overwhelmed and sensing the mounting hostility that stemmed mostly from his conflicts of interest regarding the relay, Cornell had turned to Smith, asking him to take on the task of disbursing equipment to the contractors. Smith said he would be happy to, but under no circumstances would he dispense Cornell's relay until the workmanship and reliability were vastly improved.[11] Realizing finally that Smith might be right, Cornell asked him if he knew what Page had been doing thus far for a manufacturer.[12] In response, Smith had provided him the only really useful piece of information he had delivered in the last two years: Smith would loan Cornell his (and Page's) secret weapon—the Boston instrument-maker, Daniel Davis.[13]

The catch was (and where Smith was concerned, there was always a catch), if he was to fabricate the relays, Davis insisted that Cornell pay for the materials up front. Since Cornell had few funds of his own to speak of, Orrin Wood had supplied him with a solution, suggesting that Ezra ask the contractors pay for the materials for the relay in advance.[14] This was obviously bad advice, but having reinvested his entire salary in the Magnetic Telegraph Company stock and with no cash of his own to spare, Cornell had followed it. This decision naturally had not sat well with the subcontractors, especially when Cornell afterward refused delivery until they had signed his release acknowledging intellectual ownership.[15] Faxton, in the midst of his attempt to hire Cornell away from Kendall, had acquiesced, but nevertheless reiterated his complaint to Morse about Cornell's high-handed tactics, and especially the fact that he had been asked to pay $300 for the copper wire needed for the relay in advance.[16]

With Kendall having fired him, and with Page, Vail and even Smith all arrayed against him, why was Morse still insisting on keeping him? It is apparent that Morse had been instrumental in getting Cornell's relay patent approved (defective though it was), and while Morse had begun to disparage Cornell publicly and privately, he had really done nothing to stop his sales of the relay, and in fact appeared to be encouraging them. Morse's actions had been shrewd. He needed more time to get Breguet's version of the relay patented, and also time to find a manufacturer in the United States, so for now he needed a "straw man," someone for the subcontractors to do battle with while he got his ducks in a row. Once Morse had his patent, he could then come to their

43. Organization Man

rescue with a genuine, and most importantly, patented set of "full Morse equipment." The official line was that Cornell's well-intentioned bumbling was simply a stopgap measure run amok. On January 10 Morse had written Faxton, "In reply to yours of the 8th.... Mr. Cornell's magnet may answer but it is greatly more expensive ... and vastly more cumbersome and no more efficient than my instrument which weighs a trifle over 1 [sic] pound, able to be carried in the pocket and cannot be over $20, and probably not more than $15, or even $12."[17]

Morse still had the two Breguet devices in hand, but he had no one as yet in the U.S. who could manufacture it, and he had no desire substitute Page's less elegant version in the interim.[18] Morse had noted to Vail that, as far as Page was concerned, Breguet's improvements were the "ne plus [ultra]."[19] He (Page) "seems as much pleased [by it] as if he had made it himself."[20] Any future claims Breguet might lodge had as good chance of success in America as Morse's patent claims had held in Europe; that is to say, nil. So, in several respects, in theory, Cornell's relay had been really the only remaining threat still standing in his way. All Cornell's thrashing about had accomplished one thing. If Morse had harbored any hopes of reprising his earlier role of returning from Europe like a "phenom" bearing aloft a device that would transform America, this time he had to hurry. Still, the outcome seemed inevitable, and Kendall could not help from crowing: "As to the sale of your relay magnet, I suppose the principle you lay down is correct.... But all this is immaterial, as the whole are to be immediately superseded as I wrote you, and not, as you suppose, by adopting any suggestion of yours."[21]

The delicate "watchmaker" quality of Breguet's workmanship was simply beyond the capability of most American workshops to replicate, at least at the price that Morse was willing to pay. It was not for lack of trying. Having repaired the one broken one, Morse had lent Vail the other of the devices to replicate. Vail had sent it first to Stokell in New York and then Clark in Philadelphia, but Morse had found the results unsatisfactory.[22] Vail had written, "Stokell is a slowpoke."[23] Under pressure from the subcontractors, Morse finally turned back to Breguet, ordering thirty more relays at a cost of 450 francs.[24] It would take Breguet two months to fabricate them[25] and an additional month to get them through customs. In the meantime Morse had to find some way to placate O'Reilly and Faxton.

Though Cornell may have replaced Smith as Morse's personal *bête noire*,[26] despite Vail's continued rabid exhortations to fire him, Morse remained adamant about turning a blind eye for the time being as Cornell continued to sell his "Cornell relay" to his customers right under Morse's nose.[27] So Morse had fumed and postured and belittled Cornell, Vail had fumed and postured and belittled Cornell, and in the end, neither could do really anything much about him.

Morse's decision to try to buy time by setting Cornell up as a "straw man" while he perfected his patent and lined up a manufacturer had worked initially, but the "fight club" scenario quickly wore thin. Faxton's last letter had reached Cornell at Audubon's, where he and Morse were together preparing to test the cross-Hudson line. Feeling confident of imminent success on several fronts, Cornell had assured him "the new little magnet they talk about I as yet know nothing about it, but I have confidence in them not getting any magnet that will supercede [sic] my arrangement. <u>The balance of power will be on my side</u>."[28]

What had been a game of shadows and counterpunching to this point inevitably devolved to a knock-down-drag-out. Still unable by March to supply the required number of improved relays, it seemed clear that Morse's promises had been all tap-dancing, blustering and posturing. Morse nevertheless had continued insisting that he could supply his improved version shortly,[29] blaming the delays on miscommunication and malicious interference by Cornell. In actuality, he was still having difficulties getting Breguet's shipment through customs. By the time he was able to make good on his pledge to the subcontractors, Cornell had managed to get his relay installed on the Albany-Buffalo line and on the Boston line as well, but as Kendall had warned him, it was a losing battle.

Still, with things coming to a head, if Morse did not get a patent quickly, Cornell's device might yet triumph in the field. Morse's assurances seemed like a blue plate special at the diner: one that seemed too good to be true—once you sit down and order it, the waitress politely and cheerfully informs you they have just run out. By mid–April, with Breguet's relays yet to arrive but with the patent now in place, Morse had written Faxton, "I should have had two or more receiving magnets for you long since had you ordered them. But you only made inquiries respecting them without ordering any." With that excuse logged, Morse finally came clean with regard to the real reason for the delay: "I can furnish you a dozen or more of these magnets so soon as I have received them from onboard ship."[30]

44

Crossing the Rubicon

Crossing a large body of water like the Hudson was a problem that had preoccupied Morse since the embarrassing demonstration at Castle Gardens in 1842 and perhaps earlier. Despite having dropped the problem squarely in Kendall's lap when he left in August, even while in Europe, Morse had continued struggling with the question. His voyage had reminded him of the larger implications. If there was to be a trans-Atlantic cable (as he had predicted to Secretary Spencer), this problem had to be solved. While in Europe, Morse had turned to François Arago,[1] but Arago proved to be not much help. In the past Morse would have tasked his little scientific coterie with attacking the issue, but that was already largely a thing of the past, an aristocracy that had outlived its usefulness. For now, he had placed all his hopes in Ezra Cornell, and part of the reason Morse would continue to rely on Cornell even after Kendall fired him was his confidence that Cornell would solve this problem. In any case, the problem of the crossing the North River, or as Kendall had referred to it, "crossing the Rubicon,"[2] was in the hands of a man who would shortly become Morse's bitter antagonist over the relay.

In the meantime, Vail remained camped out in Philadelphia, where he was working away on his book and theoretically waiting for Burbank to get his line from that city to Baltimore finished. Cornell had charge of running the portion of the Philadelphia line from New York to Somerville, New Jersey. The remainder was under the authority of A.C. Goell. Part of the reason Kendall had hired Goell was to act a buffer between Cornell and Vail, who by now were not talking and could not stand one another. Despite the continuing friction, running the lines in New Jersey had eventually proceeded relatively smoothly, but there had been no action on the most daunting task that faced them: crossing the river.

The initial plan had been to cross beneath the river near Chambers Street. Smith had given Cornell instructions to run a line inside the city from #10 Wall, down to the pier at the base of Chambers Street, from whence he could proceed directly to Newark.[3] It was an ill-conceived plan. The river was at its widest here. Ships were constantly plying the piers and laying off at anchor, so the likelihood one of them would snag a cable placed in the river here was magnified. By mid–July, it had been abandoned. Instead they had opted to run a line to Fort Lee.[4] Here, where the river was narrower, the plan was to suspend a line *over* the river between the towering cliffs, between the Palisades and Washington Heights.

Cornell, now faced the task of getting the line from Newark, New Jersey, to Fort Lee, and thence across the river to Audubon's mansion on the upper west side of Manhattan, where they had set up a telegraph office in the basement laundry room. The mansion, "Minnie's Land,"[5] named after Audubon's deceased wife, stood on a low bluff overlooking the Hudson. To the south was the vast Trinity Cemetery that stretched eastward, halfway to the Harlem River. At the foot of the bluff was one of the three Manhattan ferry landings for the Fort Lee Ferry. The laundry had become both laboratory for Cornell's experiments with underwater cables and the relay and his sanctuary from the smoke-filled rooms at #10 Wall, where he had to listen to the complaints of the all the contractors and Crawford Livingston's constant coughing fits.

Cornell had run the ten miles of cable up the west side of Manhattan starting at lower Broadway all the way to 155th Street, but then, instead of continuing across the river, he had appeared to just drop the project, apparently taking the opportunity of his isolation from the hubbub of the express office to attempt to repair relations on the home front and to pursue his improvements to the magnetic relay.[6] By the beginning of the last week of August, Kendall was losing patience, and was strongly suggesting to Cornell that "as soon as your contract for posts is completed ... I wish you to prepare the wire for crossing at Fort Lee ... press the whole business with all dispatch."[7]

Besides the weather, there was no real source of urgency for the cross-Hudson piece until Goell finished his part of the line, and winter was still months away. So, despite Kendall's stern directives, Cornell had been the one essentially controlling the progress of this effort and of the subcontractors as well, and Goell's "equipment shortages" were in part due to the fact that Cornell was content for the present to bide his time.

By September, the plan to suspend the wire over to Washington Heights had been abandoned. The "hurry up" school of invention was obviously not going to work here, and having little confidence in Cornell's technical acumen, Kendall had turned to Rogers and Vail (asking them to involve Charles Page) to come up with the best method for running a submarine cable.[8] Rogers's response indicated that he had already given the matter a good deal of thought and that he apparently did not require Page's or Cornell's input. His solution was a pipe of about four or five inches' diameter wrapped in several layers of canvas saturated in gum elastic (rubber) with the wires bundled inside, adding, on second thought, that one could probably do away with the pipe altogether and just wrap the wires in the gum elastic. He also suggested they have an optical telegraph set up on either shore as a backup in case the cable broke.[9] The latter, at least, was a logical suggestion. By September 14, Rogers had "perfected" his plan for a "pipeless" submarine cable.[10]

Kendall had hoped to capitalize on the competition between Cornell and Vail to get a solution in place quickly, but bringing Page into the mix had been a tactical mistake. Cornell, obviously not wishing to open a new front of contention with the Patent Office, aside from his duties on the Philadelphia line, he had spent most of September focusing on his relay and getting it ready for the Utica Fair. Following his return from Utica, Kendall had abruptly informed Cornell that he "had made an arrangement with Dr. Page for his relay magnets,"[11] which obviously was devastating news to Cornell, enclosing this under the same cover with Page's suggestions for a submarine cable.[12]

Page's solution had been to use lead pipe with a wire wrapped in canvas saturated in India rubber, tar and naptha, buried a few feet beneath the river bottom to prevent ships' anchors from snagging it.[13] This was essentially just Colt's method that he had pioneered in 1842, and it was also similar to the one English engineers Charles West and W.J. Taylor had recently adopted for stringing a cable under the English channel from England to France. West and Taylor had obtained permission from the Admiralty to test their underwater cable later in 1845,[14] but it too was essentially no different from the one Colt had used in the 1842 Castle Garden demonstration, except for the fact that it employed iron sheathing rather than lead.[15]

By October, with little progress having been made on getting across the river, Kendall had begun to grow nervous. He had resorted to cajoling and threats to get Cornell to accomplish the task of crossing the Hudson before winter filled the river with floating ice,[16] following this up with a slew of what were essentially no more than common-sense suggestions that could only have served to annoy Cornell further.[17] In the midst of this had come Page's accusations regarding Cornell's possibly having stolen his relay, but part of the reason for Cornell's inattention to the project thus far had been a bad case of poison ivy that he had contracted at Audubon's.[18]

While Cornell remained distracted by his foundering attempts at patenting and selling the relay and the deteriorating situation with MaryAnn at home, under constant pressure from Kendall, in late October, he had returned briefly to the original concept of getting a line across on masts situated on the opposing cliffs.[19] When this failed, he had turned to Page's solution, which he must have realized was essentially no different from Colt's. Colt had finished a line crossing under the East River in November. It had worked briefly and then failed. With the weather closing in, Kendall was losing all patience and, perhaps thinking Cornell unaware of Colt's problems, had written, "Colt's experiment is conclusive that the river may be crossed with insulated wire enclosed in a lead pipe.... I have all along expected and urged you to make preparation for it. I am sorry it is postponed to the last moment; but as your attention is now upon it...."[20]

O.S. Wood had already informed his brother-in-law about Colt's problems,[21] but Kendall's letter had struck just the right balance between insult in cajoling to set Cornell to work. Intent on attacking the problem methodically on his own, Cornell had sent a response to Kendall by hand via James Eddy the very same day.[22] He would spend most of the next two weeks testing various cable configurations. Kendall, with little confidence in Cornell's inventive abilities, concluded resignedly that if all else failed, "we must, for the present ... use visual telegraphs for crossing, or a boat, or both."[23] This last would turn out to be prophetic.

Cornell's solution was not based on India rubber like Page and Rogers's, but on a new substance that had just come on the market called "metallic rubber." Around the corner from #10 Wall was located George Beecher's fabric supply store. Beecher was Charles Goodyear's brother-in-law, and Goodyear had set him up as an exclusive agent in New York City for the material he had invented. Vulcanized rubber promised to be not only waterproof but tougher and more resilient than India rubber. After getting an estimate from Beecher (and also from the other major rubber supplier in the city, H.H. Day), Cornell had shipped his entire cable to Goodyear's shop in New Haven to be coated with the new product, vulcanized rubber.[24]

With Kendall breathing down his neck, and still distracted by his ongoing efforts to patent the relay, Cornell had failed to stipulate to Goodyear any time frame in which all this had to be accomplished. Also, rather than agreeing on a price per foot thereby limiting his exposure, Cornell had settled on a price per pound of rubber required, which amounted to a license to overcharge him. Consequently, not only had Goodyear taken his time, but the voucher had come as something of a shock.[25] It had stated simply for "India Rubber." Only when Kendall objected to the price had Cornell divulged the fact that he had been dealing with Goodyear's new process which was considerably more expensive.

Despite the embarrassingly high cost, it appeared Cornell had bested Vail, Rogers and Page in terms of design. Not only was the product he employed more suitable to the task, the idea to use Goodyear had been shrewd politically

44. Crossing the Rubicon

as well. Goodyear was Morse's fifth cousin and Morse's predilection for dealing with family was well known. With Goodyear's shop situated in New Haven, it might also provide an excuse for Morse to take a trip there and rub his success in his former mentor Silliman's face.

Still, the cable had to find its way into the river. By way of imposing a deadline, in a somewhat joking manner, on December 11, Kendall had warned Cornell that Vail would soon be on his way to Fort Lee, testing the entire line as he went. Kendall said, "I hope you will have the office in New York ready and will have crossed the 'Rubicon' before he [Vail] reaches it."[26] When "the Rubicon" was not crossed by December 19, obviously by then having received Goodyear's bill, Kendall had issued an ultimatum, threatening to shut off funds entirely, saying, "Your letter of the 17th alarms me ... relying upon you I have given assurances to our Trustees and Subscribers which have already turned out erroneous in point of time and are likely to do as to expense by thousands of dollars."[27] He further instructed Cornell, "If your experiment for crossing the river fails, I wish you to proceed at once to Newark and establish an office there."[28]

Cornell enjoyed working in Audubon's English basement, and hated the cramped and chilly quarters of the downtown office, where he was forced to listen to Crawford Livingston's extended coughing fits and the constant gossip.[29] Though he had been somewhat distracted by his efforts with the relay, in taking on the river project, he was endeavoring to accomplish something in a technological sense that not even the great Morse had been able to resolve. This had given him a sense of purpose and a true feeling of intellectual independence and worth. But Kendall's most recent ultimatum meant effectively that, should he fail, he would be put back under Vail's authority in the backwater of Newark as essentially a pole-setter. This was a non-starter for him. Even when Cornell had been foundering in Boston a year previous, cast adrift, giving his demonstrations for no pay and separated from the still gainfully employed Vail by a distance of three hundred miles, he had given vent, telling MaryAnn, "Vail is multiplying his enemies daily and is already in a position that renders his hatred as harmless as his conduct is contemptible. I have the brunt of the latter, but have a long train of followers."[30] Cornell would certainly rather resign rather than submit himself now to Vail's authority, and he had begun casting about for other employment opportunities.

It had already been an exceptionally cold winter. Goodyear had taken a month longer than expected to complete his task,[31] exciting Kendall's impatience and forcing Cornell to lay the cable into the river with the ice already floating in it. By the last week of December, after some interesting experiences with induction caused by tidal movement, as a result of which the line appeared to work even after the battery had been put away,³² Cornell had finally been ready to put his new "Goodyear" line into the river.³³ Ironically, it was the day he received Kendall's latest ultimatum, but he had been delayed by the weather.³⁴ On December 22, the *New York Tribune* carried the news, "The progress on the various lines of Telegraph are slow but we believe steady.... On the route between New York and Philadelphia they have succeeded in laying a wire in a lead pipe across the River and its fellow, will be put down in a day or two."³⁵

Section V
Prodigal Son

45

Audubon's Laundry

Vail, as promised, had begun his testing the Pennsylvania portion of the line the week after Cornell's cable had been put into the river.[1] Morse was particularly anxious that there be no misunderstandings and had even provided Vail with explicit instructions as to how he was to communicate with Cornell at Fort Lee once they were both ready.[2] Cornell had already demonstrated in December[3] that Morse's original in-line relay on the test line would not function on a line of this length. He had gone to the extent of retrieving the original apparatus from the Baltimore-Washington line, putting it in place at Fort Lee just to prove his point.[4] When they inevitably failed to signal Somerville, he had replaced them with his own receiving magnets, so it was Cornell's relay that was in place now for the test. When Vail received Morse's letter informing him of the substitution of the relays, he had been livid, but there was not much he could do about it from where he sat in Philadelphia.[5]

By January 12, Vail was ready to begin his testing on the Philadelphia–New Jersey section. The station at Fort Lee required two working magnets. With one of Morse's new French magnets broken at the moment, Vail had brought one of Page's relays with him for his end of the test. On the same day, Cornell had assured Morse that his line to Somerville was solid and in good working order[6] and had been fully tested and found to work, as had the cross-Hudson line from Fort Lee. Vail, who had been testing the line running from Somerville to Philadelphia all week, made the same pronouncement.[7] Finally, the independent and concurrent tests were evidently successful, as Kendall had written glowingly, if somewhat over-solicitously, "I am happy to hear that our line is likely to work well both from you and from Mr. Vail."[8] All that remained was to test the line all the way through from New York to Philadelphia.

The papers with the latest London commodity reports were being carried aboard the steamer *Oxford*, due in New York from Liverpool on the fifteenth of January. Part of Morse's present anxiety stemmed from the fact that he had, somewhat over-optimistically, promised the superintendent of the Philadelphia Stock Exchange that he would be telegraphing the reports from the London commodity exchange to Philadelphia from New York, before they could get there by express.[9] He had arranged to have the newspapers messengered up to Audubon's once the ship arrived at the dock in New York, but when he arrived at Audubon's on the morning of the fifteenth, he had been informed by Cornell that apparently some of his assurances had been somewhat premature. The line from Fort Lee to Somerville was still not working properly. Morse had instead dispatched the commodity news by courier to Somerville, where it had been telegraphed to Philadelphia, reaching the exchange at half past twelve on the fifteenth, an hour and a half before the express packet from New York.[10] Certain "powers that be" had opposed this, and Morse's success in this case had not made everyone happy.[11]

Cornell's failure, it turned out, had nothing to do with the relay. Some of the wires near Somerville had been stretched too tight and had broken. Crews were dispatched to repair the breaks in the line and finally, the next day, all stood in readiness again. A nervous Morse sat ensconced in Audubon's bucolic but ice-encrusted house overlooking the Hudson. Cornell was perched, overlooking the other side of the frigid river from the Palisades at Fort Lee. The two relays on Cornell's design[12] were connected to the Philadelphia line on one side and on the other to the black cable snaking its way down to the river where it disappeared in the grey choppy waters, reappearing just below Audubon's porch across the river.[13] Cornell's expectations were high and his excitement at a fever pitch; the critical cross-Hudson piece had already worked well now for twenty-five days without interruption, and there was no reason to suspect the test today would not work.

As arranged, at precisely 10 o'clock, a frazzled Morse took his seat beside the register in Audubon's basement, impatiently straining for the sharp "tack tack (space) tack tack" signaling that the line was open. A half hour passed and nothing—the only sound the swishing of the bare branches of the great, graceful elms in the yard as the wind worked them into a tizzy ... then an hour ... then another half.... Morse changed the battery configuration—still nothing. The cross-Hudson link had failed,[14] probably severed overnight by the fresh ice floes in the river. Morse had immediately blamed Cornell, writing Vail, "The trouble is all here, and you may judge of my anxieties ... surrounded by obstacles the nature of which you can conceive."[15]

Still, despite the failure, Morse had not fired Cornell, and continued counseling Vail to "keep cool" with respect to him.[16] There were other areas where, despite the present friction, Cornell had shown himself indispensable for the moment. For one thing, the stubborn substation question still had not gone away. With Morse away in Europe, Faxton had gone on his merry way all that fall, stringing the line with telegraph. He had relay stations situated at Little Falls, Utica and several other intermediate points under the mistaken assumption that Morse had a plan for seamless transmission through the substations which for some reason was just being withheld by Vail. This, as events had shown, was not the case. In fact, Morse was well aware and considerably embarrassed by this problem, which Cornell had purported to have solved. He was still facing this exact same problem on the Philadelphia line. Messages received at the courthouse in Newark had to be hand-copied and then retransmitted on the Philadelphia circuit. Cornell had been insisting to Morse that using his patented "independent circuit" configuration would solve this problem, but Morse had remained skeptical. In the wake of the failed cross-Hudson test, Cornell's assurances had become even more suspect. Morse had followed his own advice to Vail and kept his tongue in check.

The afternoon following the failed test at Audubon's, Morse had received another letter from Faxton in Albany inquiring urgently about how to set up the intermediate substations on the Albany-Utica line.[17] If Cornell's idea was correct, his relay would not only solve the current problem with the Newark office but solve Faxton's substation problem for the contractors as well. With Cornell remaining at Fort Lee to try to ferret out the problem with the cross-Hudson line, Morse had gone on to Newark,[18] where, that evening, using Cornell's recently patented "independent circuit," he reconfigured the telegraph apparatus.[19] Somewhat to Morse's chagrin, it worked like a charm. The messages from Fort Lee could be forwarded to Philadelphia without having to copy them over and retransmit them.

Finally, Morse had an answer for Faxton, but unfortunately it was an answer for which someone else (Cornell) presently held a patent. While Page had managed to get the main magnet excluded from Cornell's patent, the independent circuit configuration had remained in. Nevertheless, Morse had sent all the information on to Faxton, still stewing in the meanwhile about how he might elbow Cornell out of the picture. For a few days he considered simply acknowledging the source of the contribution, but had put that aside as his anger at Cornell over the continued failures mounted.[20] Despite his usefulness, Cornell was now clearly in the enemy camp and no quarter would be given. Morse would find a way to fold Cornell's circuit under his authorship just as he had the relay— by combining it with the relay. By the end of February, Morse had Cornell's independent circuit configuration diagrammed and duly witnessed by Vail— all ready to be shipped to the Patent Office along with one of Breguet's relays.[21]

Though no longer officially an employee of the company, at Morse's urging

Cornell had continued his experiments with the Goodyear cable, still obtaining inconsistent results. Despite the mishap due to the ice, Cornell's basic approach still seemed viable.[22] His cable remained in the river after having been hauled out, the sheathing repaired and replaced.[23] Once again the results were inconsistent. Despite the delays, Morse evidently still had enough confidence in this solution and hope for its future regarding the Atlantic cable that he had till now ignored Vail's hue and cry for Cornell's removal.

As for Cornell's submarine cable, as per the pattern established previously with the relay, Morse had already, behind his back, asked Charles Page to come up with an improvement on whatever Cornell had devised. Whereas Cornell had incurred a great expense in time and money by shipping the entire cable to Goodyear, Page suggested they might instead be able to accomplish this themselves by using another of Goodyear's products: metallic rubber impregnated fabric. Also, in what was a more radical departure, Page recommended dispensing with the lead pipe altogether, substituting a "rope of four twisted strands coated with rubber and then encased in rubber fabric."[24] Morse was desperate. It was, however, too late in the season to test Page's idea properly. Kendall had already made the decision to continue using the ferry for now. They would try crossing the river with a wire again in the spring—going back for a time to Cornell's original suggestion of suspending the wire above it on masts.

There was, under the circumstances, always suspicions that the problems had been intentional. The day before the test at Audubon's, Morse had noted ominously to Vail, "C[ornell] has ordered an immense quantity of wire wound for instruments of some kind ... he designs to go into the business wholesale."[25] Though the cross-Hudson line was not working, thanks to Kendall, there was at least a backup plan in place: they would route messages to Jersey City. Despite the various failures and feuds, the telegraph service was officially opened for business from New York to Philadelphia by the 26th, only incorporating a delay of ½ hour, long enough for a boat to row across the river with the messages and transfer them to Newark.[26] Jersey City was supposed to be merely an expediency. In the spring, a line would be constructed between Fort Lee and Newark, but the problems with the cross-Hudson piece were starting to look for now insoluble.

Cornell, at this point, could either abandon his efforts or continue working for no pay. If Morse wanted so badly to keep him on, Cornell this time was not going to make it easy. Vail had already informed Morse that someone besides Cornell should head the New York office. Cornell, in advance of Vail's visit to New York, had thrown out all of Morse's receiving magnets there and replaced them with his own (or at least so he told Vail).[27] Pride was one consideration, but Cornell had worked for "pride" before; and while pride may have been good for breakfast, it was not as tasty at dinner. By the end of February, Cornell had somewhat guiltily admitted to his son Alonzo that he was ready to throw in the towel on crossing the river. Alonzo had helpfully suggested to his father that he try stringing a line on masts from Washington Heights to Fort Lee,[28] but as the elder Cornell now understood, his whole utility to Morse had rested on his success with the submarine cable. Barring that, he was no longer "the indispensable man." It was time to cut his losses and move on.

46

Tit for Tat

By February, Colt had added his voice to the growing chorus of criticism against Cornell, saying, "We are still bothered with Cornell's magnets."[1] In March, with Cornell finally out of the picture,[2] Morse was ready to proceed with patenting Breguet's relay, along with Cornell's "independent circuit," under his own name. Though Morse was listing it as an "improvement" to his 1840 patent,[3] the relay obviously represented to him far more than a just peripheral add-on. His preoccupation with it over the preceding year was more than enough evidence of that fact.

The reason was Morse's financial interest in his original patent had been diluted by the percentages he had doled out over the years. His scientific reputation was being chipped away at by Henry and others with charges that his work was simply derivative, and with the Bain telegraph system now being adopted by O'Reilly, there was the possibility he might be entirely eclipsed. The relay, as had been amply demonstrated, represented a critical component, one for which he, with Page's help, could now claim sole and undisputed intellectual authorship. It provided him leverage over the other patentees that he no longer had with the original patent. If Morse succeeded, his technological perspicacity and mastery of the domain of the telegraph would finally be unbreachable (or so he hoped).

When Vail had first suggested a conspiracy between Cornell, Smith and Livingston, Morse had pooh-poohed it.[4] Having brought precious little insight into the brouhaha between Morse, Page, Vail and Cornell, Kendall was still certain of one thing: he was sticking with Morse "come hell or high water," so wherever he might land on the relay issue, Kendall would be right there with him. The relay embodied portions of Henry's research that had been overlooked or discarded by Morse in his earlier efforts.[5] The 1846 patent was cast essentially as an effort to play catch-up. Most of these stemmed from Henry's discoveries regarding the role of resistance in circuit configuration, but as Cornell had demonstrated, besides the simple "Intensity/Quantity" distinction, there was also the utility of what Morse and Cornell had come to call a "marginal" or "independent" circuit that was critical for seamless retransmission through the substations. A third element was something Page had developed as an improvement on Henry's 1835 invention of remote switching: incorporating a solenoid as an actuator.[6] All three of these concepts were now about to become "improvements" to Morse's original patent.

The economic landscape, in the meanwhile, was changing as rapidly as the physical one. New alliances were forming, old ones withering. Smith and Livingston's New York and Boston Magnetic Telegraph Association, like Kendall's company, had been headquartered at #10 Wall Street amidst the chock-a-block express companies.[7] The other neighbors at #10 were two expresses owned by Livingston and Wells, and on each side of them were concerns owned by Livingston or one of his competitors.

While O'Reilly remained sulking along with Vail in Philadelphia, he was sending out not-so-subtle overtures about creating a competing company, running from New York to Philadelphia to Wilmington, with Kendall at its head.[8] Colt had recently been cozying up to Smith in Boston, and a partnership seemed in the offing. David Burbank, running the Baltimore-Philadelphia line, had made some noises early on about acquiring the entire original patent, but he seemed content now to limit his interests. Livingston and Wells, with interests in both the New York-Buffalo and Boston-New York lines, remained the greatest potential threat to the dominant Magnetic Telegraph Company, and like O'Reilly, they had begun looking into abandoning the Morse system altogether, using House or Bain's setup instead.[9]

The fact Morse and Vail sided with him against Cornell had not placated O'Reilly in the least, and in fact had the opposite effect. If Cornell could get away with opposing Morse with no apparent retribution, then perhaps he (O'Reilly) could too; perhaps Morse was a "paper tiger." House and Bain both had come up with competing schemes that seemed to work as well as Morse's.[10] O'Reilly, over the winter, began exploring these as alternatives as well. The fly in this ointment was that all these schemes still relied on electromagnetism, so Morse might well claim they infringed his patent.[11] Anticipating such a response from Morse, O'Reilly would take the first shot, leveling public charges of "monopoly" against Morse in the newspapers.

The atmosphere at #10 Wall Street that winter, the nexus between Morse and Kendall's Magnetic Telegraph Company and Smith and Livingston's New York and Boston Magnetic Telegraph Association, had been so highly charged as to be enough almost to power both the lines without resorting to batteries (something Cornell had found he could actually do). Vail had attempted to inflame Morse's suspicions by accusing Cornell of conspiring with Smith and

Livingston to take over the company.¹² Cornell was caught between new and old benefactors and malefactors, and it was increasingly difficult to distinguish between them. Faxton was criticizing him behind his back to Morse, while attempting to hire him. Both sides continually plied him for information about the other. His most recent stock purchase had been in part a means to reassure Morse and Kendall where his true loyalties lay,[13] but he had been keeping his options open with both Smith and Faxton as well.

Knowing how heartily Morse distrusted him, when inevitably pressed for details regarding his communications with Smith, Cornell presented "extracts" of Smith's letters (just as Smith had done for him when Kendall was first contemplating firing Cornell—in December).[14] Smith, in his usual fashion, had continued to play both sides of the fence, building his empire with other people's money and inflaming Cornell against whomsoever he could. It was from Smith, not Kendall nor Morse, that Cornell first learned the details of what Morse had brought back from France, a "relay magnet not bigger than your fist," one which Kendall had let him know was intended to replace and supplant his.[15] But when offered the rights for Cornell's relay for the Boston line, Smith's canny response had been to offer Cornell stock in exchange for the rights, and further, only providing Cornell furnished him a valid copy of his patent (probably knowing full well that the existing patent covered only the independent circuit and not the relay itself).[16]

Morse's inexplicable tolerance had only served to fuel Vail's ire, and Vail, for his part, had no intention of "keeping cool." Instead he had sent semi-anonymous letters to newspapers with the outlandish claim that Cornell's relay patent was stolen from Colt.[17] The letter had been signed "O.H. Hicks" but there was no doubt who it was from, and Cornell's rebuttal closed with the lines, "Mr. Colt may (knowing that independent circuits may be established), ultimately succeeded in hitting the proper arrangement ... but his failure thus far ... even with the assistance of 'Mr. O.H. Hicks,' proves conclusively that there is no foundation in truth in the assumptions." It also proved somewhat conclusively that Cornell had been something of a fool in this case, unnecessarily making an enemy of Colt rather than admitting to problems with the relays.

47

An Indispensable Plague

Morse's "meritocracy of the intellect" had always been at root a kleptocracy, and he still needed innovative and inventive subordinates that he could later bully into silence. This was also what had been at the root of all of Morse's warnings to Vail to "keep cool."[1] The real question now was why Morse insisted, despite the evident friction, in keeping Cornell on. Why didn't Morse just adopt Colt's solution? Colt had been faced with much the same problem Cornell faced in crossing the Hudson River with his Offing line. Morse had used Colt's submarine cable once before—why did he not just utilize it again? Part of the reason was that his former collaborative relationship with Colt was by now one of outright competition. The last thing he needed was an infringement suit from Colt on the use of his cable design. Where Page and Rogers's solutions were essentially nothing more than a variation of Colt's submarine cable from 1842. Cornell had solved that with his innovation of introducing the Goodyear rubber instead of India rubber. So Morse had prevailed on Kendall, despite Kendall's having fired Cornell, to come up with some face-saving way to keep Cornell on the project. Kendall's solution was to put Cornell directly under the authority of the board of directors, but without any salary.[2] Understanding the basis of Morse's objection, Kendall also began openly advertising in the newspapers for someone (anyone) to replace Cornell.[3]

Morse's method of dealing with subordinates who had, in his view, too liberally availed themselves of the title of "inventor," was by now all too clear. While he might tolerate devices like the cable layer, or even advances in peripheral technology, like Cornell's wire-winder and glass insulators, anything impinging directly on his work was subject to ruthless (if not always immediate) intellectual confiscation. Avery's telegraph key and Vail's "code" (despite their later stated objections) as well, had been folded in initially without so much as a peep.[4] Any and all of these innovations bearing on or relying on his telegraph apparatus would inevitably somehow be incorporated into the great

Morse juggernaut, swept up as by a snowball rolling down a mountain, and as Cornell was about to learn, "woe betide him" that opposed it.

By January 1846, Morse had taken to referring to Cornell privately in terms as he had formerly reserved for Smith, calling him not only a marplot but "shrewd"[5] and "the plague,"[6] warning Vail while waiting for the first transmission from Philadelphia, knowing that Cornell stood between them on line, "Say nothing by telegraph you would not wish Mr. C. to see," and promising, "We will have a reckoning with him should he be found on my ground."[7] By February, the temporizing had proceeded to undisguised antagonism, with Morse saying, "Mr. Cornell has so bewitched and befouled everything he has touched ... that I have hard work in restraining my indignation at his officious meddling."[8] When Cornell had not abandoned his plans to patent his full version of the relay, Morse had written ominously, "And for this thing he expects to take out a patent. He can't get it ... measures shall be taken to stop his further interference on my ground."[9] Elaborating on the same disparaging language he had used with Faxton, Morse told Vail with regard to the relay, "There is one thing in respect to Cornell in which you are in error ... what he calls *his* is but a clumsification of mine."[10]

For all his saber-rattling, Morse had been the one keeping Cornell on board, ostensibly for the reason that he needed him to get a line across the river.[11] The cross-Hudson effort was dying a slow death. The line remained in the river but was only working intermittently. When Cornell had not succeeded by January 26, Kendall, along with the notice announcing the opening of the telegraph, had publicly acknowledged the failure, writing a long letter to the *Washington Union* promising "liberal compensation" to anyone who "can and will take us across the North River."[12] When the reasons could not be effectively determined, Kendall decided they would retry stringing a line on masts over the river when the weather broke; but for now, they would continue to rely on the ferry.[13] The submarine cable approach had to be dropped and would not be retried until two years later.

This essentially ended Cornell's utility to Morse. In late March 1846, Cornell went to Albany to meet with Faxton regarding the position on the New York–Albany line. Faxton had been making overtures since January, but when Cornell was fired from the Magnetic Telegraph Company, he had preferred remaining an independent contractor/inventor for the time being, so he might continue selling and perfecting his relay. He had been kept busy in the interim, with jobs for Livingston and Smith on the Boston line, and the role of contractor relieved him from any further accusations of conflict of interest regarding the relay.

By mid–February, despite continued allegations of disloyalty and Vail's unabated and rabid urging to do away with Cornell, Morse had still not fired

47. An Indispensable Plague

him and was still advising Vail to "keep cool."[14] In the meantime, Morse was taking concrete measures to ensure Cornell could not proceed with his relay patent nor leverage his position with Faxton to place any more of his relays in the field.[15] Morse told Vail to "see Mr. Kendall and ... prepare the preliminaries of some arrangement, by which our instruments shall have my or y[ou]r stamp of approval. All will work right with a little consultation, after throwing overboard that marplot [Cornell] who has confused all our arrangements."[16]

By April 11, Morse had his patent for the relay in hand and Cornell's efforts were doomed. A by now somewhat confused Wood had belatedly assured Cornell that both his device and Page's were installed and working satisfactorily on the Albany-Buffalo line and that he and Faxton would vouch for that fact should any questions arise.[17] But with the six devices that Vail had Clark fabricate in Philadelphia, plus the fifteen Morse expected to receive momentarily from France, this would be enough to temporarily satisfy Faxton and O'Reilly. It would not, however, be until the following October that Morse would be able to find a competent American manufacturer who could satisfactorily emulate Breguet's design at the price he had promised.[18] When at last he did,[19] this finally ended the present flap with the subcontractors.

Morse's public relations campaign to cast Cornell's device as clumsy and amateurish had succeeded, largely due to the inherent defects in Cornell's design, incompetent assistance with his patent, poor workmanship in manufacturing (contrasted with the elegance of Breguet's engineering), and Morse's psychological hold over Page in the Patent Office. All these combined had resulted in things devolving exactly as Kendall had bluntly predicted to Cornell in January[20]: Morse's elegant new relay would triumph in the marketplace, and Cornell would be forced to drop all attempts to patent or put his version into use. Faxton would, in October, hire Cornell on to oversee construction of the vital New York to Buffalo link.[21] Having effectively dealt with Cornell's challenge to his intellectual hegemony, Morse would eventually be able to brag to Faxton, "Nothing has been invented heretofore like my telegraphic system. No improvement has been made in it by any other person since I have had it in operation."[22] The age of the amateur scientist was officially over, partly thanks to a man that Joseph Henry himself had classed as an amateur.

48

The New York–Offing Line

Young Samuel Colt had been by far *the* most successful offshoot of Hamilton's grand scientific/manufacturing experiment at the base of the Passaic Great Falls, SEUM.[1] Organized in 1835, the Patent Arms Manufacturing Company manufactured the world-famous Colt revolver. The company, though, had not sprung out of thin air. The site between Mill and Boudinot Streets had been purchased by Colt's father and two of his uncles in 1813 from the SEUM corporate entity. They operated it first as a nail and rolling mill plant before Samuel had taken over operations in 1836 to manufacture handmade revolvers. Tearing down the old nail factory, Colt had built a more modern facility in its place. (Eventually, when Colt's company failed, the site would become the Allied Textile Printing Company).[2]

In 1842 Colt was suffering personal problems concerning his brother, John, who had been convicted of murder, and then committed suicide while in jail waiting to be executed. Colt had attributed the failure of his company to his reliance on hand-labor (boring and milling) that had led to an inconsistent product, but clearly his depression over his brother had a hand in it as well.[3] By the time Colt's company folded, Morse had already been turned down by the government in the first of his several attempts to fund the telegraph, and nothing had come of efforts to patent it in Europe either. Interest in his device seemed to be nil. By 1842 these failures had combined to dampen even Morse's enthusiasm for the project.

After closing Patent Arms in 1842, Colt had turned his efforts to the submarine battery, getting his former best customer, the government, to underwrite them. The general idea was that, with it, the Navy could essentially mine major harbors without jeopardizing friendly shipping by connecting the explosive to a detonator on land via an electrical cable. This had led to a demonstration at the American Institute Fair of 1842 that was successful in part, due to his assiciation with Samuel Morse. Even before meeting Colt, Morse had long been preoccupied with the problem of crossing large bodies of water with an electrical signal. This, he knew, was a critical technological hurdle facing the telegraph, and one for which regard he had himself undertaken several experiments, including an attempt to electrify the Delaware River (using it as the conductor to operate a telegraph on the opposite shore), all of which had failed miserably. When Colt, using a borrowed run of Morse's cable, succeeded in detonating one of his new watertight underwater mines at Castle Garden off the Battery, this had assured a skeptical Morse that what he had seen as his greatest obstacle was in fact surmountable.[4] This had set a reenergized Morse back on the path to seeking government funding, pursuing his own idea—a national telegraph.[5]

Following the success of the Test Telegraph and the formation of the Magnetic Telegraph Company, Samuel Colt had partnered with New York book dealer, William Robinson, to create the New York and Offing Electric Telegraph Association. The idea was for it to serve the New York Mercantile Exchange, bringing them the overseas commodity news well ahead of the express packets.[6] The plan was to row out and obtain the papers with the reports of European commodities markets from inbound steamers anchored off Sandy Hook, New Jersey. This would then be relayed to the Mercantile Exchange from Coney Island via telegraph, arriving well ahead of the expresses.[7] Colt, like Smith and Livingston and the other subcontractors, was counting not on the public or the government this time, but rather on dedicated commercial customers, investors, and private news organizations.

Colt was clearly the exception to the rule when it came to the subcontractors. Not only did he have personal history with Morse, like him, he was an experienced inventor, a businessman and a recipient of government monies for scientific projects. He was one of Ellsworth's original New England "golden boys." Ellsworth had invested $500 of his own money in Colt's Patent Arms before becoming Patent Commissioner. Both he and Morse clung tenaciously to the idea of the government as their eventual primary end consumer.[8]

Colt, with the backing of William Robinson, a prosperous New York City dry goods merchant, had obtained the contract from the Kendall's Magnetic Telegraph Company to run the cable between the New York Mercantile Exchange and Coney Island in Brooklyn (which at the time was actually still an island).

Colt succeeded getting a submarine cable from Coney Island and across the East River at the Fulton Ferry in late October,[9] but by mid–November the line had failed.[10] After investigating other means, including setting masts on each shore near the ferry terminal, he had succeeded by March with a second cable set in the East River at Hell's Gate. The New York and Offing Company opened for business on April 7, in Post's new fireproof building. The public was invited to visit free of charge for one week.[11] Though the line itself was not all that financially successful, Colt's business model would prove robust. In just a few years, the major news services all would have their own dedicated lines up and down the eastern seaboard. When Morse had aggressively pursued the same customers Colt sought, striking deals with the major newspapers, edging Colt out of the Philadelphia Exchange,[12] the relationship had grown less collegial.

The question posed above—as to why Morse did not simply wait for Colt to succeed and then adopt his methods instead of allowing Cornell to struggle along on his own—has actually several answers. First, though East River currents were more treacherous, the distances involved were far less than at Fort Lee. Secondly, Cornell had already proven himself a master at devising innovative means to solve seemingly intractable problems, and perhaps Morse had wanted to push him to see what he would come up with.

The rest of the answer probably lies in Morse's fond hopes for the Atlantic cable; it was a grand goal on which Morse had set his future sights.[13] He had said, "If I can succeed in working a magnet ten miles, I can go around the globe."[14] Colt was no "babe in the woods" like Cornell when it came to protecting his intellectual property. Nor had Morse forgotten the embarrassment of Castle Gardens. If Morse simply employed Colt's method when it came to the grand project of crossing the Atlantic, Colt might reasonably claim the invention for himself and hence "steal" the credit from Morse. Also, ever since the Tatham fiasco, Morse had become a firm believer in a "duplication of effort" approach when it came to solving tough technical issues. So, while Colt was merrily stringing his submarine cable below the Long Island Sound and across the East River, he let Cornell continue to work on devising some other means of running a line underneath the Hudson River.

While Morse's habit of creating lists of interrogatories when he is felt his self-esteem threatened is well established, it is about this same time, just when things were coming to a head with Cornell, that Cornell seems to have developed a similar obsessive tic for cataloging things. Its first appearance had been after his uncharacteristically spontaneous life insurance purchase, and it surfaced again now in a more noticeable way. Having been forced to admit to his failure in regard to the submarine enterprise, Cornell wrote a letter to his son Alonzo on February 22[15] detailing the precise cost per year for every year of life that a man who drinks will spend from age eighteen to forty. It is somewhat a standard admonishment to a son to be frugal in his habits, but its form and timing were odd and compulsive—this precise calculation of the estimated progressive cost of the use of alcohol over years. It was, however, something that would have made an insurance underwriter proud.[16] (This same obsessive tic would reemerge years when he was later running for the state legislature, albeit in a different form: that of tallying and cataloging all the existing livestock in Tompkins County.) For now, as it had been with Morse, it is simply his method of coping with stress when faced with a life change. Perhaps this was his way of dealing with his failure when it came to the cross-Hudson cable.

49

Rebirth of a Notion

A fourteen-year patent could be extended at the discretion of the Commissioner of Patents from seven up to an additional fourteen years, provided the inventor could show that he (or she) had not had ample time to profit from the invention. Morse's patent for the independent circuit and the magnetic relay, that elegant little device of Breguet's, having passed in 1846 (reissued in 1848) was due for renewal in 1860. Charles Grafton Page's help, in his capacity of chief patent examiner, had been critical at the time the original patent, allowing what was demonstrably a derivative idea to succeed.

When the most vocal opposition to the renewal had come from an unexpected source—the very person who had passed it in the first place Morse must have been shocked. By this time Page was no longer a patent examiner. Since leaving the Patent Office he had acted as an advocate for other inventors. Stricken now by a seemingly indissoluble poverty, he had perhaps come to realize the one inventor he should have helped during his tenure (and failed to) was himself. Page's arguments against the renewal had been delivered in a printed broadside distributed in an unmarked envelope to all the patent office employees as well as to Amos Kendall. Titled rather broadly "Invention of the Magnetic Telegraph,"[1] it detailed how the magnetic relay actually belonged not to Morse but to himself and Charles Wheatstone, stating, "The facts have not all transpired but I think enough to show conclusively that the invention claimed in the patent of 1846 is not Morse's and further that he is entitled to no credit whatever in this connection."[2] This was quite a stunning turnaround for the former self-proclaimed "warm advocate." Until the appearance of the manifesto, Morse and Kendall had expected more or less a cakewalk. Two days before Page's document appeared, Kendall had rather sanguinely written, "Our friends think it well to let Greenough take the other side. I will see Mason today and lead him to expect an additional fee in the event of success."[3]

By this time Ellsworth was no longer Patent Commissioner, and his successor, Philip Thomas, had decided, in what was a high-profile case, to strike down the requirement that all arguments be in written form, allowing the parties to make their case in person.[4] Morse's lawyer had complained, "I have just received information that an oral argument is to be had on the 2nd of April. This involves the necessity of my remaining here til that time and also of my making a much more thorough preparation than would be necessary to enable me to make a written presentation."[5] Page was not the only one fighting the renewal. O'Reilly, embroiled at the time in fighting Western Union's bid for the Pacific Telegraph, had submitted a brief,[6] but the forces operating against Morse were somewhat in disarray. O'Reilly and Page had not made common cause in their arguments against granting renewal, so they had no combined effect. Page's opposition was based purely on the technical assertion that Morse had lied in the original grant, while O'Reilly's was more socioeconomic, based on his old argument that Morse was in effect still trying to "perpetuate an odious and tyrannical monopoly."[7]

O'Reilly's opposition had been expected, but to Kendall, it appeared that Page had simply lost his mind. He had responded to Page's broadside with incredulity, pointing out to him that Morse's claims "once certainly, if not twice have been passed upon by you as a sworn Examiner in the Patent Office,"[8] adding, "Until shown how your course in this matter is consistent with honor and integrity [I] sign myself, Yours with *due* respect."[9] He further pointed out that Page had already disclaimed credit, stating unambiguously: "I have never claimed that invention, publicly or privately, directly or indirectly,"[10]

Apparently undeterred by any argument Kendall could make regarding the apparent irrationality of his stance, Page had responded, "Whatever doubts I may have entertained ... were long since dispelled by the firm conviction that Morse is entitled to no claim or credit whatsoever in connection with the receiving magnet or with the use of main and local circuits." He added archly, "In regard to my 'motives,' my 'honor' and my 'integrity' pardon me for saying that I shall endeavor to take good care of them."[11]

Page had made what seemed like another tactical error—making no mention whatsoever of Cornell's contribution to the relay. As Page well knew, Cornell's prior claim subsumed at least part of what Morse's had ridden roughshod over to gain the original patent. As Kendall and Morse were both no doubt aware, if any voice should have been raised in opposition to a renewal, surely it should have been that of Ezra Cornell's. One can only speculate on the reasons why Cornell did not wish to appear at the hearing, but clearly he had other irons in the fire that were far more important financially. At the time his current partners, Hiram Sibley and T.S. Faxton, along with Cyrus Field and Norvin Green, were just now in Washington lobbying on behalf of Western Union for the Pacific Telegraph Act, which seemed almost certain to pass.[12] Field, Green and Cornell had all been staying at the Willard Hotel. Field and Green remained at the bustling Willard Hotel for months to see that the bill passed.

Though the relay might have been a matter of pride for Cornell, the fate of the telegraph bill held far greater economic consequences for him than the outcome of any relay renewal battle. It would award the contract for constructing the telegraph line from St. Louis to California to a single bidder.[13] The principals of Western Union were at the very least hoping for a lack of public opposition from Morse, and challenging his renewal did not seem the best way of accomplishing this.[14] Green had written Morse (obviously unaware of the renewal battle), "If you can come to Washington a week or ten days hence ... you might render very material service."[15]

Cornell had remained in Washington only through March, leaving for home before the passage of the bill and before the relay battle was joined.[16] Cornell was no longer really interested in his legacy as an inventor and saw his future more in the shaping of the arena of public education. He had by now set his sights on creating the People's College, which would open in September (in what was to prove to be its first incarnation) as the Ovid Agricultural College. He had written MaryAnn from the hotel, "I have got quite tired of staying here, not because the objects of interest are exhausted, but because I am unexpectedly detained and had other plans for the employment of my time."[17] Nevertheless, he had taken the opportunity to revisit Mount Vernon, where he had trekked with Avery in 1843 while working at the Patent Office: "It was 17 years since my visit to the tomb [sic] of Washington and dilapidation and decay had wrought a change in that period. The house very much decayed. So much so that rough props had to be set up, to support the piazza from falling ... the late owner, though a direct descendent of the Good and Great Washington ... has done nothing to preserve the estate from the most galloping consumption, and but for the ladies movement, a few years would have obliterated the last traces of the residence of the father of his country."[18]

Morse's initial response to Page was to gear up his famous "list-making" indictment, but he had only got to #1 when he abandoned it.[19] Instead of rebuking Page publicly, Morse had instead chosen to write Page a long private letter rehearsing (his version of) the entire series of events, saying that Kendall was now suggesting that somebody had paid him off. With no other logical explanation, Morse said he tended to believe it: "As a solution to your strange conduct it was suggested to me that [you] were perhaps [acting] as the legal counsel of my opponents. This ... would explain the sudden change from friendship to opposition."[20]

The fact that Page's sudden attack of conscience may have emanated from some profound moral transformation apparently occurred neither to Morse nor Kendall. Both had reason to regard Page's morals as laughable. In offering Morse his "warmest advocacy" back in 1845, Page had already breached the solemn duty of his office. Page had dismissed the charges of self-interest out of hand, saying he was acting only "with a view of eliciting the truth and for this purpose alone."[21] A by now exasperated and probably somewhat amused Kendall had declared in his spidery-scrawl hand, "Professor Chas. G. Page ... seems to see no dishonor ... in attempting to prove himself a perjured Examiner in the Patent Office.... My correspondence with him is closed."[22]

Following Page's attack, Kendall had advanced Morse the rather considerable sum of $3,000 to ensure that the renewal process went favorably.[23] Morse had used this money to pay for Leonard Gale to come and testify as an expert witness and in addition the services of the prominent patent attorney and jurist, Charles Mason. Despite all this, Morse would not only almost lose the renewal battle but the entire underlying patent as well, and not thanks to Page or O'Reilly, but due to his own paid witness, Leonard Gale.[24]

There were by now over 40,000 miles of telegraph line in the United States and no fewer than four distinct major companies; three of them using, to a greater or lesser degree, other telegraph systems than Morse's. O'Reilly's men were stringing networks parallel to and almost within shouting distance of Morse's crews. Other efforts had made use of the Bain telegraph. The Supreme Court's decision that Morse was not entitled to royalties on these competing incarnations of the telegraph meant that whatever equipment his competitors might employ in place of his for the sending/receiving apparatus, neither he nor the other patentees would get a penny from it. No one, however, thus far had found a way to circumvent necessity of the relay, so, aside from the impact on his legacy and reputation, Morse was dependent on the successful renewal of the relay patent if he intended to profit in any manner from those competing projects.

Morse had been preparing his side of the case since early March,[25] but clearly Page had thrown a monkey wrench into his plans. The renewal process was perhaps the most subjective of all the functions of the Patent Office. Charles Jackson's old canard objecting to Morse's claims had long fallen by the wayside, but the renewal process opened him up to attack by other antagonists who had lain dormant, and others who remained his enemies. O'Reilly's charges of monopoly had been dealt with by the Supreme Court in 1855, but the hearing gave him the opportunity to raise them again.

Page's manifesto, aside from omitting Cornell's part in all this, contained several other factual errors that it should have seemed impossible for Page to make; for example: "The first step towards the introduction of a wire for the receiving magnet smaller than the circuit wires ... *was made by myself while Prof. Morse was in Europe.*"[26] Page's version of the relay that had shrunk the size of Cornell's electromagnet tenfold had been made and delivered a month before Morse's trip, and indeed specifically *for* Morse's trip. It had been Page's device that Morse deposited with the American consul in Paris (as Morse points out in his rebuttal), so, as Kendall noted, Page's claims seem extremely odd.[27]

Gale's appearance on Morse's behalf, however, had resulted in an issue that was far more serious. In an excess of zeal, he had testified that Morse conceived and tested marginal (independent) circuits as early as 1838, but had chosen not to implement them on the test line for various reasons. This statement was highly problematic and potentially devastating, not only to the present case, but also to the underlying patent. All extant improvements to art that the inventor was aware of at the time of filing were required to be included in a patent. The relay had been listed as an *improvement* to the telegraph. If Morse had willfully omitted the marginal circuit, as Gale now claimed, from the patent of 1840,[28] this was grounds not only for dismissing the renewal, but for revoking the underlying telegraph patent.

Patent Commissioner Philip Thomas had noted this[29] and then allowed Gale to "clarify" his testimony, which Gale promptly did, claiming he had only been referring to secondary circuits within the main line, not marginal circuits at the receiving/sending station.[30] Gale had been so concerned about his slip-up and the prospect that he might have inadvertently voided the entire patent that he had personally visited the Commissioner to sort things out the Friday after the evidence had been closed.[31] This seemed sufficient to satisfy Commissioner Thomas' objections, and the renewal was granted on April 10. An overjoyed and obviously relieved Morse had requested Gale print up 500 copies of the decision to distribute.[32]

As has been pointed out, part of the reason this renewal had been of such paramount importance to Morse was that, unlike his interest in the original telegraph patent of 1840, which had been diluted to a little over a half through distribution of partial ownership to Vail, Gale and Smith, Morse was the sole owner of the 1846 patent. The thirty-two-page decision was by far the most extensive and complex of any issued by that office to date, and while it may not have occurred to Morse that the decision was important for other reasons, Gale would write about it to Morse: "It is without doubt the most important question [case] that has ever come before the Commissioner of Patents."[33] In a sense it was; it represented Morse's final hurdle to intellectual and financial hegemony over the entire American telegraph. But the case was important for other reasons as well; the actions of Page had raised another issue: "Was it possible

49. Rebirth of a Notion 253

for a patent examiner to later impeach him [or her] self after leaving office?" If so, this could undermine the whole function of the Patent Office.

Whether this was said to justify Gale's hefty consulting fee of $1,056,[34] and to further massage Morse's ego, it is difficult to say; but Gale was unquestionably right in his assessment regarding the importance of the case. His confusing testimony at the patent hearing though in fact been only the latest in a long line of sycophantic obfuscations. In testimony to the Supreme Court in 1854, Morse had stated that the conversion to the single-wire configuration had occurred to him "at some point" during the erection of the Baltimore-Washington test line, but was not fully implemented until probably 1846.[35] The diagrams sent by Morse to Vail in April of 1844 still clearly show the presence of separate lines for sending and receiving.[36] The Test Telegraph had been a five-wire system with two redundant dependent circuits.[37] The major improvement to the lines after they were erected was to employ an external pole-to-ground wire as a ground wire, substituting for the fifth wire inside the conduit.[38] Gale's deposition, issued in another, earlier case, had been reintroduced now specifically to support these spurious claims.

The same held true for the question of "independent circuits." The division of main and auxiliary (marginal) circuits was patented by Cornell and not incorporated by Morse until 1846. Gale had claimed they were part of the original conception, even predating the erection of the test line, and this was what had gotten him into hot water now[39]:

> Before lines of telegraph were set up it was anticipated that in long lines the ordinary current of electricity might not be strong enough to work the magnet at such distance so as to write but would be so strong as to open and close a side or a local circuit as suggested by Professor Henry. This mode of using one electric circuit and magnet to open and close another electric circuit either for extending the main circuit to greater distances or to operate any local circuit although not in the machine when I first saw it was discussed in an early part of 1837 before any lines had been constructed.[40]

So, in stating this again before the Patent Commissioner, Gale had only been reiterating false claims he had made earlier that had slipped by. Thomas had been the only one technologically savvy enough to fully understand their implications.

Amos Kendall had later lectured the Supreme Court that the Morse patent cases on the whole were "transcendently important in the principles of patent law."[41] This was true. Morse's patents had almost singlehandedly transformed the Patent Office. His device would result in a flood of follow-on applications, improvements related to electrical circuits and to the telegraph itself, that would engulf the Patent Office over the next thirty years and transform its operations.

Morse may have triumphed again, but he was leaving an increasingly long trail of ill-feeling and damaged reputations in his wake. Fisher, after leaving the telegraph, had been forced to leave New York to take a position as principal at a Philadelphia grammar school. O'Reilly, the perennial thorn in Morse's side, after having served in the Union Army, would eventually die penniless and in disgrace. Smith, whom Morse had characterized as a friend turned "fiend," would become so universally hated in his native Maine that the cement for his tombstone would be adulterated with newsprint. As it deteriorated, the population of Maine were treated to the news of 1876 for decades after, as section after section fluttered off into the cool New England breeze. Page would live to see his greatest achievement, the transformer, attributed to and named for another scientist, Ruhmkorff, and likewise die in penury and semi-disgrace. Morse alienated Colt and then excluded him from the waterproof cable effort also poaching his prospective clients for this New York and Offing line. He had thoroughly alienated Henry through his participation in Vail's book. (The Smithsonian regents would over the years make attempt after attempt to realign things and restore at least in part Henry's intellectual authorship with regard to the telegraph—though to little avail.) Morse had tried to destroy or at least damage Ezra Cornell's reputation as well, but he had eventually come to tolerate him, if not as an intellectual, at least as a financial equal. Cornell's name would in any case become more synonymous with the great institution named after him moreso than any of his accomplishments on the telegraph.[42]

How much Morse himself actually contributed to the world of ideas and the reputation of American science as a whole is difficult to assess. Five months after his death, an oration was given for the opening of that year's American Institute Fair, delivered by Professor Frederick Barnard, then president of Columbia University. It was an attempt to properly assess what was by any measure a remarkable but challenging and troubling legacy[43]:

> Ancient science obtained by deductive reasoning was singularly inexact and therefore failed. Modern science was not only true but it was fruitful for the Industrial Arts were born of it. In the laboratory we have these arts in embryo; in the workshop we have science in application. In the concourse of industries which have arisen in this way, I am proud to affirm, that America holds an honored place. The cotton gin, without which the machine spinner and the power loom would be helpless, is American. The power shuttle which permits an unlimited enlargement of the breadth of the web, is American. The planing machine is American. Navigation by steam is American. The mower and reaper are American. The rotary printing presses are American. The hot-air engine is American. The sewing machine is American. The machine manufacture of wool cards is American. The whole India rubber industry is American. The band saw originated in America. The machine manufacture of horse shoes is American. The sand blast of which the large capabilities are yet to be developed is American. The gauge lathe is American. The only successful composing machine for printers is American. The grain elevator is American. The

49. Rebirth of a Notion

artificial manufacture of ice, exhibited here two years ago under the name of the Carre process was originally by Professor Alexander S. Twining, an American. The electro magnet was invented and immediately after its invention was first practically applied in transmitting electrical signals by Prof. Joseph Henry, an American. The telegraphic instrument introduced a few years later into public use, which has since obtained universal acceptance, was invented by Prof. Samuel F.B. Morse, late one of the regents of our Institute, an American.

All the attempts to diminish his reputation by enumerating his various crimes would fall largely on deaf ears. By now Morse was enshrined as an American icon and that was that. Even the reporter for the above event was inclined to joke in closing; "After this burst of patriotism everyone amused himself as seemed good in his own eye."[44]

Coda: King Edward of Kalamazoo

On November 19, 1855, a day before T. Romeyn Beck's untimely and suspicious passing, MaryAnn Cornell had penned a few poignant lines to her absent husband who was, as usual, away from home attending to telegraph business: "I have never seen so many sudden deaths as I have this fall. It teaches us a powerful lesson on the uncertainty of human life."[1] If anyone did not need a lesson in the uncertainty of human life, it was Ezra Cornell.

By the end of 1860, with the Pacific Telegraph Act having finally passed the Congress, Ezra Cornell's concerns and focus had turned temporarily back to his family and his farm, and to his new project called "The People's College." The telegraph had brought him wealth and independence, but had also shown him his own limitations and failings. Though his aspirations now veered toward the ideal of public education, he still fancied himself in some degree an inventor. The one he was working on now was not likely to "open the eyes of the world," as he had once written of the magnetic relay; it was not some astounding leap of technology, but rather a humble rain gauge. He had written Joseph Henry, hoping for some encouragement, but Henry's reply had been somewhat dismissive: "The subject of self-registering instruments ... has long occupied the attention of persons interested in investigations of this kind ... an account of the number proposed and actually constructed would fill many volumes."[2]

Three weeks before receiving this reply from Henry, he had received a rather more disturbing communication in the form of rambling six-page missive from his brother Edward in Michigan, from his sprawling farm in Pine River, Gratiot County. It described recent events on the farm and clearly suggested Edward's imminent re-descent into insanity.[3] Ezra had learned from other sources, the farm was a mess. Within a month Edward would be committed the new Kalamazoo Insane Asylum. So by 1861 the "black ship" had reappeared

256

in full sail on the horizon as the specter of familial insanity edged closer to the otherwise blissful shores of Lake Cayuga—it was disturbingly familiar as Edward's concerns seemed to focus on his wife's infidelity and the fear was that it would lead to the same gruesome outcome as had Alvin Cornell's obsession with the loyalties of his wife.

What is interesting for us is that these two separate eruptions of insanity—Alvin's and that of his brother, Edward—coincide and neatly bookend the beginning and end of Ezra Cornell's involvement with Morse and the telegraph. Whether or not we believe mental illness to be hereditarily transmissible, it is clear that the majority of the population of the United States in the mid- (even to the late) 19th century believed it to be. By the time of Edward's episode, all that was clear was that the accumulation of instances was now dangerously close to putting a blot on the family name. The stigma that mental illness carried adhered not only to the persons afflicted but their entire families as well. Even vast wealth or reputation, of which Ezra by now had surfeit of both, was no antidote.[4]

It had fallen to Ezra, therefore, the by now unquestioned patriarch of the family, to address this situation.[5] Whether or not such a sad fate or eventuality, was in the cards for him personally, the long shadow had already found fertile ground in the lives of some of those closest to him. He had felt and seen and contemplated its strange fruit firsthand, in all its pathetic and frightening aspects. There had been no need to publicly acknowledge it in the case of Alvin, but as he grew older and more prominent, indeed even harboring political aspirations, it was proving increasingly difficult to ignore or escape. This time, with Edward, it had struck closer to home.

Both cases seemed to hark back to an earlier one in the Cornell genealogy; the familial murder that had occurred in the Plymouth Colony of Rhode Island.[6] To summarize the circumstances of that case: in 1673, in Portsmouth, Rebecca Cornell, a prominent member of the Quaker colony, was brutally beaten and burned to death, an event at first ascribed somewhat oddly as an accident, but one which led eventually to the indictment of her son Thomas Jr. for the act. In March of that year, charges were brought against Thomas by his Quaker brethren. Despite there being only circumstantial evidence, the tribunal had been led to the somewhat suspect, but certainly more expeditious conclusion that Thomas had been possessed by devils. The only known cure available at the time for this condition was unfortunately fatal to the patient. Thomas, upon being found guilty, was promptly and unceremoniously hanged.

Having been in the asylum since January, by May, Edward was anxious to get back to his farm: "It is with pleasure I am enabled to announce to you my health is better than it has been any time since I left Ithaca.... I wish to leave by the thirteenth for home."[7] In early June, Cornell had received another letter from his brother-in-law Hiram Robertson in Albion, Michigan. Having run out of funds, he was requesting a loan to complete construction of his new brick home on Spectacle Lake.[8] Ezra had already built a home for his parents. An urgent plea for release from the asylum from Edward had been sent under cover of this letter from Hiram.[9] Edward's letter was also a strange mix of accounting concerns and rambling emotional pleas. In cordial, fraternal tones he thanked Ezra for settling the outstanding long overdue bill[10] for board at the asylum and went on to ask plaintively that he intervene with his doctor, Edward van Deussen, to get him released. After stating definitively he was completely cured, he somewhat disingenuously acknowledged the duty to go home to his farm in Pine River to check on his wife. But then, in somewhat less welcome terms, he restated his oft-expressed desire to visit his dear brother in Ithaca, once the chores were done at the farm: "I wish to leave the asylum now and visit Ill., Coshochton, Ithaca, N.Y. City, Providence and then home again. Will you try and spare me the means to travel."[11]

This letter alluding to Edward's desire to revisit Ithaca, his old stamping grounds, could not have been welcome news to the upright, now wealthy burgher, and civic booster of the Ithaca Fire Department (and soon to be member of the New York Assembly), Ezra Cornell. One may entertain the only marginally surreal notion that the now powerful Ezra had somehow arranged to have the county seat of Gratiot, Michigan, renamed "Ithaca" expressly so that his confused and disoriented brother might think himself already having arrived at "Ithaca," therefore avoiding what inevitably would have proven an embarrassing and awkward family reunion. Almost anything was available for a price.[12]

In any case, something had clearly precipitated Edward's sudden angst, and his urgent desire to be released. It was also evident that there was little if any credence to be accorded his claim that he was cured. He had written two other letters, one to his terminally ill father and the second to brother-in-law Hiram, stressing the view that unless he could return home posthaste, his marriage was in danger of dissolving.[13] Portions degenerated into gibberish and incoherence, but he at last let slip the point upon which all this sudden urgency devolved, what had agitated him to the point of distraction: it was a letter received from his wife, Angeline Mosher. Though he denied it, it was clear she had inadvertently, perhaps through some otherwise innocent chatter, fomented in him an impotent jealousy, an unshakeable conviction now lodged in his tormented psyche that while he remained locked up, his wife was making plans to leave him. Hiram informed Ezra of this: "As we rec'd a letter [from] Edward this last week, begging to be released from the Asylum, I thought it best to forward it to you [thinking] that if he is as well as he states that he may get his discharge immediately."[14] To Ezra, who was well aware of the tendency, the case was taking on increasingly ominous tones, and once again some disturbingly reminiscent of earlier situations in his family.[15]

Edward van Deussen, superintendent of the Kalamazoo Asylum, where Edward was confined, was a former student of T.R. Beck's, and so he would be at this point the third[16] of Beck's disciples to exert some strange power over Ezra Cornell's life. To add yet another strange and haunting parallel, another ripple on the pond of this series of sad and tragic rondels, as it turns out, Angelina Mosher, Edward's wife, was a direct descendent of Hannah Mosher, wife of the grandson, Steven (the son of the murderer Thomas Cornell, Jr.), of the murdered woman in the Plymouth Colony, Rebecca. Ezra, Edward and Alvin therefore were all descended directly through this match between Steven Cornell and Hannah Mosher.[17]

The Civil War was now on, and though Ezra had originally planned to go to Washington with medical supplies for the Ithaca Regiment in June, he first arranged a somewhat ad hoc side trip to Albion by train, ostensibly to speak directly to Edward's doctor in person. He arrived in Kalamazoo in mid–June, clearly hoping to get matters sorted out with some degree of finality. He was successful in persuading van Deussen (probably with the help of a substantial donation to the hospital) to release his brother (presumably on Hiram's recognizance), but then mysteriously disappeared before the release could be effectuated. No doubt the train ride back to New York was not a happy one. His sanguine expectations for Edward's recovery had been thoroughly dampened by Superintendent van Deussen's sobering evaluation.

Ezra had also taken some pains to ensure that the other part of his brother's plan would not succeed. Before leaving Kalamazoo, he had tasked the relatives in Albion with ensuring Edward's direct passage home to Pine River, a task which, not surprisingly, turned out to be not all that easy. His filial conscience satisfied, he turned once again to the strains of philanthropy, hoping to drown the familial contretemps amidst the larger insanity of a fratricidal war.

Receiving no further news of Edward by the time he was ready to depart finally for Washington, he proceeded on the mission of mercy. He traveled south along with his somewhat wayward son Oliver Hazard Perry (first stopping at Brooklyn to take care of some real estate issues), bringing the hospital supplies to the Ithaca regiment. This would land both father and son at the First Battle of Bull Run where, with the specter of the dead and dying fresh before him, at least for now perhaps dimmed the troubling echoes of the recent visit to Kalamazoo. Following the battle amidst the influx of dead and wounded to Washington City (and a harrowing nighttime scramble back to the safety of the Willard Hotel that included a peremptory and hasty exit from the barn of a Southern sympathizer). There he whetted his rhetorical skills, conveying his thoughts in perhaps a less poetical but equally scathing form in a letter to the *Ithaca Journal* decrying the performance of the Union troops at the battle.

By July, Ezra had hand-forwarded the requested funds to Hiram for the completion of the house on Spectacle Lake. On July 12, Hiram sent the acknowledgment for receipt of $360 and a debtor's note for $300. He began his letter with the expected but still unsettling news: "Edward has gone to Gratiot [County]." Clearly the plans to get Edward back to the farm had gone awry in Ezra's absence. Sister Mary had written, "Your request from Kalamazoo I attended to to the best of my ability. Edward went to MacClureVill[e]."[18]

Most likely Cornell had not received this letter until his return from Washington in late August, so at least he was spared the suspense of Edward's wandering misdirections. By now well acquainted with his brother's propensities and being himself largely responsible for his release, Ezra probably lived in constant dread the rest of that summer in expectation of receiving a letter perhaps detailing the gruesome death of his sister-in-law, Angeline Mosher, a humiliation and a tragedy for which he would have had to bear the lion's portion of public and private blame.

Edward Cornell in fact reached and remained at home without further incident, writing himself to Ezra on July 14.[19] It is likely this letter, like Hiram's, only reached Ezra after his return from Washington, but in any case it could only have exacerbated Ezra's fears. Though full of chatty plans for the wheat and potato crop, it portrayed an agitated state of mind, nervously repeating phrases, going on at length about clearing the land, something he was clearly in no condition at that time to undertake. From the scrawl of the hand and the obsessive repetition, it was clear his mental state was deteriorating rapidly. The letter also reiterated Edward's desire to visit Ithaca.

On July 16 there was another letter, this time from Edward's wife, Angeline.[20] Its tone is determinedly rosy as she dealt with the events and flood of conflicting emotion surrounding Edward's return and Angeline's hopes of reestablishing a normal household. It seems to indicate, according to her, despite the prior evidence in Edward's own hand to the contrary, that he has improved considerably. This, of course, proved a vain hope, and shortly after his arrival a second, more honest and clearly agitated letter followed, from Hiram's wife Mary, confirming Edward's progressively deteriorating condition.[21]

As for Ezra, he had witnessed that year, with the coming of the war, a nation built on rational principles and good intentions, entirely and almost without warning descending into general madness, transmuted from a sunny disposition of self-confident optimism and thrust into a dark, schizophrenic, self-rending passage. He now had firsthand personal experience of a similar polar transformation in the form of his brother's onrushing mania. Having built a life on rational principles and self-restraint, he could not have helped but wonder: if the nation itself was capable of being propelled on such a mad, self-destructive, and disastrous course—when so possessed, could one individual be exempt?

Ezra Cornell no doubt at several points had felt the chill hand of madness on his own shoulder, and heard a ghostly, disembodied voice whose ministrations (real or imagined) he had sought to banish through clean living and the invigorating purgative of frequent manly exercise, and more recently, the tabulation of livestock; all this was intended to invite into the frame of his intellect the cleansing light of reason. Despite all this, he too was no doubt a somewhat unwilling subscriber to the prejudices of the day regarding insanity, and had felt its plaintive tug at his breech-coat and heard its dark whisper first hand. Perhaps for a time, as events further revealed how far the red tide of madness had penetrated the bloodlines of his family, he lived in fear that no matter how much fresh air and exercise he obtained, one day it might emerge from his own otherwise well-regulated Quaker psyche and wreak havoc on all he had so painstakingly built.

Two parallel eruptions of familial madness neatly bookended Cornell's career with the telegraph. That of cousin Alvin that had devolved from spousal murder; and Edward, who thankfully, had only contemplated it. If one believes in ghosts, one might say that over each of these events, as with the luncheon at the van Rensselaer Manor in 1823, somehow hovered and presided the disembodied spirit of the brilliant Albany doctor and educator, T. Romeyn Beck and the poor victim, Rebecca Cornell.

Afterword

Morse is often portrayed as a *naïf*, an almost childlike figure, with his floppy hat and bushy beard, a bumbling artist buffeted by a maelstrom of greed and ambition, at the mercy of far shrewder and more calculating men. Nothing could have been further from the truth. We have seen how, in the case of Fisher, he could be both ruthless and self-serving, eager to sacrifice anyone despite their having been a colleague and friend. We have seen how, in the cases of Smith, Page, Cornell and others, when threatened, he would not shrink from demonizing them, and then afterward reshaping the truth to suit his purposes and self-image. This in itself falls far short of heinous and could be ascribed simply as "business," but in the process he also sought to arrogate to himself the trappings of a superior moral nature. Where he does not possess any particular leverage, as in relation to Henry, he attempts to manufacture it by obfuscation and misdirection, using his social connections to influence and protect both his moral stature and his business interests. Of the two men who remained loyal to him throughout his career, Vail would finally realize that neither profit nor fame would follow from his long association with Morse; that the "God" of the telegraph will tolerate no others before or even beside himself. Henry O'Reilly had figured out early on, that fortune might be found only in opposing him. In the terms he had once applied to Ezra Cornell, it was Morse himself who most often "bewitched and befouled" things, turning an otherwise truly unique and remarkable legacy into something at times sordid, and this was what overshadows his claim to be the American Leonardo. (Unlike Leonardo, however, it seems clear his efforts in the artistic field were at best second-rate.)

While his accomplishments as an inventor can and have been called into question, there is one score on which Morse's achievements are unquestionable and undeniable—that of shaping the enterprising American spirit of scientific accomplishment and competitiveness. No other contemporary American scientist, not even Colt, certainly not Henry, none would or could have conceived or executed a grand experiment on the scale that Morse did at the time. Those who were by nature and custom used to proceeding by incremental and careful steps were taught by example the value of the bold leap. The scale of his ambitions seems somehow commensurate with the scale of the new country as it was taking shape. When it came to bravado and showmanship, Morse was unrivaled. This was perhaps thanks in part to his early career as an artist. No one but an artist could have accomplished such an intuitive leap, and it is for this, the final transformation of the American psyche, the translation of the "can do" spirit onto such a vast scale and canvas, that Morse will, in the end, be most remembered, and for which we most have to thank him.

By contrast, his impact in the arena of practical invention was further reaching than even Leonardo's; as the first true pioneer of the information age, Morse was at the nexus of change. Indeed, the telegraph represents the lynchpin, the starting point of the entire modern electronic age. Today, when we have Facebook and Twitter, the telegraph seems a quaint artifact of a bygone era, but we should remember that in 1860 it took a month and a half for news from California to arrive in New York. By 1861 it would take a minute and a half. Many times he rejected the "typing telegraph" in favor of his code. Morse code, a non-analog form of human communication composed of two elements—dots and dashes—while not the first binary code in human history, was certainly the first electronic binary code. As with the computer with its binary language of ones and zeroes; just two elements that could be grouped to render any conceivable meaning.

So the modern era owes much to this singular genius despite the fact that his claims to invention seem, in many cases, either spurious or concocted ex post facto. To the 19th-century individual, what the telegraph represented was no less than the virtual annihilation of space and time, and these were the terms that the newspapers used in speaking of it. With the laying of the Atlantic cable, communications across the globe would become virtually instantaneous. It was not only the speed of communication, but the entire conceptual realignment resulting from the advent of electronic communication, so that we today are still surfing the aftershock.

This book is based on the premise that what had formed in the past an interesting side-note, the contributions of Joseph Henry and Ezra Cornell to the development of the telegraph, in a technological sense were in fact central to the ultimate success of the project. It was not the highly educated and well-connected Morse, but two individuals who started out in life as day laborers (Henry and Cornell), who consistently pointed the way past what seemed to be insurmountable hurdles. It also forms an instructive and dramatic gloss on events to see how public ambition, private angst and greed and acquired wealth, at the right point in history will all seem to conspire to set in motion these great leaps of social progress, but that once they are in motion, little obeisance need be administered to the truth of matters. The curious set of circumstances linking Henry, Cornell and T. Romeyn Beck, which we purposefully introduced as a kind of subterranean echo, hopefully lends a further resonance to these events. We find in them a relentless tendency for events to conspire in producing what appears to us as the singular "man of fate," in this case Morse, as a vessel chosen to accomplish great ends.

Ezra Cornell, it can be argued, was certainly, in every way as much as Morse, fate's man on the scene, though perhaps in a different sense. The telegraph made both of them immensely wealthy beyond their wildest dreams. What distinguishes the two above all was Cornell's refusal to elevate himself at he expense of others, while Morse eagerly clambered atop the prostrate form of anyone in the vicinity who posed an inconvenience. This is where the two differ. When Fog Smith encouraged Cornell to trample the temporarily eclipsed Alfred Vail, despite Vail's evident antipathy and contempt toward him, Cornell declined. When Page first asserted a claim to the invention of the relay, Cornell's response, though his stance was taken as disingenuous, was to put both devices before his customers and allow "the better man to win." Whatever his ability to exercise justice or equity at the moment, Cornell, unlike Morse, remained dedicated to the ideal of fairness. Cornell eventually paid off his debts to Beebe and his other backers in Ithaca, but there remained a metaphysical debt owed by Cornell to T.R. Beck that, for reasons which will become apparent later (after this story ends), was to remain unacknowledged.

Both Morse and Cornell, however, each in his own way, viewed themselves as first and foremost supremely moral men, possessed of a hidden moral core that elevated them above the common run. While this may be an admirable trait, it is also the classic formula and setting for hubris. It becomes clear, in their desire to stage-manage their own posterities (whether personal or public), the depths they would sink to accomplish this. It is this desire that provides the tension in their individual moral sagas, in which their true personalities sometimes emerge more clearly than in the polite utterances to which history affords us access. There remain tantalizing threads we may pull at to reach this deeper stratum; for instance, the list-making tic that we see first in Morse in times of stress, and which emerges in Cornell as well, though at a later stage in his life. The desire to tabulate and enumerate as a way of dealing with profound emotional stress or challenging situations seems to possess them both, as if listing things were the corrective for a potentially disordered universe.

This is the classic definition of hubris: the desire of the "moral man" to rectify and order things that the gods have left in disarray through inattention, which often ultimately results in their downfall. Miraculously, instead of striking them down, the gods instead chose to elevate both Morse and Cornell, and did so time and again. It is Shakespearean in scope (one imagines the trials of Lear). It is as if the appearance of madness brought on by numerous afflictions had become so convincing and sincere that even the gods could not help but be enthralled and amused. But in this version Lear emerges triumphant as perhaps the greatest con-man who ever lived. As noted late in the book, madness at the hand of fate, feigned or otherwise, can prove less amusing to us mere mortals. So in the end, the question of whether Morse was truly the great genius that history tells us he is, or merely the greatest con-man who ever lived, that judgment, in some respects, is beside the point.

Appendix A:
Morse's Deposition

I further state that the combination of machinery in constructing my telegraph as put in operation in 1844 was different from that originally contemplated and described in my first patent in the following respects viz. The combined circuits of my first patent were the combination of two or more circuits as links in a main line for the purpose of renewing the power and propelling forward indefinitely the electric current in such volume as to render the power more available at the distant point and to charge an electro-magnet with sufficient magnetic force to work a register or move the lever of a relay magnet suggested by the probability indicated by my own experiments and the experiments of scientific men that sufficient magnetic power could not be obtained from the electric current through a very long circuit to make a mark of any sort. This difficulty the undersigned proposed to obviate by means of two or more circuits each with a battery coupled together and broken and closed by means of the same principles as the receiving magnet now used these links of one main line are to be made so short as to secure the necessary magnetic power. The register was to be placed not in a short circuit as now arranged but on a link in the main line. But this arrangement was liable to the practical inconvenience that it would always require two lines of wire both always in order because the receiving magnet would work only in one direction. While preparing to build the line from Washington to Baltimore I ascertained by experiment upon one-hundred-and-sixty miles of insulated wire and sometime previously upon thirty-three miles of wire that magnetic power sufficient to move a metallic lever could be obtained from the electric current of a circuit of indefinite length and that there was no necessity for combining two or more circuits together for the purpose of renewing the power at short intervals on the main line. I then devised the present combination which enables me to work the same wire both ways dispensing with one of the two wires originally supposed to be necessary under all circumstances. This combination consists of one main circuit connected by the receiving magnet with as many short office circuits as may be desired upon which respectively, are the requisite registers and not upon the lines of the main line as originally contemplated. Any of these office circuits may be separated from the main line without affecting its efficiency whereas the breaking of a link in the chain of circuits originally contemplated would interrupt all communication. In that combination the battery at each station was to perform the double purpose of working the register and breaking and closing the next circuit in the main line. In the present combination, the purpose of the battery on the main line is to close and break the short independent office circuit which works the register. This new combination of parts was a most valuable improvement upon my first plan. A part of this improvement was used on the experimental line between Washington and Baltimore for the first time in May 1844 and the whole of the improvements in the year 1846.

Source: Shaffner, *The Telegraph Manual*
(New York: Pudney and Russell, 1859), 418–19.

Appendix B:
Questions Prepared for
Professor Henry by Morse, 1839

Question:
Have you any reason to think that magnetism cannot be induced in soft iron at the distance of a hundred miles or more by a single impulse or from a single battery apparatus?

Answer:
No.

Question:
Suppose that a horseshoe magnet of soft iron of a given size receive its maximum of magnetism by a given number of coils around it of wire or of ribbon and by a given sized battery or number of batteries at a given distance from the battery, does a succession of magnets introduced into the circuit diminish the magnetism in each?

Answer:
No.

Question:
Have you ascertained the law which regulates the proportion of quantity and intensity from the voltaic battery necessary to overcome the resistance of the wire in long distances in inducing magnetism in soft iron?

Answer:
Ohm has determined it.

Question:
Is it quantity or intensity which has most effect in inducing magnetism in soft iron?

Answer:
Quantity with short, intensity with long wires.

Chapter Notes

Preface

1. There is an interesting anecdote that occurs following the success of the telegraph. A commission for painting the capitol dome became available when the originally chosen artist died. Morse submitted his name. When he was not chosen, he took the opportunity to settle some old debts he had "overlooked" to fellow artists. See circular letter to Fink and Harvey, and their subsequent replies in which Fink bemoans Morse's loss to the brotherhood of artists: Samuel Morse to Frederick Fink and George Harvey, March 16, 1846, MPLOC.

Introduction

1. Edward Lind Morse, ed., *Samuel F.B. Morse: His Letters and Journals*, vol. 2 (New York: Houghton Mifflin, 1914), 258.
2. Ezra Cornell to Mary Ann, January 1, 1844, 1/8, ECPKL.
3. Elaine Forman Crane, *Killed Strangely: The Death of Rebecca Cornell* (Ithaca: Cornell University Press, 2002). This occurred under disturbingly similar circumstances. Thomas Cornell was convicted of the murder of his mother, Rebecca.
4. Cornell seemed to suffer from a recurring bowel complaint.
5. Cornell would conveniently forget the bitter conflicts that forced him out of the American Electro-Magnetic Telegraph consortium. Professor Goldwin Smith of Cornell would represent him at the banquet given at Delmonico's Steak House on December 29, 1868, presided over by Supreme Court Justice Salmon Chase, who himself had been Morse's antagonist in the patent lawsuit.

6. Alfred Vail, *The American Electro Magnetic Telegraph with the Reports of Congress* (Philadelphia: Lea and Blanchard, 1845). Morse, who was away in Europe when Vail's original volume was published, claimed to be ignorant of the contents of the ensuing pamphlet, which is unlikely since he had a financial interest in it. More likely he was concerned that Henry could use it to challenge his patent than that Vail would reveal trade secrets. (Hence the entire exclusion of Henry's contributions that led to the feud with Morse and Vail.)

Chapter 1

1. It's interesting to note that even by the time of Morse's second visit in 1845, the only American foreign corresponding member of the French Academy of Science at that point was Benjamin Rumford (b. 1753-d. 1814). Rumford was actually British-born. He had laid the foundation for the equivalency of "work" and "heat."
2. Robert Bruce, *The Launching of Modern American Science* (New York: Knopf, 1987), 26. Quoting Benjamin Gould here. Benjamin Apthorp Gould was an astronomer who later headed the longitude division of the U.S. Coast and Geodetic Survey.
3. George Daniels, *American Science in the Age of Jackson* (Tuscaloosa: University of Alabama Press, 1968), xii.
4. Ron Chernow, *Alexander Hamilton* (New York: Penguin, 2004), 372, 374.
5. The "Report on Manufactures" was eventually delivered on December 5, 1791. While it was not initially adopted by the Congress, it is widely regarded as having provided the economic framework for manufacturing in this country.

6. Chernow, *Alexander Hamilton*, 373. Alexander Hamilton to William Duer, April 20, 1791.
7. Robert Jones, *King of the Alley* (Philadelphia: APS, 1992), 159.
8. Peter Onuf, ed., *The New American Nation*, vol. 2: *Establishing the New Regime* (New York: Garland, 1991), 158–9.
9. Some may prefer the "marriage of the dismal and non-dismal sciences."
10. Some may object here (and with some justification) to characterizing SEUM as fundamentally an intellectual organization rather than a political one. While Hamilton unabashedly presented it to everyone as part of his economic scheme for the country and supportive of tariffs, it seems clear he also intended it to encourage a sustainable technological dialog, and more importantly, one not focused solely on agriculture.
11. Bruce, *The Launching of Modern American Science*, 12.
12. This dual ideal, later embodied in the Smithson bequest, would form the foundation of the Smithsonian Institution.
13. Hamilton's concept was of a semi-autonomous free trade zone similar to Hong Kong in its relationship to mainland China.
14. Alexander Hamilton to the Board of Directors of SEUM, April 14, 1792. Historical Society of Pennsylvania, Philadelphia. The plan was partially to fund SEUM directly from import tariffs.
15. Gordon Wood, *Empire of Liberty* (Oxford: OUP, 2009), 102.
16. John Larson, *The Market Revolution in America* (Cambridge: CUP, 2010), 20.
17. Douglas Irwin A., "The Aftermath of Hamilton's 'Report on Manufactures,'" *Journal of Economic History* 64:3 (2004): 800–821.
18. It has been pointed out to me that this is a reading of 20th century values into what was then standard and accepted science. I will politely stick with my assessment. I don't think, for instance, slavery should be accorded any contextual moral slack just because it was the accepted practice at the time.
19. Theodoric Romeyn Beck, *Eulogium on the Life of Simeon Dewitt* (Albany: Skinner, 1835), 20–1.
20. ALS Peter Collinson to John Bartram, July 10, 1739. *William Darlington Memorials of John Bartram and Humphry Marshalls* (London: Lindsay and Blakiston, 1849), 132.
21. Albert E. Moyer, *Joseph Henry: The Rise of an American Scientist* (Washington D.C.: Smithsonian Institution Press, 1997), 226.

22. Such as when Amos Eaton replaced Beck as head of the mineralogical survey because he agreed to focus more on van Rensselaer's pet project, the Erie Canal.
23. Originally called SPAAM, the Society for the Promotion of Agriculture, Arts and Manufactures; later changed to SPUA, the Society for the Promotion of the Useful Arts. For ease of reading, I use them interchangeably though the name change did not officially take place until 1800.
24. Tammis Groft, *Albany Institute of History and Art* (Albany: Hudson Hills, 1998), 16.
25. In a sense, this very lack of coordination would become a crutch for some and a means of justifying some of the wilder and more difficult to justify claims to inventive priority, Morse included.
26. POJH, vol. 1, 64.
27. Jon Meacham, *Thomas Jefferson: The Art of Power* (New York: Random House, 2012), 389.
28. George Dangerfield, *Chancellor Robert R. Livingston of New York* (New York: Harcourt Brace, 1960), 254 (fn). For various reasons, this did not occur.

Chapter 2

1. For a diverting description of this trip, see Andrea Wulf, *Founding Gardeners* (New York: Knopf, 2011), 84–94.
2. Referring to a letter of Livingston's of December 10, 1790, regarding using mercury to reduce friction on grain grinding millstones. http://www.loc.gov/resource/mtj1.013_0183_0191/?sp=1, last accessed October 2015. See Jefferson's reply to Livingston, February 4, 1791. *The Works of Thomas Jefferson*, vol. 6 (New York: Cosimo, 2009), 187.
3. *Ibid.*, letter of Livingston of December 10, 1790. Livingston had tried and failed to obtain the cooperation of local farmers in testing this theory; however, he did obtain a patent for it, one of the first issued in the U.S. (Livingston, Robert R., *Diminishing Friction of Spindles*, Clermont, NY, 4 AUG 1791). Obviously the uncooperative farmers were exhibiting far greater common sense than Livingston, as mercury is poisonous. This fact, however, was not generally known at the time, even by physicians who administered it as a remedy for social diseases.
4. Meacham, *Thomas Jefferson*, 259. While Freneau at first declined, Jefferson was eventually

successful, and his *National Gazette* would become a populist, anti-Federalist mouthpiece for the next two years.

5. Dangerfield, *Chancellor Robert R. Livingston*, 250–1. Jefferson and Madison met again later with Livingston at Clermont, but the substance of those discussions is not recorded.

6. SEUM, The Society for Establishing Useful Manufactures.

7. *Ibid.*, *Works of Thomas Jefferson*, 189–92. This is the bill that would establish the Patent Office. It was introduced in 1791 and reintroduced in 1792. It was meant to replace the Patent Act of 1790 that, according to Jefferson, set too high a bar for granting a patent.

8. *Transactions of the Society for the Promotion of the Useful Arts*, Part I (Albany: Childs and Swaine, 1872), iii.

9. Paul Anbinder, ed., *Albany Institute of History and Art* (New York: Hudson Hills, 1908), 16–18.

10. Robert R. Livingston, "Essay on Sheep," in *Hudson River Panorama*, ed. Tammis Groft (Albany: SUNY Press, 2009), 31.

11. Alexandra Oleson, *The Pursuit of Knowledge in the Early American Republic* (Baltimore: Johns Hopkins Press, 1976), 21.

12. "Elkanah Watson Commonplace Book," no. 66. Elkanah Watson papers on deposit at the New York State Archives. (Cited in William B. Fink, "Stephen Van Rensselaer, the Last Patroon," Ph.D. diss., Columbia Teacher's College, 1950), hereinafter, "Fink, SVR."

13. Philip Schuyler, "Suggesting a Plan to Introduce Uniformity into the Weights and Measures of the United States of America," in *Transactions of the Society for the Promotion of the Useful Arts*, vol. 2, ed. John Barber (Albany: Webster and Skinner, 1807), 39–64.

14. *Transactions of the Society for the Promotion of Agriculture, Arts and Manufactures*, vol. 1, 2nd ed. (Albany: Webster, 1801), Article VI of the Rules and Regulations governing SPAAM, ii.

15. *Ibid.*, 173.

Chapter 3

1. Groft, *Hudson River Panorama*, 19.

2. Some may feel I am painting with too broad a brush here. Marc Rothenberg, noted Henry scholar, contributed the following probably more accurate description. "A natural philosopher was what we would now call a physicist. By the 19th century, chemistry, mineralogy, and geology were all clearly distinct from natural philosophy. And no botanist would call himself a natural philosopher. He might call himself a natural historian. The Yale catalog for 1825 listed one professor for chemistry, pharmacy, mineralogy, and geology (Silliman) and one for mathematics and natural philosophy (Olmsted). There is also a professor of botany for the medical school. Specialization occurred, but not as narrowly as today." In response I would say that while these distinctions may have been maintained within the walls of academia, it was fairly common for even a medical doctor, such as T.R. Beck, up to the 1830s to class himself as a "natural philosopher" when it suited him, due to the catholicity of his interests ranging afield from his official specialty. By 1831, Lewis Beck was taking pains to distinguish the field of chemistry from that of natural philosophy. He writes, "Chemistry borders closely in many instances upon Natural Philosophy but the distinction can be easily drawn. It is the office of Natural Philosophy to investigate the sensible motions of all bodies, whereas chemistry studies the constitution and qualities of these bodies. The Natural Philosopher contemplates whole masses and ascertains their properties, while the chemist notices the operations of their particles, observes their reciprocal actions and seeks to discover all the changes that may occur." Lewis Beck, *Manual of Chemistry*, 13. In any case, the idea of a specialist was only slowly dawning, and Lewis Beck also referred to himself alternately as a botanist, mineralogist, zoologist or chemist, whenever it seemed fitting or convenient.

3. Oleson, *The Pursuit of Knowledge in the Early American Republic*, 123. Gives the ratio of agricultural to non-agricultural subjects. Also J.E. DeKay, *Natural History of New York* (Albany: Thurlow Weed, 1842), 171.

4. T. Romeyn Beck, "Annual Address for the Society" (Albany: Websters and Skinners, 1813). It was re-presented to the legislature in April. *Documents of the Assembly of the State of New York*, vol. 4 (Albany: Carroll and Cook, 1843), 17.

5. *Transactions of the Society for the Promotion of Useful Arts*, vol. 3 (Albany: Webster, 1814). List of officers.

6. Edward van Deussen, *A Biographical Sketch of T. Romeyn Beck* (New York: Holman and Grey, 1856). Unfortunately, there is really no good extant biography of Beck, who was unquestionably a seminal figure in American

science and jurisprudence, both in respect to the treatment of the insane as well as more generally. The closest is the sketch created as a eulogy by Edward van Deussen.

7. Simeon Dewitt's "Theory of Music," unpublished. *Ibid.*, "Eulogium."

8. Ezra Cornell to Mary Ann, July 6, 1845, 2/9, ECPKL. Apparently, Beebe bought 4 acres of the Dewitt estate in 1845. How Cornell acquired that parcel and the rest is not known.

9. POJH, vol. 2, 91.

10. *Transactions* 1, Appendix 25–6. (cited Fink).

11. S.H. Sweet, *Documentary Sketch of New York State Canals* (Albany: Benthuysen, 1863), 102.

12. Chancellor James Kent was not only an esteemed jurist and educator, he was an early and active member of SPAAM, delivering the first annual address of 1796. For a discussion of Kent's role in SPAAM, see Brian Murphy, *Building the Empire State* (Philadelphia: University of Pennsylvania Press, 2015), 202–3. He was also a longtime trustee of the Albany Academy. POJH, vol. 1, 51 (fn.)

13. *Quarterly Journal of the New York State Historical Association* 23, Alexander Clarence Flick ed., NYSHA Association (1942).

14. The project had faced stiff opposition from those in the Southern Tier and Delaware Valley who felt it inordinately benefitted upstate interests. The eventual response would be the proposal of a state road from Albany to Binghamton (now Route 86).

15. Munsell, Annals, 124.

16. The Albany Academy was eventually funded through grants from the Regents of the State of New York, from the Albany City Corporation Council, and the largesse of private donors such as Stephen van Rensselaer (as well as through tuition and lecture fees). See note below.

17. The Albany Female Academy was founded by Ebenezer Foote. The Troy Female Seminary, founded by Emma Willard (cousin by marriage of Romeyn Beck's protégé, Sylvester David Willard), would follow seven years later and would offer women the first real opportunity to study science, history and higher mathematics.

18. *Transactions* 3, 258.

19. Oleson, *The Pursuit of Knowledge in the Early American Republic*, 124.

20. *Ibid.*, 125.

21. Daniel Barnard, *A Discourse on the Life, Services and Character of Stephen van Rensselaer* (Albany: Hoffman and White, 1839), 71 (cited Fink).

22. *Messages from the Governors*, vol. 2 (Albany: J.B. Lyon, 1909), 957–59. The New York Agricultural Society did not come officially into existence until 1832, whereas the Board of Agriculture had been formed in 1819. Its charter was extended to 1825, at which point it was disbanded.

Chapter 4

1. Paul Grondahl, *Mayor Erastus Corning* (Albany: SUNY Press, 2007), 77.

2. Joel Munsell, *Annals of Albany*, vol. 6 (Albany: Munsell, 1855), 119.

3. Beck was appointed principal in 1817. POJH, vol. 1, 41.

4. POJH, vol. 1, 51(fn).

5. Dekay, *Natural History of New York*, 173.

6. "Memoirs of the Board of Agriculture" (Albany, Southwick, 1828), Misc. Papers. (Cited Fink, 5.)

Chapter 5

1. Ethel McAllister, *Amos Eaton: Scientist and Educator* (Philadelphia: University of Pennsylvania Press, 1941), 298.

2. George W. Clinton, *Journal of a Tour from Albany to Lake Erie by the Erie Canal in 1826* (Buffalo: Buffalo Historical Society, 1910).

3. Fink, SVR, 200.

4. Eaton had been given a life sentence for forgery in 1811. He was pardoned by Governor Tompkins in 1815. Governor Dewitt Clinton was the one to invite him to Albany to lecture. POJH, vol. 1, 124 (fn); Fink, SVR, 200–201.

5. Frederick Merrill, *Description of the State Geologic Map of 1901* (Albany: SUNY Press, 1902), 5.

6. See letter of Samuel Mitchell to T.R. Beck, December 6, 1820, and footnotes. POJH, vol. 1, 52–3. The logic in describing Eaton as head of the 1820 and 1821 surveys, though widely repeated, is certainly an example of (to paraphrase Sheldon Cooper) *propter hoc ergo ante hoc*.

7. The sequence of the reports tells the tale here:

- A *Geological Survey of the County of Albany NY*, by Amos Eaton and TR Beck 8vo 1820.
- A *Geological and Agricultural Survey of*

Rensselaer County NY Taken under the direction of the Hon Stephen Van Rensselaer, by Prof Amos Eaton, 8vo Albany 1822.
• *A Geological and Agricultural Survey of the District adjoining the Erie Canal in the State of New York Taken under the direction of the Hon Stephen Van Rensselaer,* by Prof Amos Eaton, 8vo Albany 1824.
8. Merrill, *Description of the State Geologic Map,* 5.
9. Stephen van Rensselaer III to Amos Eaton, August 30, 1822, Eaton Papers, NYSL, 1/3.
10. Amos Eaton, *A Geological and Agricultural Survey of the District Adjoining the Erie Canal* (Albany: Benthuysen, 1824).
11. It seems evident that even by 1822 Eaton had become the recipient of some of the monies no longer going to the Albany Institute.
12. Fink, SVR, 151? (page last digit obscured).
13. *Ibid.,* 201, 205.
14. Rensselaer School Prospectus, 1827, quoted in Palmer Ricketts, *History of Rensselaer Polytechnic Institute* (New York: Wiley, 1914), 51.
15. David Hackett, *The Rude Hand of Innovation: Religion and Social Order in Albany, New York* (New York: Oxford University Press, 1991).

Chapter 6

1. Beck's letter is not found, but van Rensselaer's reply is. Stephen van Rensselaer III to T.R. Beck, February 21, 1823. T.R. Beck Papers, ms. collection, NYPL.
2. "Minutes of the Albany Academy," May 1823. POJH, vol. 1, 70 and 72.
3. It seems the Board was unwilling to let SVR off the hook that easily. Since the Lyceum constituted the "2nd department of the Albany Institute," the second vice presidency of that department was accorded to SVR as well as the presidency of the Institute (in absentia). "Transactions of the Albany Institute," vol. 1 (Albany: Webster and Skinners, 1830), 44.
4. Memorandum dated April 23, 1827. Archives of the Albany Institute.
5. Letter in the possession of the Albany Institute, 1823, Richard Varick DeWitt, Lewis Caleb Beck, J.T. Cooper.
6. James F. Dana to Lewis Beck, September 17, 1823. Rauner Special Collections Library, Dartmouth. (Not to be confused with James Dwight Dana, who was also a noted geologist.)
7. POJH, vol. 1, 61.
8. See bill for board and tuition dated October 25, 1824, to SVR signed by Beck. NYSL, Manuscript and Special Collections, van Rensselaer papers, boxed household receipts, SC7079. 9/7, 1820–24.
9. Joel Munsell, Annals, 275.
10. Whatever polite interchanges and favors that still accrued, it was clear that van Rensselaer had, for whatever reasons, become less enamored of Beck. Aside from the choice of Eaton to head the mineralogical survey and the new Rensselaer School, when van Rensselaer was tapped to help prepare a report for the Board of Regents regarding the future of the Columbia College of Physicians and Surgeons, he would recommend that the medical men on its Board of Trustees (which at the time included Beck) be replaced with non-medical academics. Fink, SVR, 191.
11. Amos Eaton to T.R. Beck, March 13, 1822, Beck papers, ms. collection, NYPL, index #293. There was as yet no physical laboratory at the Institute, so the demonstration had actually taken place in the Troy Court House. This in essence constituted a direct challenge to Henry, who would eventually respond in spades.
12. See the author's *Donderburg's Pumpkin Vine.*
13. *Encyclopedia Americana,* vol. 9 (Philadelphia: Blanchard, 1857), 589.

Chapter 7

1. Samuel to Lucretia Morse, August 29, 1823, MPLOC.
2. Fink, SVR, 126.
3. *Ibid.,* 128.
4. Van Rensselaer took up residence at Peck's in 1822 along with Martin van Buren, Rufus King and Henry Warfield. *Congressional Directories,* ed. Goldman and Young (New York: Columbia University Press, 1973), 127, 139.
5. It appears the Reverend Morse had a longstanding admiration for inventors. See his addendum to the letter from Eli Whitney. "Keep this for autograph of Eli Whitney." Eli Whitney to Jedidiah Morse, January 2, 1823, MPLOC.
6. Kenneth Silverman, *Lightning Man: The Accursed Life of Samuel F.B. Morse* (New York: Knopf, 2003), 16–17.

7. Morse was doing well enough in this period to send $800 home to his parents to settle his bills. S.F.B. Morse to Jedidiah, year date only, 1821, MPLOC.
8. Anna Rutledge, *Artists in the Life of Charleston: Through Colony and State* (Philadelphia: APS, 1949), 138-9.
9. Carleton Mabee, *The American Leonardo* (New York: Knopf, 1943), 84-5.
10. Silverman, *Lightning Man*, 64.
11. *Connecticut Journal*, New Haven, January 28, 1823. Reprinted *Connecticut Herald*, February 11; *Boston Weekly Messenger*, April 24, 1823.
12. This is interesting from the standpoint that it represents Morse's first foray into inventing and his first attempt to deal with the Patent Office. The machine was unworkable due to the weight of the marble; besides, someone else had already patented a similar machine. A letter sent to the Secretary of State just before his departure to Albany attempted to lodge a "caveat," which is interesting also because caveats were not really instituted in U.S. patent law until 1836.
13. S.F.B Morse to Lucretia, August 10, 1823, MPLOC. "I presume I shall see Dr. Silliman on his return." Silliman and his wife had proceeded Ballston Spa to "take the waters."
14. *Ibid*. Also see Richard Morse to Samuel Morse, August 7, 1823, MPLOC.
15. S.F.B. Morse to Lucretia, August 16, 1823, MPLOC.
16. *Ibid*., and August 27 and September 3, 1823. In this last letter, Morse says he will be returning home the following day; he has not found full employment and is "feeling miserable in doing nothing," MPLOC.
17. SVR had taken over the seat in March the previous year.
18. "Van Rensselaer (Rensselaerwyck) Manor Records, 1630-1899," Ms. Archives Collection NYS Library, Box 80. Van Rensselaer had recently completed the addition of two octagon wings to the manor house. See two bills for lead pipe dated August 1823 (burnt materials).
19. S.F.B Morse to Lucretia, August 10, 1823, MPLOC. "General van Rensselaer called on me yesterday and invited me to call and see him."
20. S.F.B. to Lucretia Morse, August 16, 1823, MPLOC.
21. Jedidiah did provide his son a letter of introduction but it was to Elkanah Watson, who had served on the SPUA Board. Certainly not the most prominent individual in Albany with whom he was acquainted (though possibly the most level-headed).
22. "I have just seen Mr. Stephen van Rensselaer whom you know was at college with us ... he ... calls on me every day while I am painting." ALS Morse to his brothers, May 17, 1812. Edward Lind Morse, ed., *Samuel F. B. Morse: His Letters and Journals*, vol. 1 (New York: Houghton Mifflin, 1914), 73. There is no record of SVR IV ever having attended Yale, which makes this passage rather strange. It may be Morse had mixed him up with his brother, Alexander, who had been expelled from Yale (Fink, SVR, 247). Morse goes on in the same letter to refer to another Yale classmate named Ralph Ingersoll, who had graduated two years earlier in 1808. SVR IV was in London at the time of Morse's residency there and became engaged to a woman (*ibid*.). As regards London, there is further proof in the form of a letter from Baring Bros. indicating an enclosure from SVR IV to SVR III (which is not found). Letter: Baring Brothers, London, to Stephen van Rensselaer, Albany [N.Y.], August 18, 1812. NYSL, Manuscripts & Special Collections. Van Rensselaer Manor Papers, SC7079 77/6. NYSL, Manuscripts & Special Collections. Also, two other letters seem to indicate that SVR IV was doing the "Grand Tour" in the style of a European gentleman of the day. Stephen van Rensselaer [IV], Rome [Italy], to Stephen van Rensselaer, Albany [N.Y.], October 14, 1812. Van Rensselaer Manor Papers, SC7079 77/5. NYSL, Manuscripts & Special Collections. Stephen van Rensselaer [IV], Paris [France], to Stephen van Rensselaer, New York, February 9, 1813. NYSL, Manuscripts & Special Collections, Van Rensselaer Manor Papers, SC7079 77/6.
23. *Ibid*. There is some understandable confusion here, as there were several Stephen van Rensselaers, but this clearly must have been SVR III (1764-1839) and supposes Morse's classmate was SVR IV (1789-1868). As noted elsewhere, despite this clear evidence, Yale has no record of SVR IV attending; and except for Morse's letter, most histories have him graduating Princeton. Also see Samuel Irenaeus Prime, *The Life of Samuel F.B. Morse* (New York: Appleton, 1875), 45. Letter of Morse's, dated June 15, 1812. Prime's biography was commissioned by Morse's family and is generally regarded as being inaccurate in many details. It is, however, a good resource for original source material.

24. S.F.B. Morse to Lucretia, August 16, 1823, MPLOC.
25. *Ibid.* Which son is not exactly clear.

Chapter 8

1. S.F.B Morse to Lucretia Morse, August 24, 1823, MPLOC.
2. He was required to step down due to his age. Kent was an avid admirer of Hamilton and wrote a memoir of his acquaintance with him that is included in *Memoirs and Letters of James Kent*.
3. Morse apparently had been told by Moss Kent that Chancellor Kent and his niece were traveling together. Other accounts say Kent was with his son and nephew[?]. Frederick Kilbourne, *Chronicles of the White Mountains* (London: Forgotten Books, 1916; reprinted 2013), 107.
4. S.F.B. Morse to Lucretia Morse, August 10, 1823, MPLOC. Although there is no passenger manifest listing Morse and the Sillimans on the same steamer, the evidence of Morse's hasty departure from New York suggests that the reason he wished to travel north at that time was to be on the same steamer as the Sillimans.
5. S.F.B Morse to Lucretia Morse, August 24, 1823, MPLOC.
6. August 29 postscript to August 27, 1823, letter Samuel to Lucretia, MPLOC.
7. Kane's presence in Albany was likely also due to the machinations surrounding the upcoming presidential election that had brought Poinsett there.
8. John Horton, *James Kent: A Study in Conservatism* (New York: Appleton, 1939), 266–7.
9. August 28 postscript to August 27, 1923, letter S.F.B. Morse to Lucretia, MPLOC.
10. *Ibid.*, and August 29 postscript.
11. *Ibid.*
12. Silliman was apparently still in Albany on the 24th as Morse notes Silliman's reaction to the death three days earlier of the state librarian John Cook, "the man who sneezed so singularly," as Morse describes him in this letter. *Rochester Telegraph*, September 2, 1823. As a side note, Morse's old teacher William Allston also passed through Albany on the way to New Haven. S.F.B. Morse to Lucretia, August 24, 1823, MPLOC.
13. Prime, *Morse*, 121–2.

Chapter 9

1. Stephen van Rensselaer III to Reverend Samuel Blatchford, November 5, 1824. RPI Institute Archives and Special Collections (online).
2. David Hackett, *The Rude Hand of Innovation: Religion and Social Order in Albany New York* (New York: OUP, 1991), 114.
3. Palmer Ricketts, *History of the Rensselaer Polytechnic Institute* (On demand: Forgotten Books, 2015), 7.
4. *Medical and Surgical Journal and Review* 1, ed. E. Geddings (Baltimore: Lucas, 1833), 34. In the article on suspect suicides he refers to the "censure of ... an obnoxious religious sect." He probably only felt comfortable expressing his views on the subject in a medical context, for obvious reasons. He was probably not so much an atheist as a pietist like his colleague Amariah Brigham, who wrote on the ill effects of organized religion on health and mental stability. Amariah Brigham, *Observations on the Effects of Religion on Health and Physical Welfare* (Boston: Marsh, 1835). (Note: there was a T. Romeyn Beck, DD, of New York. This is a different person, but he was an acquaintance of Morse.)
5. *Ibid.*
6. Eaton, in his book, specifically notes that it is intended as a textbook for the student of geology. Joseph Henry would participate in the "expedition" in 1826. Samuel Rezneck, *Education for a Technological Society* (Troy: RPI Press, 1933), 34. This seems to be part of Eaton's unsuccessful attempt to woo Henry away from the Albany Institute to the Rensselaer Institute at Troy. POJH, vol. 1, 124–7.
7. T.R. Beck to Amos Eaton, September 10, 1822, T.R. Beck papers, ms. collection, NYPL. Eaton would only respond when the request was reissued by van Rensselaer himself.
8. Son of SPUA member Dr. Jonathan Eights.
9. Bill from Rawdon Clark and Co. to Stephen van Rensselaer for 1,000 copies of a map dated August 6, 1824, NYS Library Ms. and Archives Division. Van Rensselaer papers 1820 to 1824. The bill was for $685, indicating these were high quality lithos. Eights would also end up illustrating the resulting report (dedicated to van Rensselaer) with his picturesque scenes from along the canal-way. Lithography was relatively new at the time.
10. As Reingold notes, Jabez Delano Hammond represents another strong and interesting

link between Joseph Henry and S.F.B. Morse. Morse and his cousin were both occasional guests at Hammond's house in Cherry Valley. Hammond, as one of three State Road commissioners, was the one who had likely chosen Henry to participate in it. Hammond's neighbor was S.F.B. Morse's cousin James O. Morse on Otsego Lake. Hammond would remain close friends with Henry, and Henry would send him reprints of his scientific articles. Ibid., POJH, vol. 1, 99 fn. T.R. Beck had gone to great lengths to get Henry, placed with the Road Survey. Ibid., 100, fn, mentioning Beck's letter to Martin van Buren of March 10, 1826. (The State Road was a sop to the opponents of the Erie Canal in the Southern Tier.)

11. Ibid., POJH, vol. 1, 117–20. Presumably the offer was not made.

12. The Lancastrian School had been chartered in Albany in 1812 along with the Albany Academy and built by the Corporation of the City of Albany in 1815. It lasted until 1834. In 1839 it reopened as the Albany Medical College. While the Albany Academy catered to Albany's elite, the Lancaster school had been created to serve the less fortunate students. Lancastrian schools sprang up across New York State, including Ithaca.

13. Rensselaer School prospectus, 1827, as quoted in Palmer Ricketts, History of the Rensselaer Polytechnic Institute (New York: Wiley, 1914), 51.

14. Maclure was an avid proponent of the Swiss Pestalozziann system and a contemporary of Eaton. Along with Robert Owen, he was one of the founders of New Harmony. Markes E. Johnson, "The Parallel Impacts of William Maclure and Amos Eaton on American Geology, Education, and Public Service," Indiana Magazine of History 94 (June 1998).

15. "Rensselaer School Extended Circular, 1830," http://archives.rpi.edu/rensselaer-school-extended-circular-1830-transcript, last accessed 10/25/2015.

16. "In addition, the students had the use of two farms in the manor of Rensselaerwyck at their disposal for their practical agricultural study." "Journal of Exercises in Shop and Manufactures ... in the Fall Term of 1826," Amos Eaton Papers, NYSL. Fink, SVR, 208.

17. Lives of Eminent American Physicians and Surgeons of the Nineteenth Century, ed., Samuel Gross (Philadelphia: Lindsay & Blakiston, 1861), 689–93. Beck had published the Gazetteer of Illinois and Missouri (Albany: Webster, 1823), based on his travels in the Midwest, and contributed two articles on botany to the American Journal of Science in 1826 (vols. 10 and 11).

18. In the Constitution and Bylaws of 1826, L.C. Beck is ranked as professor of botany, mineralogy and zoology. Harlan Ballard "Amos Eaton," Collections of the Berkshire Historical and Scientific Society, Pittsfield, 1897, p. 238; "Biographical Record Rensselaer Polytechnic Institute," p. 132. Beck would have his revenge, albeit mostly posthumously. When an official statewide geological survey was commissioned in April of 1842, it was Beck who was appointed to head it. Beck, having worked for the survey from 1835 to 1841, had published his work as part of the Natural History of the State of New York in 1842. Eaton passed away in May of 1842.

19. Vermont Historical Gazetteer 3 (1877), 521. Lewis Beck would remain in the role of assistant to Eaton in the fields of botany, mineralogy and zoology until 1828, when he was hired on by Rutgers University as professor of chemistry and head of the Department of Natural History. Romeyn Beck, though nominally an officer of the Rensselaer School, would continue to focus the bulk of his efforts on the organization he had headed and virtually built, the Albany Academy.

20. Within a few years SVR would attempt to dissociate himself from the school entirely. By this time, Eaton, like Beck before him, had been put into the position of periodically begging for funds to run the institution. See letter, Amos Eaton to Stephen van Rensselaer, November 3, 1826, Gratz Collection, Historical Society of Pennsylvania (cited Fink).

21. William Demarest, A History of Rutgers College (New Brunswick: RUP, 1924), 275 (cited Fink).

22. Ibid., 295 (cited Fink).

23. Allen Robbins, A History of Physics and Astronomy at Rutgers (Louisville: Gateway, 2001), 27.

24. Ibid., 158. When Henry discussed what he had learned from Dana, it was mostly about how to tilt a galvanic cell so that his students could better observe the contents during experiments.

25. POJH, vol. 1, 190.

26. Henry maintained a lifelong interest in geology and taught the subject at Princeton, but I think it safe to say it did not form the crux of his intellectual passion.

27. Moyer, Joseph Henry, 63, notes by now even Silliman in New Haven had heard of

Henry. The 1828 *Transactions of the Albany Institute* refers to "our popular lectures."

28. Oleson, *Pursuit of Knowledge in the Early American Republic*, 148.

29. POJH, vol. 1, 201, 216 (fn), 217. Moyer, *Henry*, 69–70.

30. Wollaston had been the first to come up with the idea, but Henry, by substituting hydrogen as the basic unit, was able to create what was essentially a working slide rule for chemical analysis. *AJS* 14 (July 1828): 202. There were attempts to market this device, but apparently no attempt was made to patent it. POJH, vol. 1, 195, excerpt from Beck's autobiography.

31. POJH, vol. 1, 214.

Chapter 10

1. Botanist Asa Gray recalls, on visiting the Albany Academy, being greeted by Lewis Beck and a "grave-looking man who I was told was Professor Henry." *Letters of Asa Gray*, ed. Jane Gray, vol. 1 (New York: Houghton Mifflin, 1894), 15.

2. Ten Eyck claimed to have been a boyhood friend of Henry's. (See Hamilton Literary Magazine citation below.)

3. *Albany Chronicles*, ed. Cuyler Reynolds (Albany: Lyon, 1906), 439.

4. Here, it seems, is the making of a mystery, as there is a puzzling year-long lacuna at this time in the otherwise minute scholarship of Henry's life.

5. It is not clear exactly whose children he was tutoring, those of van Rensselaer Sr. by his second wife Cornelia Paterson, or those of Harriet van Rensselaer, wife of SVR IV.

6. Large ad placed on August 29, 1823, in the *Albany Argus*.

7. POJH, vol. 1, 28.

8. Albert Moyer, *Joseph Henry* (Washington D.C.: Smithsonian Press, 1997), 25. Moyer also notes Henry's exposure to Gregory's *Popular Lectures on Experimental Philosophy, Astronomy and Chemistry*, which a young Henry chanced on when he was ill, and which had apparently been left on the kitchen table. *Ibid.*, 21. I think it only fair to note here that a noted authority on Henry, Marc Rothenberg, disagrees with my reading of Moyer and the conclusions I come to regarding poverty in Henry's upbringing and some specific events therein.

9. Munsell, *Annals of Albany*, 93.

10. *Albany Chronicles*, ed. Cuyler Reynolds (Albany: Lyon, 1906), 440. It is still in use today as one of the premier authorities in cases of medical jurisprudence.

11. *Ibid.*

12. *Albany Chronicles*, 462.

13. Alternately, Quimby. Aaron B. Quimby was an expert in the field of stationary steam engines whom Beck had hired as a teacher in 1822 through his association with Columbia College. *Ibid.*, POJH, vol. 1, p. 59. This is undoubtedly same the A.B. Quimby who contributed several articles to *AJS*, vol. 7, 1824, 316–23, and vol. 12, 1827, 344–58. As Reingold points out, there is very little known of Quimby except for this article, but the article portrays someone with advanced knowledge of mechanical engineering and the mathematical bases of motive power.

14. *Henry's Canal Journal*, POJH, vol. 1, 137–54.

15. Henry had only been a full professor for a few weeks, but it seems clear that as far as Eaton was concerned, his status was that of a student, not an instructor. POJH, vol. 1, 135, 155.

16. Torrey was a professor at West Point and Dana was on the Board of Visitors, which had been called in specifically for this event.

17. Henry's "Journal of a Trip to West Point and New York," POJH, vol. 1, 155–61.

18. Henry offers no reason for the abrupt departure from Palmyra and the visit to West Point, saying simply, "Palmyra. I left the canal at this place." POJH, vol. 1, 153.

19. As a side note, it was on this same trip, apparently, that he discovered another new pedagogical device of which he would make great use: the blackboard. *Ibid.*, 157 (fn).

Chapter 11

1. Elaine Forman Crane, *Killed Strangely*. The account of the parallel nature of two murders, one in 1673 by Thomas Cornell, and the other in 1843 by Alvin Cornell, form the major premise of the book. Ezra Cornell's association with Morse and the telegraph is bookended by two eruptions of familial insanity, in both of which T.R. Beck had exerted direct or indirect influence. The beginning of Cornell's history with the telegraph coincides with his cousin Alvin's heinous act in Jamestown, New York, and ends coincidentally with the committing of his brother Edward to the Kalamazoo Asylum for the Insane. In the former case, it was Beck that influenced the governor to commute

Alvin's sentence from death by hanging to incarceration by appealing to Governor Bouck. In the latter it was Beck's disciple Edward van Deussen who had Edward in his care, and to whom Cornell petitioned for his release in 1860. In 1863 another of Beck's protégés, Sylvester David Willard, laid the foundation for Cornell University by convincing the legislature to assume the mortgage for Cornell's first failed educational venture, the Ovid Agricultural College, which became the Willard State Asylum for the Chronic and Indigent Insane. For us, even in the absence of definitive knowledge, this brace of murders lays a foundation for insight into the formative elements of Ezra Cornell's psychology, as he certainly was aware of both instances of madness in his patrimony. Unlike the case of his cousin Alvin, whose sentence was commuted thanks to the efforts of T. Romeyn Beck, there was in the earlier instance no one to intervene at the last minute, no loving sister inclined toward Christian mercy, no commutation based on the recommendations of a pillar of the medical community, and no last-minute reprieve for Mr. Thomas Cornell Jr. from the powers that be.

2. See Elaine Forman Crane, *Killed Strangely*.

Chapter 12

1. Fisher, *Silliman*, 290–1. Silliman to Professor J.L. Kingsley. October 9, 1819.

2. *Ibid.*, 293. Chancellor Kent's "thank you letter" to Silliman, October 14, 1820.

3. George Fisher, *Life of Benjamin Silliman* (New York: Scribner, 1866), 307–9. Silliman was clearly suffering both ill-health and depression at this time.

4. Amos Eaton, *A Geological and Agricultural Survey of the District Adjoining the Erie Canal* (Albany: Benthuysen, 1824), 5. Silliman would not have passed up this opportunity unless he was indeed very ill. (See my article, "The Shivering Stones of Orange County," in the *OCHS Journal* #41, 2012.)

5. Alexander Fisher (b. 1794, d. April 22, 1822) was a mathematician, physicist, poet and musician who entered Yale at the age of 15. His article on musical temperament had appeared as the first article in the first issue of Silliman's *AJS*, vol. 1, 9–34.

6. Professor Fisher had perished on the schooner *Albion*, shipwrecked off the coast of Ireland in April 1822. The depth of Silliman's affection and respect for Fisher was made apparent in his eulogy. He was so distraught following the event, he had asked another professor to take over his teaching duties for the upcoming term at Yale. Mortimer Blake, *History of the Town of Franklin, Mass.* (Franklin: 1879), 152–4. Morse had painted a posthumous portrait of Dr. Fisher (see illustration).

7. Richard McKay, *South Street* (New York: Haskell, 1934), 147.

8. Anna Rutledge, *Artists in the Life of Charleston*, vol. 39 (Philadelphia: APS, 1949), 199.

9. Fisher, *Silliman*, 368.

10. This may be an assumption, but it is a reasonable assumption. That Kent was an avid reader of the *Journal* is evidenced by his letter to Silliman of August 7, 1819: "I am obliged to interrupt your studies for a moment to thank you for the pleasure and instruction you have given me in another work of yours which has lately been presented by you and to which I have become a subscriber. I have read attentively ... your Journal of Science." William Kent, *Memoirs and Letters of James Kent* (New York: Little, Brown, 1898), 246. Silliman several times expresses admiration for Kent's extensive library.

11. James Dana, "Galvano-Magnetic Apparatus of Prof. Dana," *AJS*, vol. 6 (1823): 330–1. Henry would later adopt the same idea presented in the article of an overwound insulated coil, which was in fact crucial to his massive electromagnets that he would begin demonstrating in 1828.

12. It can be argued that Schweigger's insulation preceded Dana's, but the purpose was really entirely different. Dana specified "a steel wire five inches long, around which passes about seventy-five turns of brass wire ... wrapped in sewing silk." Unfortunately, the accompanying diagram does not show the brass coils clearly enough to see if they are contacting each other or overwound, which is possibly why it was overlooked in contributing to the idea of an overwound electromagnet. *Ibid*.

13. B. August 14, 1777, d. March 9, 1851. Danish.

14. B. April 8, 1779, d. September 6, 1857. German.

15. B. May 22, 1783, d. December 4, 1850. English.

16. *Transactions of the Society of Arts, Manufactures and Commerce*, vol. 43 (London: Flindell, 1825), 37–52 (and Plate 3). Sturgeon is credited with having invented the first workable electromagnet. What is perhaps equally

interesting is that the article describes what is in essence a primitive electrical motor (Plate III, Fig. 7).

17. Silvanus Thompson, *Lectures on the Electromagnet* (New York: Johnston, 1891), 17–22.

18. Either multiple batteries or a single battery with multiple compartments. (See quantity vs. intensity discussion.) The question of quantity vs. intensity has to do with the circuit resistance, and also whether the conductors are connected in parallel or in series.

19. To avoid confusion, I am using the term "overwound" to indicate that charged wires can be wound over themselves several times, thus requiring them to be insulated to avoid a short regardless of the number of galvanic sources. The question of the number of galvanic sources is dealt with below in the chapter "Quantity vs. Intensity."

20. Though the work was first published in Silliman's in 1831, it was shown as part of a lecture at the Albany Academy in 1827. Dana, while at West Point, had instructed Henry on exactly what kind of battery and coils to use for his classroom demonstrations. Moyer, *Joseph Henry*, 46, 56–7.

21. Henry only credits Schweigger. However, from Dana's description it is clear he anticipated Henry by four years. However, if Dana was destined to remain obscure, it was partly his own fault. The compass diagram had failed to show an overwound coil, and the text does not indicate the reason for the insulation. Also, he did not realize or failed to measure the increase in magnetic potential created by overwinding the coil, and so the accomplishment was easily swept under the rug until Henry explicated and exploited it more fully in 1828.

22. Another way Morse and Henry may have met earlier than either claim is that both likely attended Dana's lectures at the New York Athenaeum in 1825; we know for certain Morse attended them based on his article in Shaffner's. The fact that Henry attended was asserted by Henry's daughter after his death but Henry himself never acknowledged this. Moyer, *Joseph Henry*, 59. It appears, just as Morse charged, that Henry did have a problem acknowledging Dana's influence. In a letter to Lewis Beck dated September 21, 1827, he describes Ampere's more primitive version of Dana's device, adducing it as evidence of the magnetic nature of earth's auroral arch. POJH, vol. 1, 196–99.

23. Samuel Prime, *Morse*, deposition of Matilda W. Dana, 162–4. It is not clear who sought out whom, but according to his wife, Dana was a frequent visitor to Morse's art studio during 1826.

24. James Wilson, *The Memorial History of the City of New York*, vol. 4, part 2 (New York: New York History Company, 1893), 424. Morse was obviously so impressed with Dana that he kept verbatim copies of these lectures that were reproduced in Shaffner's *Telegraph Journal* in 1855.

25. In Shaffner's, vol. 2, Morse quotes at length from Dana's lectures to show that he was well aware of the technique of overwinding at the time, 67–73. Morse quote found on 71.

26. Silverman, *Lightning Man*, 288.

Chapter 13

1. That is done amply elsewhere by Hochfelder, et al.

2. It could be argued that the concept of self-induction should be included in these, but Henry himself was extremely cautious of this claim because of the parallel work of Michael Faraday.

3. In the context of the patent case, Henry would adduce the eyewitness testimony of geologist James Hall that he had created, in essence, a working telegraph by 1832. POJH, vol. 1, 442 (fn). Other accounts have him transmitting signals over a distance of a mile and a half by 1831. Moyer, *Joseph Henry*, 69–70. Henry later would claim "over a mile." Joseph Henry, *Scientific Writings of Joseph Henry*, vol. 2 (Washington, D.C.: Smithsonian, 1887), 434.

4. Brian Bowers, *Sir Charles Wheatstone, FRS* (London: Institution of Electrical Engineers, 2001), 124.

5. Moyer, *Henry*, 67. Schweigger initially called it a "multiplier" or "doubler," but it became known as a "galvanometer."

6. Schweigger published his discovery of the multiplier in the November 1820 issue of *Literary Gazette* in Germany. Henry discussed its bearing on his work in "On the Application of the Principle of the Galvanic Multiplier to Electro-Magnetic Apparatus," *American Journal of Science and Arts* 19 (January 1831): 401, reprinted in *Scientific Writings of Joseph Henry*, vol. 1 (Washington, D.C.: Smithsonian Institution, 1886), p. 38. First published in *Transactions of the Albany Institute* 1 (ibid., reprinted in 1830), in which he notes his paper was first delivered in 1827.

7. I.B. Sebring, *Life of Lewis C. Beck, M.D.* (Schenectady: privately printed, 1934), 8.

8. Mary Ferris, "Catherine Elizabeth van Cortlandt," in *New York Genealogical and Biographical Record* 26, no. 1 (1894): 141.
9. Lewis C. Beck, *A Manual of Chemistry* (Albany: Webster and Skinners, 1831), 102. This comprehensive work, containing (among other things) a detailed description of Henry's ideas and experiments, appeared virtually simultaneously with Henry's article in the *American Journal of Science*. In it, it is interesting to note, Beck still conceives of electricity as a fluid, as would most scientists even to the late 19th century.
10. *Hamilton Literary Monthly* 26, no. 8 (Hamilton College, April 1892), 331–2. Ten Eyck was a graduate of Hamilton College. This article gives Ten Eyck the credit for obtaining the length of over 1,000 feet of copper wire and actually of conducting the bell-ringing experiment that Henry cites later on as evidence of his (Henry) devising a telegraph. The article, interestingly enough, insinuates that Henry and Ten Eyck were boyhood friends. Obviously the "force at a distance" concept laid the foundations for the telegraph.
11. *Scientific Writings of Joseph Henry*, vol. 1 (Washington, D.C.: Smithsonian Press, 1886), 37–8. Even from the use of silk as an insulator, it is clear Henry's improvement in creating an overwound coil was remarkably influenced by Dana's.
12. Mary Ferris, *Catherine Elizabeth van Cortlandt* (privately printed). Courtesy of Albany local and Albany Academy Historian John McClintock. Perhaps it was to get his poor nieces out of this chore that Lewis Beck suggested the alternate method.
13. Ibid., *Transactions of the Albany Institute*, 1828. Henry notes he had created a magnet that could lift fifty times its own weight using a magnet with insulated overwound wire.
14. Henry's persistent poverty at this point in time (in my opinion) is conveyed plaintively in a postscript to a letter to Silliman in which he confesses, "Please excuse this scrawl as I am writing just after daybreak without a candle." POJH, vol. 1, 317. Henry to Silliman, December 28, 1830.
15. "Die galvanische Kette, mathematisch nearbietet," Dr. G.S. Ohm (Berlin: Riemann, 1827). This was the consolidation of his work with circuits, but the papers had been published earlier.
16. It was the preeminent scientific publication of the day and featured articles from all branches of science, from astronomy to zoology. *AJS*, vol. 19 (1831): 400–404. See Henry's letter to Silliman of December 10, 1830, asking to be included in the next issue. The article was added as an appendix. In the same issue was a long article titled "Electro-Magnetic Experiments" by Moll, 329–337.
17. Gerard Moll, b. 1785, d. 1838.
18. Ibid., Silliman's, 402. In saying this, clearly Henry meant only the bell-ringing and compass deflection experiments, but he does not make clear that the subsequent experiments were conducted by him alone with Beck "kibitzing."
19. "Most of the results in this paper were witnessed by Dr. L.C. Beck and to this gentleman we are indebted for the several suggestions and particularly that of substituting cotton well-wrapped for silk thread.... Dr. Beck also constructed a horse shoe of round iron." Ibid., *AJS*, vol. 10, 407–8.
20. Henry may have been circumspect on this point because, according to Ten Eyck, it had actually been he and not Henry who had conducted the famous bell-ringing experiment. See note above from the *Hamilton Literary Monthly* 26, no. 8. Silliman does not include Ten Eyck's name in the title and apologizes if this is not Henry's intent. POJH, vol. 1, 322.
21. In a letter of August 24, 1832, Professor John Maclean of Princeton suggested to Henry that he inform Beck of their negotiations, but apparently Henry had chosen not to. POJH, vol. 1, 443. The brief notice of Henry's departure, given to the trustees at a hastily called meeting on October 11, is further evidence of Beck's pique. Ibid., 460 (fn).

Chapter 14

1. Ames's studio was located at 41 South Pearl Street. Morse was renting rooms at 94 North Pearl, a few blocks away.
2. Anarcharsis, "Travels in Greece." S.F.B. Morse to Lucretia Morse, August 27, 1823, MPLOC.
3. Morse sometimes referred to the picture as *The House of Representatives*, and sometimes as *Congress Hall*.
4. William Dunlap, *A History of Design*, vol. 3 (Boston: Goodspeed, 1918), 58. Formed in 1821 by Poinsett, Morse and sculptor John Cogdell, it was essentially a sales venue for established artists like Rembrandt Peale and John Vanderlyn, who had studios in affluent Charleston, and a conduit for garnering Morse portrait work. It would last only two years.

5. Postscript of August 29, 1823. Samuel Morse to Lucretia August 27. "Mr. Poinsett of S.C. is also in town."
6. Poinsett's mission had been highly confidential and sensitive. America had no formal diplomatic relations with Mexico at the time, so he had been personally tapped by James Monroe in 1822 to evaluate the legitimacy and popular support for Don Augustin Iturbide, who, enjoying the support of the Mexican military, had proclaimed himself emperor. Poinsett, on his return, had predicted Iturbide's subsequent overthrow, which in fact occurred just a few weeks later. Charles Stillé, "The Life and Services of Joel R. Poinsett," reprint of article in *Philadelphia Magazine of History and Biography* (1888): 31.
7. The architect William Jay; the sculptor John Cogdell; Charles Fraser, who had painted the portrait of Lafayette; and others. (Morse would later paint a portrait of Lafayette as well.)
8. Morse had a portrait studio in South Carolina on and off since 1816. Kenneth Silverman Research Collection on Samuel F.B. Morse; MC 173; 1/1819–209; NYU Archives, NYU Libraries. Also, Carolina Gazette, February 24, 1821, re: auction.
9. There is no evidence of correspondence between the two at this point.
10. Fink, SVR, 127–8.
11. Ibid., 126, 130.
12. *Papers of Henry Clay*, vol. 3, ed. Robert Seager (Lexington: University of Kentucky Press, 1963). David Wood to Henry Clay, August 27, 1823, ibid., 475–7, 512 (fn). Wood was also a board member of the Albany Academy.
13. There is no direct evidence that this was the reason for Poinsett's visit. Neither is there any written evidence for the existence of the "corrupt bargain" itself, but it is generally accepted as fact by most historians. The sequence of events here is far too abrupt and coincidental for it to be anything else than the striking of some political deal with SVR regarding the upcoming election.
14. Fink, SVR, 128.
15. Ibid., 130–1.
16. Ibid.
17. Samuel Morse to Joel Poinsett (undated), images 15–17. Bound volume, 9 December 1823–9 February 1828, Morse Papers LOC. Refers to the meeting with Poinsett in Albany.
18. In fact, they were suffering the aftereffects of the same economic depression that had afflicted the U.S. in 1819. In a sense this was the first worldwide depression.
19. Whether this is a deliberate deception of Poinsett's or he himself had been duped is hard to say, but suffice it to say, embracing Morse's expertise in art could not have been detrimental to his political mission, and he probably would have made a good traveling companion.
20. Morse, *Samuel F.B. Morse: His Letters and Journals*, vol. 1, 247.
21. S.F.B. Morse to Lucretia, August 27, 1823, MPLOC.
22. Samuel. Morse to Stephen van Rensselaer (III) (apparently) September 4, 1823, MPLOC, image 19–20 of Bound volume, 9 December 1823–9 February 1828. Letter is marked with pencil "December 1823," but accompanying postmark on the cover bears a date of September 4. (Edward Lind dates it the day previous, September 3.) Morse had written his wife that he intended to pack up and leave the following day (the 4th) but it is clear from other correspondence (note: van Schaick) that he did not leave for New York until several weeks later at the earliest. It seems that he left the *House of Representatives* picture in Albany after he left, as there is a notice of it showing on State Street in an October notice in the newspaper. *Albany Argus*, October 7, 1823.
23. From a circular letter to Poinsett and van Rensselaer, undated but marked with pencil "December 1823," image 21–23 of Bound volume, 9 December 1823–9 February 1828. The postmark/address on the back is to Samuel Morse, artist, Albany, September 4, indicating this was misfiled. This form of a set of tabulated interrogatories used in times of emotional distress is one shared by Ezra Cornell, and is largely the same format he would use preparing for his interview with Henry in 1842. Henry's replies were as succinct as Poinsett's. Pencil notation by Morse (probably) "Tuesday, dine at 3 o'clock."
24. Silverman, *Lightning Man*, 69.
25. This apparently was a temporary political sop, as Poinsett by 1825 had been made minister to Mexico and traveled there, but without the benefit of Morse's company.

Chapter 15

1. Mabee, *American Leonardo*, 299. Samuel Morse to T.S. Faxton, March 15, 1848.
2. David Hochfelder, *Joseph Henry: Inventor*

of the Telegraph? Accessed February 20, 2015, http://siarchives.si.edu/oldsite/siarchives-old/history/jhp/joseph20.htm.

3. In Reingold there is close to a one-year hiatus in materials for the year 1823. Albany historian John McClintock, however, uncovered one document that clearly places Henry as librarian for the Albany Academy in the academy's spring term. Undated note indicating "Principal [Beck] to Joseph Henry, librarian ... February 1, 1823," found in Archives and Collections of the Albany Academies.

4. Reingold notes that some biographers put the beginning of Henry's service as a tutor in the van Rensselaer home as early as 1821–2 instead, but that he regards this as suspect, stating that the dates were most likely 1822–3. "It seems more probable that Henry tutored the van Rensselaer children after leaving the Academy, let us say 1822–1823 or 1824." POJH, vol. 1, 64.

5. Memorandum indicating "The Principal [Beck] to be Librarian Dec 8, 1826 Joseph Henry, to be D[itto] February 10, 1823," Memorandum #86, Archives and Collections of the Albany Academies, also courtesy of John McClintock.

6. There is the letter from an R.L. Williams that seems to clearly indicate that Henry was still a member of van Rensselaer household at least as late as January 1824. The letter, dated January 10, 1824, requests that Henry use his connections to introduce a musical composition of Williams's to Harriet van Rensselaer. POJH, vol. 1, 64. The song was "Kind Robin Loves Me" (fn).

7. O'Reilly v. Morse, 56 U.S. 15. An appeal of the decision of the District Court of Kentucky. In what was a controversial decision, the court and Chief Justice Taney would vindicate Morse on all but one (the last) claim (8): that his patent covered any device using electromagnetism, which obviously was a reach.

8. Reingold notes Henry's visit to the National Academy of Design in 1830 but discounts the possibility that the two met at the time, though Morse was then president of that organization. POJH, vol. 1, 281.

9. "It will be perceived that Prof. Henry has mis-recollected the circumstances of our first acquaintance, making the date 1837, in New York, instead of 1839 in Princeton." Ibid.

10. *Shaffner's Telegraph Companion*, vol. 1, ed. Tal Shaffner (New York: Pudney & Russell, 1854), 17. Appearing in the second issue of *Shaffner's* (1855) is Henry's deposition, given originally in 1849 in Boston, regarding Morse v. O'Reilly (*ibid.*, vol. 2, no. 1, 97–112). The deposition was introduced in several subsequent patent cases, including *O'Reilly v. Morse*, 56 U.S. 15 How. 62 (1853). In Morse's footnotes to Henry's deposition he clearly stops barely short of accusing Henry of outright lying: "A memory defective in an unusual degree is the most charitable construction for Prof[essor] Henry that can be put on such manifest aberrations from fact attempting to recollect what was said by another."

11. *Ibid.*, 8.

12. Morse had written, "I think that you have pursued an original course of experiment and discovered facts of more value to me than any that have been published abroad." Morse to Henry, April 24, 1839. Prime, *Morse*, 420.

13. Samuel Morse to W.W. Boardman, August 10, 1842. This would have been a highly unusual statement for Morse to make to a virtual stranger. It was the kind of comment he usually reserved for letters to his brother Sidney.

14. Leonard Gale to Morse, December 29, 1843, and Joseph Henry to Morse, January 24, 1844, MPLOC. When Morse looked to Gale for advice on whether or not to go to an overhead cable solution, Gale had gone straight to Henry. Gale proposed running two test lines that winter between the Patent Office and the Capitol: one above ground, one below. It was Henry's comments regarding the unavoidability of short circuits in underground cables that caused Morse to finally abandon the underground cable approach. There can be no doubt that Morse's eventual falling out with Henry and Fisher pained Gale greatly. Gale would devote most of his portion of the congressional memorial to Morse to the conflict between the two. *Memorial of Samuel Finley Breese Morse* (Washington: GPO, 1875), 15–19.

15. Alonzo Cornell, *True and Firm* (New York: A.S. Barnes, 1884), 89–90.

16. William Taylor, *An Historical Sketch of Henry's Contribution to the Electro-Magnetic Telegraph* (Washington: GPO, 1879), 97.

17. *Ibid.*

18. The public break between the two is usually traced to the publication of Vail's *History* in 1845, in which Vail virtually excludes Henry from any credit. Morse claimed entire ignorance of the contents of Vail's book, which appears highly disingenuous based on the circumstances, and the correspondence between he and Vail, and the fact the Morse held a

1/4 pecuniary interest in the volume. It seems clear the book was not the sole cause of the falling-out between Morse and Henry, but only the "straw that broke the camel's back." It didn't take Vail's book for a man as perspicacious as Henry to realize he was being accorded no credit. Seeing this in print was what had set him off.

19. There are several extant theories, all of which seem to contradict the other. There is the one presented above by the author, that despite their mutual disclaimers, the two in fact did meet in 1823 at the Van Rensselaer Manor but it was a meeting which, for various reasons, neither wished to acknowledge publicly. Then there is Henry's daughter, who claimed (after his death) that Henry attended Dana's lectures at the New York Athenaeum, which we know Morse attended as well. This too might have significance if it occurred, as it is in these lectures that Dana elaborated on the idea first presented in the 1823 Silliman's article of an overwound insulated electromagnet. We are forced to ask the question here: why is Morse so adamant about excising any reference whatsoever to Henry's influence prior to the date he claims to have conceived the telegraph aboard the *Paquebot Sully* since he then freely admits the influence of Dana and even adduces a model electromagnet Dana constructed in the court case? Henry had asserted no pretensions to invention or rights to economic advantage, yet Morse would attack him viciously. Whether it can be traced to 1823 or not, Morse clearly seems to be operating on a guilty conscience on this score. Samuel F.B. Morse, "The Electro-Magnetic Telegraph. A Defense Against the Injurious Deductions Drawn from the Deposition of Prof. Joseph Henry," *Shaffner's Telegraph Companion* 2, no. 1 (January 1855): 6–96.

20. Samuel Morse to Horatio Hubbell, June 12, 1854, http://atlantic-cable.com/Article/Hubbell, last accessed, November 20, 2015.

21. As Morse's biographer Carleton Mabee put it. Morse described it as a "bolt of lightning."

22. *AJS*, vol. 19 (1831): Appendix, 400–08.

23. October 1832.

24. Ibid. *AJS*.

25. In fact, many of Sturgeon and Barlow's early attempts had involved compass deflection, so, while he could claim ignorance of the article, to claim it was not suggestive of a telegraph because the receiving indicator was a compass was merely ridiculous. The key here is that Henry proved that the pulse was detectable at a far greater distance than had been conceived to be possible.

26. Samuel F B. Morse, "A Defense Against the Injurious Deductions Drawn from the Deposition of Prof. Joseph Henry," *Shaffner's Telegraphic Journal* 2, no. 1 (1855): 15(fn). (Emphasis in the original.)

27. When the feud between Morse and Henry heated up, Henry appealed to geologist James Hall to corroborate his version of the facts. Hall had been a student at the Rensselaer School at the time. Moyer, *Joseph Henry*, 70; POJH, vol. 1, 442 (fn).

28. It seems, whenever Morse lies (except in the case of Fisher's firing), he enlists Gale or Vail to swear to it. This would eventually backfire, as we shall see, in the context of the 1860 (improvement) relay renewal battle, in which Gale's enthusiasm for Morse's version of the truth almost lost Morse the entire underlying patent. Gale would restate this as the final lines of his eulogy to Morse. "Memorial of Samuel Finley Breese Morse" (Washington: GPO, 1875), 19.

29. Silverman, *Lightning Man*, 159–60. This does not at all jibe with the fact that Morse claimed to have attended all of Dana's lectures, in which this technique was specifically addressed in detail.

30. Moyer, *Joseph Henry*, 25.

31. The Horatio Alger–type story was not accorded the same romantic gloss then as it is by a more modern sensibility.

32. *Annual Report of the Board of Regents of the Smithsonian Institution* (Washington, D.C.: Harris, 1858), 85–88.

33. Ibid., 89.

34. We later delve into this in far greater detail, but suffice it to say that the cost of enlisting Page's cooperation in undermining Cornell and his relay, and for buying Page's complete silence on the matter thereafter, was about $65. Unlike poor Page, who would die in poverty and semi-disgrace, Cornell in a sense would have the last laugh, serving as a pallbearer at Morse's funeral.

35. Charles Page to Samuel Morse, July 28, 1845, MPLOC.

36. Cornell had actually been expelled from the Quaker meeting (church) for marrying MaryAnn Wood.

Chapter 16

1. *Shaffner's Telegraph Companion*, 2, 85.
2. William Taylor, *An Historical Sketch of*

Henry's Contribution to the Electro-magnetic Telegraph (Washington, D.C.: GPO, 1879), 22.

3. Either as a single fluid or Coulomb's double fluid flowing in opposite directions (to account for positive and negative charge).

4. Brian Bowers, *Sir Charles Wheatstone, FRS* (London: Institution of Electrical Engineers, 2001), 124. Even the use today of the word "current" is a reflection of that venerable prejudice.

5. It was not until 1832 at the earliest that Faraday began to criticize the "two-fluid" version of electricity. Frank James, *Michael Faraday* (Oxford: OUP, 2010), 63.

6. Henry was to go so far as to propose that, conversely to Barlow, the electromagnetic force was inversely proportional to the speed of the current (*ibid., AJS*, vol. 19, 403).

7. Barlow published his results in the *Edinburgh Philosophical Journal*, Sturgeon in the *Transactions of the Society of Arts*, both of which Henry likely would have had access to through the Albany Academy library. (Beck was fanatical about keeping abreast of European developments.) Henry's article in the next note is addressed to Sturgeon's failures.

8. Joseph Henry, "On some Modifications of the Electro Magnetic Apparatus," *Transactions of the Albany Institute* 1 (1830): 22–24. As Henry notes, this was delivered as a lecture on October 10, 1827. This is found in the 1828 *Transactions*.

9. 1789–1854. German.

10. Michael Schiffer, *Power Struggles: Scientific Authority and the Creation of Practical Electricity before Edison* (Boston: MIT Press, 2008), 45. Faraday and Wheatstone had also been experimenting with several different battery configurations based on Ohm's Law at the time. This was the Cruikshank battery.

11. Not that it was some other kind of electricity, but rather that the circuit was capable of being manipulated by means of restructuring the individual components.

12. Joseph Henry, *Scientific Writings of Joseph Henry*, vol. 2 (Washington, D.C.: Smithsonian, 1887), 431–3.

13. *Ibid.*, 433–4. "Quantity" roughly corresponds with amperage, "intensity" with voltage or potential. Henry also referred to voltage as "tension" (see letter to Fisher of April 17, 1843), which has resonance with the modern terminology "high-tension wire." Henry had first introduced this distinction in his lectures at the Albany Academy in 1829. *Albany Chronicles*, 531.

14. Moyer, *Joseph Henry*, 69–70. Henry's later accounts put this at ⅕ mile (per an article in Silliman's). Most likely he continued the experiments, eventually reaching a mile and a half (see previous footnote in this regard).

15. Henry himself later gives us a clue as to why he never pursued the economic implications of the telegraph, saying that he almost immediately thereafter became intellectually consumed by the problem of electrical self-induction. Henry, *Scientific Writings of Joseph Henry*, vol. 2, 435.

16. POJH, vol. 1, 195–199. Torrey of West Point had pointed out one of the errors with respect to the atomic weight of sulphate of copper.

17. Moyer, *Joseph Henry*, 75.

18. *Ibid.*

19. *Ibid.*, 342–3.

20. There is good reason for confusion as to Henry's state of mind regarding this as his actions seem inconsistent. He did seek on occasion to profit from his inventions, viz. his electromagnetic ore separator and logarithmic scale of chemical equivalents, but he just was not good at it. Why Henry did not bring a suit against Morse in the case of the telegraph has been discussed ad nauseam. As Carleton Mabee, Morse's Pulitzer Prize–winning biographer, succinctly and humorously put it, "Morse was more interested in cash where Henry was more interested in credit."

21. Morse, in discussing his thinking claims to have realized the utility of the "quantity" and "intensity" distinction by 1836 or 1837 (hence the benefits of a relay), and claims elsewhere he simply chose not to implement this until late 1844 or 1845. Gale, in the following testimony, backs him up on this claim in one of the more remarkable sets of interrogatories that no doubt inspired the Smithsonian regents to dub this piece pure sophistry. *Shaffner's Telegraph Companion* 1, 74–5.

22. *Ibid.*, 9. Morse claims "a part of this improvement" was used in May 1844. In fact, by the time Morse left for Europe, the in-line repeaters were still in place.

23. Prime, *Morse*, 344–5.

24. See section below on the 1860 relay renewal application. Gale had asserted this in his deposition and then quickly retracted it when the Patent Commissioner informed Morse this was grounds for invalidating the entire underlying 1838 patent, if it was apparently intentionally omitted.

25. See 1860 relay renewal discussion chapter below.
26. Ibid., 409–10.
27. Ibid., 344–5.
28. *Shaffner's Telegraph Companion* 1, 74. Usually, if Morse would either lie or obfuscate regarding the date when he first conceived the marginal circuit, Gale would be the first to swear to it.
29. As noted above, the author surmises it was probably Gale, not Morse, who framed these interrogatories. Prime, *Morse*, 341–2.
30. George Iles, *Leading American Inventors* (New York: Holt, 1912), 142.
31. Prime, *Morse*, 422–3; Joseph Henry to Samuel Morse, February 24, 1842.

Chapter 17

1. The patent, U.S. #1320, was owned by Ambrose Barnaby. Cornell had bought the rights for Georgia and Maine, obviously the two least valuable and hence probably least expensive territory from any standpoint.
2. Both dated January 21, 1843: one Jeremiah Beebe to Nathaniel Tallmadge, and one to Christopher Morgan, both Whig members from Beebe's district. 1/6, ECPKL.
3. Smith was the editor of the *Eastern Farmer*, which merged with the *Maine Farmer and Mechanic's Advocate*, around March of 1843. *The Cultivator* 10, no. 3, 6.
4. The biography *True and Firm* states Cornell went to Georgia in the fall of 1842 and walked the entire way from Washington, D.C. This is obviously an exaggeration based on Cornell's own account here of traveling by steamboat. Ezra Cornell to MaryAnn, February 15, 1843, 1/7, ECPKL.
5. Bill from Denmeads and Daniels. February 5, 1843, 1/7, ECPKL.
6. Ezra Cornell to MaryAnn, February 10, 1843, 1/7, ECPKL. This is datelined Plymouth and appears to be in someone else's hand.
7. Ezra Cornell to MaryAnn, February 26, 1843, 1/7, and October 29, 1843, 1/8, ECPKL.
8. Ezra Cornell to MaryAnn, February 15, 1843, 1/7, ECPKL.
9. Ezra Cornell to MaryAnn, February 26, 1843, ECPKL.
10. Morris Bishop, *A History of Cornell* (Ithaca: CUP, 1962), 14.
11. Ezra Cornell to MaryAnn, February 26, 1843, ECPKL.
12. Ibid.
13. Theodoric Romeyn Beck and Amariah Brigham, "Analysis of the Testimony on the Trial of Alvin Cornell for Murder and of the Subsequent Proof which led to the Commutation of His Punishment," reprinted in *Transactions of the Medical Society of the State of New York* 6 (1846).
14. Gilbert Hazeltine, *Early History of the Town of Ellicott* (Jamestown: Journal Printing, 1887), 369.
15. He had been staying with a cousin, John, in Plymouth, but mentioned to MaryAnn he intended to head for Georgia.
16. The date here has an ink blot on it (which is unusual and may have been intentional), but appears to be April 2. Cornell says he just received the letters of March 12 and 16. Ezra Cornell to MaryAnn, April 2, 1843, 1/7, ECPKL.
17. The gruesome murder had received wide coverage in the papers. It seems improbable that Ezra Cornell would have been unaware of Alvin's story by now.
18. S.F.B. to Sidney Morse, February 23, 1843, MPLOC.
19. William Thue, *Electrical Power Cable Engineering* (New York: Marcel Decker, 2005), 1–2.
20. Samuel Morse to Louis McLane, April 11, 1843, MPLOC. "I have this morning advised Mr. Smith of his appointment."
21. Samuel Morse to F.O.J. Smith, July 16, 1842, MPLOC. Notated on the back (obviously later) by Morse: "Showing that Prof. Henry saw my Telegraph for the first time in July 1842!!!" (Emphasis in original.)
22. Henry was wrong about this. Nathan Reingold even apologizes for Henry's misinterpreting Gauss's experiments on atmospheric magnetism. POJH, vol. 2, 302 (fn).
23. Joseph Henry, "Contributions to Electricity and Magnetism," extracted from the *Transactions of the American Philosophical Society* 6, no. 2 (1849): 47. Ironically, it will be Henry's later (also incorrect) conclusions about the inevitability of short circuits in long wires underground that will be the prime motivation for Morse's switching back to an overhead solution. Once again Henry's merest speculation would be taken as gospel by Morse when he was in a pickle.
24. Even the perspicacious and eminent Faraday would not dare to publish anything substantial on atmospheric magnetic induction until the mid-1850s.

Chapter 18

1. This is evidenced by the fact that his main intellectual preoccupation for this period seems to have been fighting with the university over the draftiness of the rooms he had rented and keeping a daily record of the weather at Prince and Crosby Streets. Weather reports and bill of damages to the university: Bound vol., 28 November 1835–18 April 1838, images 89–93, 98. Apparently this argument with the university lasted for his entire occupancy.

2. Washington Allston to Samuel Morse, February 27, 1837, MPLOC.

3. This was the *Journal of Commerce* of August 26, 1837. It is referred to in a note Morse wrote: Bound vol. 28 November 1835–18 April 1838, image 131, MPLOC.

4. Edward Lind Morse ed., *Samuel F.B. Morse: His Letters and Journals* 2, 49; letter to Catherine Pattison quoted.

5. Ellsworth, in a letter of September 1, offers congratulations on the *Journal of Commerce* article and offers advice on the caveat. Henry L. Ellsworth to Samuel Morse, September 1, 1837. First mention in the *Journal* usually refers to Morse's letters appearing in the issue of September 8 (see following note). However, it appears Morse's letters were in response to mention of his alphabet in an article on "animal magnetism" (mesmerism) that had appeared in the *Journal of Commerce* issue of August 26, 1837. See also letter to Catherine Pattison, August 27, 1837, mentioning enclosing the article. MPLOC. This is probably the origin of the rather sarcastic attempt to attach an amendment to the 1843 appropriation to support "mesmerism."

6. Taylor, *An Historical Sketch of Henry's Contribution*, 43–5.

7. Benjamin Homans, *Army and Navy Chronicle*, vol. 6–7 (Washington, D.C.: Homans, 1838), 25; letter of Samuel Morse to Levi Woodbury dated September 27, 1837. By this time, notice of Wheatstone's telegraph in England had appeared.

8. Samuel Morse to Catherine Pattison, August 27, 1837, MPLOC.

9. Homans, *Army and Navy Chronicle*, vol. 6–7, 23; Samuel Morse to Levi Woodbury dated November 28, 1837.

10. http://www.morrisparks.net/speedwell/tel/tel.html. Last accessed February 17, 2016.

11. Henry L. Ellsworth to Samuel Morse, October 6, 1837, MPLOC.

12. Contract between Samuel Morse and Alfred Vail, postdated September 23, 1837, MPLOC. The contract conveys one-quarter interest in the patent in exchange for all expenses of developing the Washington presentation. It is amended to three-sixteenths on March 2, 1838. It was reduced further to one-eighth share by the time the patent was filed.

13. Catherine was the daughter of Elias Pattison, a Troy merchant who had partnered with George Vail (not Alfred's brother but definitely a relative) in several business ventures, including the silkworm craze of the 1830s, which had ruined Alvin Cornell. There are rumors that Morse had asked her to marry him but that her father had opposed it.

14. Alfred Vail to Samuel Morse, October 14, 1837, MPLOC.

15. Homans, *Army and Navy Chronicle*, vol. 6–7, 23; letter of Samuel Morse to Levi Woodbury dated November 28, 1837.

16. Henry L. Ellsworth, receipt to Morse for $20 for patent caveat, October 6, 1837. While not exactly a patent, a caveat did offer an inventor some legal protection. The law regarding caveats had been enacted a year earlier in 1836.

17. Silverman, *Lightning Man*, 167, 169.

18. Given Smith's propensities and Morse's later actions, it is likely that Morse was seeking a far more modest sum that Smith encouraged him to inflate.

19. Report of the Commerce Committee, House of Representatives, April 6, 1843, MPLOC.

20. Silverman, *Lightning Man*, 169.

21. Prime, *Morse*, 340; Samuel Morse to F.O.J. Smith, February 22, 1838. Morse's estimate here comes to $26,000, and $4,000 was tacked on by Smith for "incidentals."

22. *Ibid.*, 342.

23. H.W. Greene, *Letters Addressed to F.O.J. Smith* (privately published, 1839), 22.

24. Prime, *Morse*, 370.

25. Smith was probably aware he was about to lose his seat anyway given the reaction of his Maine constituents to his having lavished on himself the funds for construction of a massive mansion, most of it probably at their expense. When Smith offered to accompany Morse on a trip to Europe to help him obtain a European patent (which efforts failed), the cost in exchange for underwriting the trip had been a quarter share of his patent, more than what Morse had awarded the other partners, Alfred Vail, Leonard Gale and George, Alfred Vail's brother, together. The fact that George was in fact a legal part owner of the patent is found in

vol. 30, December 1845–8, December 1846, image 253, MPLOC. Distribution of shares was now Morse 9/16, Smith 1/4, Vail (split) 1/8, and Gale 1/16.

26. *The Maine Farmer*, inaugurated by Ezekiel Holmes in Winthrop, was purchased by Russell Eaton and moved to Augusta in 1843. Source: http://www.kennebechistorical.org/histevents.htm. Last accessed, November 16, 2015.

27. Morse's complaint to Alfred Vail, dated February 23, 1843, MPLOC.

28. Silverman, *Lightning Man*, 170–2. The salary issue is referenced in the later letter to Sidney re: monies Smith owed him.

29. *Ibid.*,184. Morse would still be trying to unravel Smith's outstanding financial obligations for this trip as late as 1844. See Morse's letter to Smith, February 29, 1844, MPLOC.

30. Prime, *Morse*, 344. "The property in the invention was divided into sixteen shares of which Mr. Morse held nine, Mr. Smith four, Mr. Vail two and Professor Gale one. In ... foreign countries the proportions were not the same. Professor Morse was to hold eight, Mr. Smith five, Mr. Vail two and Professor Gale one." Alfred Vail's brother, George, was awarded his interest as a silent partner. In Europe the proportion was slightly different, with Morse holding only a ½ share and Smith 5/16th. Before leaving for Europe, Morse had asked Ellsworth to delay issuing the patent on the pretext that it would harm his chances in Europe. In fact, he probably regretted already having been so generous with Smith and was seeking a way around it. Consequently, the patent would not be issued until 1840 (U.S. Patent #1647, issued June 20, 1840). *Ibid.*, Silverman, 172.

31. *Ibid.*, 423. Prime calls this period in Morse's life "the dark ages." In his copy of his letter to Smith of December 5, 1841, Morse had added the postscript, "If this is a little Blue, forgive it."

32. *Ibid.*

33. Receipts to Cooper, Coles, Dunlap, etc., December 1841 total $250, or half. The remainder apparently was paid out over several years, as there is a receipt dated 1846, MPLOC.

34. *New York Mirror* 19, no. 42 (1841): 335. Article of October 16, 1841.

35. Joseph Henry to Samuel Morse, December 5, 1841, MPLOC. Apparently Morse was peeved because he received back only a little more than eight miles rather than the ten miles he had lent Henry.

36. Prime, *Morse*, 420–3. This obviously must have been a very emotional experience for Morse as we see him adopting the "list making" tic that he adopts whenever he anticipates a life-altering or especially trying circumstance, such as when he was contemplating the trip to Mexico, when he was firing Fisher, and when he was in the fracas with Bartlett. Interestingly, it is a trait he shares with Ezra Cornell. These questions are highly technical in nature and not at all characteristic of Morse, so it is likely (in the author's opinion) that he had Professor Leonard Gale prepare the questions for him.

37. *Ibid.*, 421–2. Joseph Henry to Samuel Morse, May 6, 1839. (See appendix.)

38. Then called the University of the City of New York.

39. Luther Harris, *Around Washington Square* (Baltimore: JHU Press, 2003), 20–1.

40. *Ibid.*

41. Samuel Morse to the Chancellor of the University of the City of New York, February 12, 1842. NYU University Archives.

42. Rent receipt dated March 29, 1843. NYU University Archives, courtesy Emily Chapin, Adjunct University Archivist, NYU. Gale was listed as "rent free" on the third floor, possibly in one of the five rooms Morse rented.

43. Francis Markoe Jr. to Samuel Morse, October 12, 1841, MPLOC.

44. Prime, *Morse*, 422–3. Joseph Henry to Samuel Morse, February 24, 1842.

45. *Ibid.*

46. Silverman, *Lightning Man*, 209.

47. Samuel Morse to the chancellor of the University of the City of New York, February 12, 1842.

48. November 30, 1841 (copy), to Isaac Coffin. Morse refers to two previous letters he had received from Coffin offering to lobby, MPLOC.

49. Original not found. See below Henry's reply of December 5, 1841, referencing both, MPLOC.

50. Philip Lundeberg, *Samuel Colt's Submarine Battery Smithsonian Studies in History and Technology*, 29 (1974): 31.

51. "I am deterred from giving up on the matter as desperate ... for the consideration that those who ... lent their aid ... would suffer." Copy of letter apparently to F.O.J. Smith, December 3, 1841, MPLOC.

52. Morse had lent Henry one reel of wire and spooled it onto two reels, promising in his December 5 letter to restore it to one reel, which apparently he did not do. It is more likely that Morse rather than Colt was the main benefactor of this arrangement. Though Colt had declared bankruptcy for his Patent Arms Man-

ufacturing company in Paterson, he had since been the recipient of a grant from the War Department in the amount of $6,000 to investigate the submarine cable (see Lundeberg, *Colt*, 21). Morse was at this time living in genteel poverty.

53. Samuel Morse to F.O.J. Smith July 16, 1842 (transcr.). Prime, *Morse*, 437. (Emphasis in original.)

54. *Ibid.*, 22.

55. *Shaffner's Telegraph Companion* 1, 73. Gale's deposition. By 1843 Morse was already broaching the subject of an Atlantic telegraph to the Secretary of the Treasury. Morse to Treasury Secretary Spencer, August 10, 1843, MPLOC.

56. Morse had asked for a more modest appropriation of $3500. Samuel Morse to Hon. W.W. Boardman, August 10, 1842. See Boardman's reply in the negative just two days later, Prime, *Morse*, 434–6.

57. Colt had set off an underwater explosion as part of the July 4 festivities in New York Harbor. Lundeberg, *Colt*, 25.

58. Samuel Morse to F.O.J. Smith, July 16, 1842, MPLOC. In what is obviously a later added postscript on the back, Morse writes, "S.F.B. Morse to F.O.J. Smith, Showing that *Prof. Henry* saw my Telegraph for the first time in July 1842!!." He thus adduced his own words to prove that Henry had never before seen a telegraph. In view of Henry's own statements elsewhere, this is obviously a construction on something Henry said.

Chapter 19

1. Schiffer, *Power Struggles*, 139.
2. Mabee, *The American Leonardo* (New York: Alfred A. Knopf, 1942), 250–1. Citing the *New York Herald* article, October 19, 1842.
3. Lundeberg, *Colt*, 34.
4. Jeffrey Kroessler, *New York, Year by Year* (New York: NYU Press, 2001), 85. Colt had actually been aboard the revenue cutter *Ewing*. The test was successful, but the attempt to sell the device to the navy was not. The demonstration was conducted off Castle Garden in New York Harbor, where the hulk *Volta* was mined and blasted into oblivion from a galvanic battery on board the U.S. Naval brig *Washington*. A year earlier Morse, probably encouraged by NYU professor John Draper, had run for mayor of New York under the "Nativist" or anti-immigrant, anti-Catholic party, but was defeated by Robert Morris. Also Silverman, *Lightning Man*, 216. (The cutter *Ewing* figures centrally in this author's novel *monoville*.)

5. Lundeberg, *Colt*, 35.
6. Lundeberg, *Colt*, 34.
7. American Institute to Samuel Morse, October 20, 1842, thanking him and Fisher for their help and the use of the telegraph between Governor's Island and the Battery on the 18th of October.
8. Probably —·—-, or "aye aye," Morse's code for "all is ready."
9. Lundeberg, *Colt*, 35.
10. Prime, *Morse*, 443.
11. *Ibid.*
12. Schiffer, *Power Struggles*, 139.
13. In the fall of 1844, when he tried to string the cable across the Hudson, Colt's method would fail miserably. Morse would then turn to Ezra Cornell and later to Professor Charles Page of the Patent Office to solve the problem. Interestingly, Page's solution would involve arguably one of the first uses of vulcanized rubber as an insulator, a method devised by Morse's cousin Charles Goodyear.
14. *Ibid.*, 26, citing account in the *New York American*.
15. Samuel Morse to Governor William Seward, July 1, 1842. MPLOC.
16. Samuel Morse to F.O.J. Smith, July 16, 1842, MPLOC.
17. Hochfelder notes Morse had made this claim of having New York investors twice before to Vail, and the investors had mysteriously disappeared. David Hochfelder, *The Telegraph in America* (Baltimore: Johns Hopkins Press, 2012), notes to Chapter 2, note 2. This seems to be an exaggeration. Just before Morse's first letter on this subject to Vail he has appealed to a congressman regarding enlisting another lobbyist on his behalf. Hon. Robert McClellan to S.F.B. Morse, June 24, 1841, MPLOC.
18. "Report of the American Institute, September 12, 1842." In Vail, *The American Electro-Magnetic Telegraph*, 88–89, and letter of James Renwick, corresponding secretary of the AINY, to the Congress, December 31, 1842, MPLOC.
19. Moyer, *Joseph Henry*, 242.
20. By 1842 Morse was eager to show Henry the improvements Gale and Fisher had made in the interim. Henry, at Morse's urging, had supplied the letter on his behalf that Morse would use to gain support in Congress. Joseph Henry to S.F.B. Morse, February 24, 1842. Henry politely declined later invitations to come see the new device in operation. POJH, vol. 5, 144, 146–7, 150–1.

21. S.F.B. Morse to his brother, Sidney, December 18, 1842, MPLOC.
22. S.F.B. Morse to Sidney Morse, February 21, 1843, MPLOC.
23. Mabee, *American Leonardo*, 260.
24. His only income had been a harebrained scheme to circumvent Smith's chokehold and partner up with James Fisher and Philip Schuyler to sell the telegraph rights in China. For this he had received $100 from Schuyler. See contract dated September 22, 1842, and subsequent correspondence with Fisher and Schuyler. MPLOC.
25. Samuel Morse to Alfred Vail, February 23, 1843, MPLOC.
26. Letter to Louis McLane, president of the Baltimore Washington Railroad from Morse, referring him to Smith on any matters pertaining to the trenching contract.
27. John Niernsee to Samuel Morse, August 13, 1843. Niernsee was a civil engineer who had made application several months back. While the letter is addressed to Morse, Morse marks on the back "Application for the Consideration of F.O.J. Smith." And while this is not proof that Smith had formed a separate company, it is clear he had taken over the hiring and firing. Cornell, the pragmatist, in his autobiography later generously awards Smith full credit for the device, saying he just happened to enter when Smith was drawing up his plans on the floor of the paper's offices and that he (Cornell) only built the model for him using his blacksmith skills. This is contradicted by Cornell's vigorous efforts to secure a patent in his own name. It was not Cornell's nature to try to steal another man's work. Smith, on the other hand, had no such qualms about theft, and consistently referred to it as "our" invention.
28. Not found. See Morse's reply below.
29. Samuel Morse to F.O.J. Smith, May 17, 1843, MPLOC.
30. Secretary of the Treasury Spencer to Samuel Morse, June 6 and 7, 1843, MPLOC.
31. Samuel Morse to F.O.J. Smith, May 17, 1843, *Collected Letters of Samuel F. B. Morse*, vol. 2, ed. Edward Morse (New York: Houghton Mifflin, 1914), 7.
32. Secretary of the Treasury John Spencer to Samuel Morse, June 7, 1843, MPLOC.
33. Samuel Morse to F.O.J. Smith, July 16, 1842, MPLOC. Morse had calculated the total cost per mile to be $568, with $150 allocated to "Excavating and Filling," which is the portion he had contracted to Smith and Bartlett. When Morse signed the contract with Bartlett, it was to be for this exact amount ($150). The contract specified an additional $3.00 per mile for depositing the cable, a process that would be economized thanks to Cornell's cable layer. Morse's letter to Smith says he expected the work to commence no earlier than September 15 and no later than October 1.
34. Mabee, *American Leonardo*, 209.
35. Walter Harriman, *History of Warner* (Concord: Republican Press, 1879), 469–72.
36. *Ibid.*
37. *The Country Gentleman*, Albany; *The Genesee Farmer*, Rochester.
38. *Scientific American* 2, no. 8 (1860), 118.
39. Various articles in the *Journal of Agriculture*, which was a publishing arm of the Patent Office.
40. Filed under June 10, 1843, this is the contract and actually one of Morse's many "lists." This list is misfiled here, as it outlines the relationship with Bartlett in contemplation of his dismissal from its inception to January 1844. Probably Morse had been introduced to Fog Smith via Bartlett. See letter addressed to Jedidiah Morse care of Levi Bartlett, November 19, 1822, MPLOC.
41. Morse to F.O.J. Smith, July 24, 1843, MPLOC.

Chapter 20

1. The hand of these missives is so cramped that it seems he is determined to utilize every available square millimeter of paper.
2. As noted, the following may be somewhat apocryphal, as Cornell later patented the cable layer under his own name.
3. Cornell gets the name wrong here. This was actually "Robinson," per a letter from A.B. Lincoln of August 24, 1843, 1/7, ECPKL.
4. Bishop, *A History of Cornell*, 14.
5. All of Beebe's recent letters to Cornell have been advising him to close up the plow business and sell off any inventory on the cheap. Jeremiah Beebe to Ezra Cornell, July 31, 1843, 1/7, ECPKL.
6. Ezra Cornell to MaryAnn Cornell, February 2, 1843, 1/7, ECPKL.
7. *Evening Post*, February 2, 1843.
8. *New York Tribune*, March 11, 1843. ChroniclingAmerica.loc.gov.
9. Henry Lanier, *A Century of Banking in New York* (New York: Doran, 1922), 308–9. Robert Comfort Cornell was also a direct descendent of Thomas Cornell, who, as we shall

see, also figured centrally in the first familial murder in the Cornell lineage.

10. *Charters of American Life Insurance Companies* (New York: Spectator, 1906), 174.

11. Viviana Zelizer, *Morals and Markets* (New York: Columbia University Press, 1979), 92.

12. Ezra Cornell to MaryAnn, February 2, 1843, 1/7, ECPKL. It is a habit that will come back in other, altered, but recognizable forms every time he is confronted with a stressful decision. The life insurance issue is so compelling to him that it takes up the bulk of two letters to MaryAnn: the one previously cited, and that of January 31, 1843, 1/6, ECPKL. In the second letter he urges MaryAnn to purchase an additional $1,000 as soon as she can. (This cannot but have alarmed her.)

13. *Ibid.*

14. Greene, *Letters*, 14. Applied to him by fellow Maine congressman, Leonard Jarvis.

15. The mansion was actually in Deering, Maine, not Portland, and had been purchased from William Woodbury, president of Merchants' Bank in Portland, transported and reassembled there.

16. F.O.J. Smith to Secretary of State Daniel Webster, August 21, 1842. Smith demands $2,000 for services rendered, with blank spaces supplied for future payments to be filled in at a future date. George Curtis, *Life of Daniel Webster*, vol. 2 (New York: Appleton, 1870), 278.

17. This would prove overly optimistic, but certainly savings were realized, as Cornell will point out in his deposition for Smith chiding Morse for favoring unfaithful people.

18. Obviously not the whole forty miles.

19. Secretary Spencer to Samuel Morse, May 15, 1843. Approving the Serrell contract, MPLOC. Apparently the original contract (alone out of all the contracts at this time) did not survive amongst Morse's papers.

20. Serrell, James E. 1843. "Improvement in machinery and in the process of manufacturing metallic pipes." U.S. Patent #2918, issued January 20, 1843.

21. Extract of Smith's letter of August 18, 1843, MPLOC.

22. Jeremiah Beebe to Ezra Cornell, August 18, 1843, 1/7, ECPKL.

23. Either this entire story is apocryphal or the dates are wrong. Ezra Cornell was severely ill by the 21st and could not have conducted this demonstration on the 19th. Letter of A.B. Lincoln addressed to Cornell care of "Robinson's," referring to his letter of the 21st that tells of his recent illness. A.B. Lincoln to Ezra Cornell, August 24, 1843. 1/8, ECPKL

24. *The Cornell Era* 39, no. 1 (1909): 363. All these accounts stem from Cornell's autobiography, and as noted are somewhat contradicted by the facts. Morse refers to a letter of Smith's of August 18 in which he agrees to a delay in the trenching. The letter was sent to him in New York, so Morse obviously was not in Portland on that date. The letter is not found, but Morse's comment on it is. Extract of Smith's letter of August 18, 1843. MPLOC. Apparently Serrell was already having trouble with the cable. Three miles a day was the estimate Smith had provided Morse. The actual figure would be less than half that, and that only in open country where there were no obstructions. Sections closer to existing structures like railway stations still had to be hand-dug, and problems with the cable itself would slow things down even further.

25. Note of F.O.J. Smith regarding trenching. July 24, 1843 (Morse's copy), MPLOC.

26. Alonzo Cornell, *True and Firm*, 75–7. Cornell's recollections here are notoriously unreliable, but the description seems too detailed to be totally fabricated.

Chapter 21

1. In May, two months after the events in Jamestown, Smith would be assigned by the court in Maine, along with W.P. Fessenden (later Secretary of the Treasury under Lincoln) to defend a couple charged in a love triangle homicide case.

2. While not exactly a love triangle, the Alvin Cornell case also had similar romantically lurid overtones. Alvin had become obsessed with the idea that his wife was planning to leave him, and as a result was trying to turn their children against him. This is precisely the same obsession that drove Ezra's brother Edward to be committed for insanity in 1860.

3. The mitigating testimony had come from friends of Alvin who had not appeared at trial because the state did not want to bear the cost of transporting them. When Beck interviewed them, they confirmed his agitated state of mind and other instances of bizarre behavior.

4. *Ibid.*

5. *Transactions of the Medical Society of the State of New York*, vol. 6 (Albany: Munsell, 1846), 215.

6. The creation of Cornell University will become intertwined with the birth of Willard's public asylum in Ovid, New York, the site of Cornell's first attempt, and also with his brother's committal to the care of Dr. van Deussen at the Kalamazoo Insane Asylum in Michigan. (Separate monograph by the author.)

Chapter 22

1. His communications with Secretary Spencer at the early stages of the Test Telegraph contain several references to a national and even a transnational telegraph. See Morse to Secretary Spencer, August 10, 1843. MPLOC
2. Samuel Morse to Joseph Henry, April 30, 1844, POJH, vol. 6.
3. Lewis Coe, *The Telegraph* (Jefferson, NC: McFarland, 2003), 167.
4. Richard John, *Network Nation* (Cambridge: Harvard University Press, 2010), 34.
5. John Rudolph Niernsee. Niernsee was a qualified civil engineer who had apprenticed with Benjamin Latrobe, architect of the U.S. Capitol, as a civil engineer. Morse had politely ignored his repeated applications. See letter of recommendation from Benjamin Latrobe Jr., April 22, 1843 (supposed to be hand-delivered); apparently Morse avoided, him as Niernsee writes on the 24th. Second letter from Latrobe Jr., April 24; letter of Niernsee, August 13, 1843 (referred to Smith), MPLOC.
6. Leonard Gale to Samuel Morse, March 21, 1843, MPLOC. This is obviously a pro forma letter written at Morse's behest. Gale offers to oversee the manufacture of the lead pipe. Vail proposed for himself a salary of three dollars per diem. Alfred Vail to Samuel Morse, March 21, 1843, MPLOC. March 31, 1843, letter appointing Vail assistant at a salary of $1,000 (not in Morse's hand), MPLOC. James Fisher and Leonard Gale, two of the same date to Morse, March 31, 1843, accepting position. Virtually identical. Morse letter of the same date offering the positions. Same day. Obviously hand-delivered.
7. Ezra Cornell to MaryAnn Cornell, October 5, 1843, 1/8, ECPKL.
8. Secretary of the Treasury to Morse, May 15, 1843, MPLOC. Approval of contract with Serrell.
9. Samuel Morse to Leonard Gale, September 8, 1843, MPLOC.
10. Henry had suggested adding a winding of coarse twine to further separate the conductors.

He also suggested that he might visit Fisher (per Morse's invitation) to do some tests on the wire during the school break. Joseph Henry to James Fisher, April 17, 1843. Morse's invitation is not found. The wires for the magnets were wound with silk. James Fisher to Samuel Morse, April 18, 1843, MPLOC.
11. *Ibid.*
12. *Ibid.* See Morse's note to Smith, July 24, 1843, noting "not earlier than September 15 nor later than October 1" as the start date. Agreeing to postpone to mid-October, Morse (note copy) re: Smith August 18, 1843, MPLOC.
13. Alonzo Cornell, *True and Firm*, 81–4. Cornell gives a firsthand description of his visit to the Serrell factory, where he first asserts an intellectual challenge to Vail by suggesting an alternate means of proving the pipe.
14. Morse, Samuel. 1843. Telegraph Wire. U.S. Patent #3316, issued October 25, 1843.
15. On September 22, 1843, Smith provides a Power of Attorney to Morse that he instructs Morse not to show the Secretary of the Treasury, authorizing him to sign Bartlett's name on bills. Copy in MPLOC.
16. *Ibid.*
17. Alonzo Cornell, *True and Firm*, 81.
18. *The American Agriculturalist*, vol. 2, ed. A.B. Allen (New York: Saxton & Miles, 1843), 215.
19. "Annual Report of the American Institute" (Albany: Mack, 1844), 161.
20. Ezra Cornell to MaryAnn Cornell, October 29, 1843, 1/8, ECPKL.
21. Payment to Serrell of $142.50 on October 10, 1843, $105 and receipt for lead for joints, October 16, MPLOC. Morse would continue paying Serrell into November, even after he had been fired.
22. F.O.J. Smith to Ezra Cornell, October 31, 1843. 1/8, ECPKL Indicating that 3 of 12 of Serrell's boxes, making up approximately seven miles, had been shipped by packet.
23. Steam freight receipt dated October 20, 1843, from Ericsson Steamboat Line for three boxes lead pipe, Philadelphia to Baltimore, MPLOC.
24. Louis McLane to Samuel Morse, April 7, 1843; Morse's reply April 11, MPLOC.
25. The B&O itself was under Benjamin Latrobe Jr. Latrobe had asked Morse to hire on John Niernsee in April, which request Morse had ignored, hence his reason for applying to McLane instead.
26. Samuel Morse to Secretary Spencer, October 28, 1843, MPLOC.

27. Samuel Morse to Secretary Spencer (two dates on one sheet), September 11 and October 28, 1843, MPLOC.
28. Alonzo Cornell, *True and Firm*, 82.
29. F.O.J. Smith to Ezra Cornell, October 28, 1843, 1/8. ECPKL.
30. Copy of the memorandum of F.O.J. Smith, November 25, 1843, relating to the saving on the lead pipe contract. On the back in Morse's hand, are the words "belongs to Smith $588.06" on the front. Memorandum undated, back dated November 25, 1843, MPLOC, which Morse will decline to pocket his but he will pay Smith.
31. F.O.J. Smith to Ezra Cornell, October 31, 1843, 1/8, ECPKL.
32. Bill from Ebenezer Chase for covering the "extra" ten miles of wire. October 28, 1843. MPLOC.
33. Serrell's letter and signed release (drafted by Smith, revised to include the phrase "through circumstances beyond his control," three documents signed by Serrell, witnessed by Smith), October 28, 1843, MPLOC.
34. Receipt from Ebenezer Chase for ten miles of wound cable, October 28, 1843, MPLOC. Chase had been working in Morse's office for two months (see weekly salary receipts).
35. F.O.J. Smith to Ezra Cornell, October 31, 1843.
36. Contract with Tatham Brothers, signed F.O.J. Smith, promising ten miles of pipe to be delivered by November 10 and an additional ten miles by November 15, dated October 30, New York, MPLOC.
37. Instructions to Fisher. Morse had told Fisher to expect a drum of wire 450' to 600' with a diameter of 6"–8."
38. F.O.J. Smith to Ezra Cornell, October 28, 1843, 1/8, ECPKL.
39. F.O.J. Smith to Ezra Cornell, October 31, 1843, ECPKL. Emphasis added. It seems clear based on Cornell's subsequent preoccupation with the "cable-layer patent" that Morse has already warned Cornell of Smith's propensity for over-reaching in regard to claims of intellectual property. Smith had been referring to it as "our machine" since its inception. This echoes Charles Jackson's letter claiming credit re: the telegraph to Morse, referring to it as "our machine."

Chapter 23

1. Vail's repeated attempts to get Cornell fired are covered in a subsequent chapter. For the rest see Alonzo Cornell, *True and Firm*, 83, 87–8. Cornell was not averse to biting the hand that fed him. He would refer to Ellsworth, who went out of his way to help Cornell over the winter of 1843–4, as "intermeddling." June 19, 1844, NYU Archives (Silverman Collection, Box 1, Folder 1843–4).
2. Per the letter to Fisher (referring to the new pipe) the "pipe" was ⅜" in bore diameter with four color-coded wires inside. Samuel Morse to James Fisher, November 1, 1843, MPLOC.
3. Alonzo Cornell, *True and Firm*, 83–4.
4. It is clear by now Morse was losing confidence in Vail's acumen and judgment re: things scientific. In his letter to Fisher appointing him to inspect the new Tatham cable, he gives him the option of using Vail's method of the vacuum pump, or Cornell's method, the forced air pump. Instructions to Fisher, November 1, 1843, MPLOC.
5. Alvin Harlow, *New Wires and Old Waves* (New York: Appleton, 1936), 90.
6. According to Cornell, he had warned Thomas Avery, who was in charge of the battery, about this and suggested the alternate means of testing by supplying the current to the transmission wires and testing the return wires; the latter course would have uncovered the defect in the insulation immediately, where Vail and Gale's method did not. Cornell said of Avery somewhat contemptuously, when asked to inform Vail of the results, "he dared not do so." *Ibid.*
7. Cartage bill for 500 lead joints, October 28, 1843, MPLOC. Serrell's bill of the same date is for 240 lead sleeves, suggesting these were perhaps two-piece sleeve joints.
8. *Ibid.*, Ezra Cornell to MaryAnn Cornell, October 29, 1843. 1/8, ECPKL.
9. *Ibid.*, 82.
10. The new contract specified a cost of $5.10 per hundred weight plus the annealing of the wire (actually thirty-five cents more than Serrell's). With a bit of creative accounting, Smith had presented this to Morse as a cost savings of more than $1,000, of which he proposed they each pocket half. Though Morse declined and credited his half of the savings back to the government, Smith had Morse write him a check for his half of the projected savings, $583. Receipt to Smith dated November 25, 1843, MPLOC. Also see undated summary titled "Statement of account with Smith" on Tatham's contract, showing "$534.60 still in Smiths hands." Correspondence 13 September 1843, 13 January 1844, image 300, MPLOC.

11. Morse to James Fisher of December 2, MPLOC.

12. Fisher's somewhat frail wife was due to deliver in December. James Fisher to Samuel Morse, December 9, 1843. "We are all well, but Mrs. Fisher is in hourly expectation of her confinement." Fisher's wife was Eliza (Sparks), who was a niece of the famed historian Jared Sparks. The baby apparently did not survive. Apparently Fisher's brother-in-law had been employed by Morse to wind the wire with the cotton twine (D. Sparks). Approval and Instructions from Secretary Spencer to Morse, May 12, 1843, MPLOC.

13. James Fisher to Samuel Morse, November 29, 1843, MPLOC.

14. Morse to Fisher, December 2, 1843, MPLOC.

15. *Ibid.*

16. *Ibid.*

17. S.F.B. Morse to James Fisher, December 6, 1843, MPLOC. "We are about to commence on Tatham's pipe." It appears by this date Morse was already aware of some problem with Tatham's pipe, as he instructs Fisher bluntly: "Inform me also if you have faithfully tried the wires." The still unsuspecting Fisher, in his reply of the 9th, misinterprets this to mean only the new wire he has received since. The next (undated) extraordinary document is notated as "evidence of Fisher's neglect" (image 185–6), and contains extracts of Fisher's letters up to December 9. What follows is a charade: Morse's transparent attempt to trap Fisher into obvious malfeasance through intentionally misdirecting the bank drafts, possibly with the collusion of his brother Sidney.

18. Samuel Morse to James Fisher, December 6, 1843, MPLOC.

19. See Fisher's approval for 10 miles of pipe dated November 15, 1843, MPLOC.

20. Vail's letter to his wife December 12, 1843, in *Early History of the Electro-Magnetic Telegraph*, 16.

21. Cornell says, "I accordingly went on with my part of the work expecting every day to receive orders to suspend laying the pipe. Thus matters proceeded until we had completed the laying of pipe as far as the Relay House." Alonzo Cornell, *True and Firm*, 84.

22. Obviously this is an offensive term, and Cornell would have found it offensive too, but in the context of an anecdotal usage, anyone who has been a salesman or saleswoman will immediately recognize the derivation and the circumstance and how it would have come up in the conversation.

23. Gale had written back on the 7th giving his excuses, but even after a second letter of the 6th, Fisher would not respond until the 9th.

24. *The New England Magazine* 4; Stephen Vail (son), *Early Days of the First Telegraph Line* (Boston: NEMC, 1892); letter of Alfred Vail, December 22, 457.

25. Leonard Gale to Samuel Morse, December 7, 1843, MPLOC.

Chapter 24

1. Comptroller's receipt to Morse, November 17, 1843, indicating $10,000 (in receipts) and amount due the government from the original appropriation, MPLOC. Prime, Morse, 477–8.

2. *Ibid.*

3. Morse would actually continue making payments to Serrell through January. The expense for labor and rent at the shop had been about $60 a month.

4. Lundeberg, *Colt*, 22.

5. The diagram Morse drew showed the in-line relays at a distance of 20 miles.

6. Leonard Gale to Samuel Morse, December 8, 1843, MPLOC.

7. Fisher and Gale, as "scientific consultants," were being paid $1,500 per year, whereas Vail was getting $1,000 per annum and Smith was getting whatever he could skim from the Bartlett contract. Abstract "D," Salaries. October 10, 1843. MPLOC.

8. James Fisher to Samuel Morse, November 29, 1843, MPLOC.

9. Leonard Gale to Samuel Morse, December 7, 1843, MPLOC.

10. S.F.B. Morse to Sidney Morse, December 16, 1843, MPLOC. The letter is datelined Washington, but Morse is referring to the meeting scheduled for the next day at Relay House.

11. James Fisher to Morse, December 15, 1843, MPLOC. Says he will start out for Baltimore the first of the week, which would be the 18th.

12. Fair warning to the reader: it is a side of Morse that is not pretty and which will resurface in an even more grotesque form later in relation to Ezra Cornell. His actions over the magnetic relay will also demonstrate a sly ruthlessness worthy of both Ulysses or Machiavelli, all stemming from the "minor" inconvenience

that Morse himself did not actually happen to invent it, a device that, as he by then had come to realize, was an essential component of the telegraph. Once Morse realized the importance of what Cornell called the magnetic relay, he would painstakingly recreate the circumstances necessary for him to lay what would appear to be a perfectly reasonable claim to sole authorship.

13. Samuel Morse to Secretary Spencer, December 5, 1843, MPLOC.

14. Vail, *Early History of the Electro-Magnetic Telegraph*, 16.

15. Alonzo Cornell, *True and Firm*, 85. Cornell (who was not there and not invited) is quoted as saying that Fisher and Gale were there at relay. He was getting his information second-hand from Smith (who may have lied to him, as clearly they were not there).

16. Morse evidently was still trying to conceal the real reason for the work stoppage. Samuel Morse to Secretary Spencer, December 18, 1843, MPLOC.

17. Unsigned in ECPKL 1/8, marked *National Intelligencer*, two dates on same sheet, December 23 and December 29, 1843.

18. Alfred Vail to (his wife?), December 17, 1843. Vail, *Early History of the Electro-magnetic Telegraph*, 16.

19. Samuel Morse to James Fisher, December 6, 1843, MPLOC.

20. Samuel Morse to Secretary of the Treasury Spencer, dated December, 1843 (Washington), MPLOC. On the back (in Morse's hand, in what is obviously a later addition when he was gathering evidence of liabilities to Smith on the trenching), he notes, "Showing Mr. Smiths own desire to have the wire upon posts after the failure of the pipe plan."

21. Louis McLane to Samuel Morse, December 15, 1843, MPLOC.

22. Alonzo Cornell, *True and Firm*, 87.

23. *Ibid*. The new Patent Office building began construction in 1836 after the destruction of the old Patent Office by fire.

24. *Ibid*., 86–7. There are notations on the vouchers at the end of December, all undated, relating to Fisher, disallowing room and board charges (asterisked), if he is to be paid his salary from this period. Vail, Cornell and Morse are listed on Abstract "D" of March 10, 1844 (MPLOC), as salaried employees; however, there are no accompanying receipts leading me to believe that they were not actually receiving money, as Morse was meticulous about this. Cornell's unpaid salary was probably later converted into stock.

25. Letter (unsigned) in Morse's hand dated December 27, 1843, 1/8, ECPKL. (Emphasis mine.)

26. Cornell, *True and Firm*, 87.

27. Various estimates for stations, wires, poles (all undated) all noted on the back as "Cornell's Estimate," December 1843, MPLOC.

Chapter 25

1. Leonard Gale to Samuel Morse, December 29, 1843, MPLOC. After learning of Fisher's dismissal firsthand, having traveled to Washington after the Relay conference, Gale had resorted to "prayer, meditation and fasting."

2. John Jordan, *Colonial and Revolutionary Families of Pennsylvania* (New York: Clearfield, 1911), 754–5. As stated below, there seem to be errors in this account.

3. The next source (Jordan) gave the date of 1836 for this, but the *Biographical Sketch of the Class of 1826* gives it as 1839. Selden Haines, *Biographical Sketch of the Class of 1826, Yale College* (Utica: Roberts, 1866), 36–7.

4. Courtesy NYU University Archives. Gale had taught there from the opening of the University.

5. See Henry's letter to Morse of January 24, 1844, cited above, mentioning Morse's reply to his critics in the papers.

6. Pro forma receipt to Tatham Bros. dated November 1843, Bound vol., 13 September 1843–13 January 1844, images, 165–7, MPLOC.

7. Invoice from B. Quackinbush dated September 5, 1843, MPLOC. There are no receipts for asphaltum, beeswax, resin or India rubber. The receipt (from Quackinbush above) specified only (in addition to cleaning acids) shellac, cochineal and alcohol. Asphaltum, like cochineal, was commonly used by artists of that era as a coloring and mixing agent for paints, so perhaps Morse thought he could substitute cochineal, which is far easier to work with than asphaltum and cheaper (but only the latter is waterproof). Some otherwise reliable contemporary accounts, notably the one noted below given in the "History of Baltimore," say that Morse's cable was coated with resin and tar. This is incorrect, as proven by the receipt annotated below of March 20 from Ebenezer Chase.

8. Lundeberg, *Colt*, 35. (Note: the linseed oil was a solvent to cut the alcohol.)

9. Joseph Henry to James Fisher, April 17, 1843, MPLOC.

10. By "cement," Henry here evidently meant the asphalt-beeswax mixture, not what we call cement. (Emphasis mine.)

11. Joseph Henry to James C. Fisher, April 17, 1843, MPLOC.

12. James Fisher to Samuel Morse, January 1, 1844, MPLOC. Morse had definitely seen this reply from Henry, as Fisher notes explicitly in a later letter that he had shown it to him

13. Some later concerns Henry expressed, regarding "celestial magnetism," had also thrown Morse into a tizzy. These too had proven bogus.

14. Obviously Henry must have known that static electricity would leap a gap, so his emphasis here on galvanic electricity (as being somehow different) is intentional, but still wrong. Technically Henry was also wrong about the "cement" insulators, as later versions of Colt's resin and tar solution and more advanced substances like gutta-percha or rubber would make excellent insulators on cables of any length, but strictly in terms of what was then commercially available and cheap, he was probably correct.

15. *Boston Directory* (Boston: Frost, 1820), 62. Lists Chase as having a shop at Carver Street.

16. The rope walk was a block east of Bloomingdale Square. The square ran between Eighth and Ninth and between 53rd and 57th Streets.

17. An actual remnant of Colt's cable survives. See the illustration in Lundeberg, figure 30.

18. Ebenezer Chase to Samuel Morse, March 20, 1843, MPLOC. First underline in original, second emphasis mine. On the back endorsed by Morse (presumably), "Winding wire, E, Chase first proposal, withdrawn."

19. Ebenezer Chase to Samuel Morse (rear), March 20, 1843, MPLOC.

20. May 2 and 4, 1843, from Secretary Spencer to Morse. MPLOC. The latter refers to the bid of "May 1, instant" which is missing from the Morse papers, as is Serrell's contract, which was approved on the 14th.

21. Contract with Aaron Benedict dated May 10, 1843, MPLOC.

22. Bill for silver solder dated May 22, 1843, and noted on back for Fisher at Bloomingdale ropewalk.

23. Ebenezer Chase receipt to Samuel Morse, May 17, 1843, MPLOC. As Colt might have put it, this was the "smoking gun" that Morse was to blame for entire fiasco.

24. Samuel Morse to James Fisher, December 2, 1843, MPLOC.

25. Morse to Fisher, December 6, 1845, MPLOC. (Emphasis mine.)

26. Fisher to Morse, December 9, 1845, MPLOC. c/o David Burbank's warehouse, Baltimore.

27. Morse to Fisher, December 12, 1845, MPLOC.

28. James Fisher's reply to Morse of the 12th, December 15, 1843, MPLOC.

29. Samuel Morse to James Fisher, December 20, 1843, MPLOC. Dismissal.

30. It is clear Morse was very much relying on Gale's continued services, as he asks him to undertake inspection of the faulty cable at the Patent Office at some point in the future. Samuel Morse to Gale and Vail, December 18, 1843, instructing them to "make the necessary experiments to ascertain the condition of the conductors." It is equally clear that Gale has foregone his salary for this period. See monthly Abstract "D," expenditures on account of salary. March 10, 1844, MPLOC, which lists only Morse, Vail and Cornell on salary. (Avery was being paid also, but his pay schedule was biweekly.)

31. Undated note saying Bartlett had agreed to go home and return in April. Smith would make sure this did not stand, MPLOC.

32. Desperately short on cash, most likely Morse had concocted this entire scenario with his brother Sidney to implicate Fisher and to defer payment to Tatham as long as possible until he could determine the exact problem with the pipe, with which delinquency he now upbraids poor Fisher. The whole sequence otherwise makes no sense except as subterfuge or malfeasance. Though Morse (probably with a wink) had asked Sidney in the letter of the 16th specifically to "diligently" search for the missing bank drafts; one does not just happen to misplace what amounted to a half year's salary behind a bookcase.

33. James Fisher to Samuel Morse, December 22, 1843, MPLOC.

34. *Ibid.*, second letter from Fisher the same day.

35. *Ibid.*, Morse to Fisher, December 20, 1843.

36. Silverman, *Lightning Man*, 232.

37. This is another example of Morse's list-

making tic. The entry is undated and notated as "Evidence of Fisher's neglect," images 185-6, Bound vol., 13 September 1843–13 January 1844, MPLOC.

38. Samuel Morse to James Fisher, December 29, 1843, MPLOC.

39. Ibid. (Emphasis in the original.)

40. James Fisher to Morse January 1, 1844, MPLOC.

41. Morse to Fisher, December 29, 1843, MPLOC.

42. Joseph Henry to Samuel Morse, January 24, 1844, MPLOC. The effective distance the current traveled in a one-mile section of cable was two miles, per Henry's new theory.

43. Silverman cites as evidence of Fisher's character his supposedly spurious claim for $55 room and board at the Reverend Rich's in Connecticut, which supposedly Morse investigated later and denied when he found out Fisher had been staying for free. Sliverman, *Lightning Man*, 227. This has all the earmarks of "EMFP" (evidence manufactured for posterity). This would have been entirely out of character for Fisher and not at all out of character for Morse.

44. Fisher to Morse, January 1, 1844, MPLOC. Fisher quotes Morse's letter to Gale of Saturday previous (not found in the collection) accusing him of malfeasance.

45. Ibid. (Emphasis in the original.)

46. The strain of the dismissal and other events (possibly losing his newborn son) probably caused Fisher to have a nervous breakdown. He would leave his position at NYU. The highest position he would attain in the years following the fallout with Morse would be as a high school principal in New Jersey. Eventually, with the coming of the Civil War, he would gain a posting as an Army surgeon and serve honorably.

Chapter 26

1. Samuel Morse to Vail and Gale, December 18, 1843, MPLOC.

2. Ibid. (Emphasis mine.)

3. It was highly unusual to address a letter to two parties, and there is no other instance in which I have seen that Morse did this. He probably did it to ensure that the "new company org chart" was known to all concerned, even though Gale had officially resigned.

4. Commissioner Ellsworth to Ezra Cornell, January 11, 1844. It is addressed to Cornell in Ithaca, but is obviously drafted by someone else and only signed by Ellsworth. That Vail was the one responsible is an assumption, but it is a reasonable assumption based on Vail's other actions. See Cornell's letter to MaryAnn above and the two following, all beginning frantically with queries about the misdirected Patent Office papers. While it would be difficult to prove that Vail was the one interfering with Cornell's mail, it certainly was not beneath him, and is only a logical conclusion given all the other shenanigans.

5. Alonzo Cornell, *True and Firm*, 87.

6. According to Cornell, Vail had drawn out the exact set of books he had requested before he had a chance to pick them up. Cornell, *True and Firm*, 88.

7. Samuel Morse to Secretary Spencer, December 27, 1843, MPLOC.

8. Morse to Cornell, December 28, 1843, MPLOC. Minute of instructions.

9. Ibid., Morse to Spencer, December 27, 1843.

10. Abstract "D," of salaries, March 10, 1844, MPLOC.

11. Unsigned letter datelined Washington, February 16, 1844, 1/8, ECPKL.

12. Silverman, *Lightning Man*, 230.

13. Nathaniel Philbrick, *Sea of Glory* (New York: Penguin, 2003), 334.

14. Ibid. Some are still in use today. To this point American seamen had relied almost entirely on European maps of the Pacific.

15. Ezra Cornell to MaryAnn, January 19, 1843, 1/8, ECPKL.

16. Ezra Cornell to MaryAnn, February 26, 1844, 1/9, ECPKL. This letter also contains the account of the tragedy onboard the USS *Princeton* that killed the Secretary of State Upshur and Secretary of the Navy Gilmer. It is clear from this letter that the so-called National Institute was by this time a functioning museum, open to the public.

17. Charles Munroe (Monroe) was one of the original investors in the Magnetic Telegraph Company and a witness to Morse's original patent. He would also figure in Cornell's attempts to patent the relay.

18. Ezra Cornell to MaryAnn, February 4, 1844, 1/8, ECPKL. Cornell often employs this tone with MaryAnn. His spelling is atrocious; e.g., "moline [malign] influences," "wright" instead of "write," "the Golden Rool," "geniel" instead of "genial."

19. *Report Upon the Condition and Progress of the U.S. National Museum* (Washington, D.C.: GPO, 1897), 313–22. (Correspondence between Abert, Wilkes and Ellsworth.)

20. William O. Craig, *Around the World with the Smithsonian* (Coral Springs, FL: Llumina Press, 2004), 16.
21. Ibid.
22. Cornell describes this room as being about 52' by 280' with marble floors that are kept immaculately clean. Ezra Cornell to MaryAnn Cornell, February 20, 1844, 1/9, ECPKL.
23. Richard Rathbun, *The Columbian Institute* (Washington, D.C.: GPO, 1917), 51–2.
24. In addition to establishing the term for patent extensions, the Patent Act of 1836 established the library of prior art.
25. Ezra Cornell to MaryAnn, February 4, 1844, 1/8, ECPKL.
26. Michael Faraday, *Experimental Researches in Electricity*, vol. 1 (London: Quaritch, 1839). It is not a certainty that Cornell took this particular volume from the Patent Office Library, but it is likely, as it was the most up-to-date standard reference on the subject, and certainly Page would have recommended it. It is certain that Vail was reading it at the time. Silverman, *Lightning Man*, 230.
27. Ibid.
28. As a somewhat humorous aside, Cornell had been invited by one of the naturalists, who had returned to the Patent Office to reassemble some of the bird specimens, to view his recently preserved collection of hummingbirds. As a result of the prolonged exposure to formaldehyde, Cornell came down sick with headaches and diarrhea. The naturalist involved was not Pickering, as he had left for Egypt by then. Cornell refers to him as "the gent." Ezra Cornell to MaryAnn, February 16, 1844. 1/8, ECPKL.
29. Cornell, *True and Firm*, 87.
30. Beck had evidently viewed this as dismissive. A chilly silence between him and Henry ensued that lasted for decades.
31. Ezra Cornell to MaryAnn, January 19, 1843, 1/8, ECPKL.
32. S.F.B. to Sidney Morse, January 13, 1844, MPLOC.
33. Dr. James Calvert, "The Electromagnetic Telegraph," accessed, January 2, 2015. http://mysite.du.edu/~jcalvert.
34. While Morse was in fact a professor, the nickname "the Professor" no doubt had originally carried with it a slightly derogatory taint, emphasizing his academic aloofness and impracticality.
35. Alonzo Cornell, *True and Firm*, 88–9.
36. Estimate for new #15 and #20 wire (undated, pencil marked 1843), MPLOC.
37. Burbank would become the subcontractor for the Baltimore-Philadelphia line. At one point (after the government declined to purchase the telegraph outright), he would offer to buy out Morse and the patentees entirely but this apparently falls through.
38. Samuel Morse to David Burbank, April 18, 1844, instructing him to resell the pipe, MPLOC. Once the cable was transferred to Burbank, Thomas Avery took over the task of stripping and reclaiming. David Burbank to Samuel Morse, May 10, 1844, MPLOC. Interestingly, from the receipts of the same day of Stabler and Canby and those to Burbank for cotton, it appears Morse was tarring and winding all the new cable that was going up overhead. It appears this was only the Serrell cable, as the unused Tatham cable was still in the Patent Office basement when Cornell was trying to cross the Hudson in November 1845.

Chapter 27

1. Ibid.
2. John W. Kirk, "Historic Moments: The First News Message By Telegraph," *Scribner's Magazine* 11 (1892): 654.
3. Initially Colt's funding and support came directly from the War Department. In 1842 he was awarded a congressional appropriation of $6,000 that Secretary Upshur threatened at one point to withhold. Lundeberg, *Colt*, 33–4.
4. S.F.B. Morse to Hon. Levi Woodbury. Dated only Washington 1843. Bound vol., 13 September 1843–13 January 1844, image 265, MPLOC.
5. Silverman, *Lightning Man*, 231, quotes Smith as telling Cornell, "I would take Professor Morse aside … tell him … how utterly worthless Vail is."
6. Leonard Gale to Samuel Morse, December 29, 1843, MPLOC.
7. Abstract "D," expenditures on account of salary. March 10, 1844, MPLOC. Though Gale had "resigned," as Morse would tell both Vail and Sidney, in fact he continued on the project as an unpaid advisor until Morse accumulated the funds again to rehire him officially.
8. See undated memo "estimate of carrying conductors on posts to the Capitol from the Patent Office." Bound vol., 13 September 1843–13 January 1844, image 304, MPLOC.
9. Letter of Alfred Vail, *New England Magazine* 4 (January 6, 1844): 457.

10. Clearly Morse was contemplating such a request. See the congressional vote tally sheet. Bound vol., 13 September 1843–13 January 1844, image 267, MPLOC.

11. Morse's state of mind is difficult to assess here. With the project on the brink of collapse, he devotes himself to coming up with plans for some kind of National Telegraph Agency with himself at the head. He would seem to have to be either delusional or absurdly optimistic given the circumstances. Bound vol., 13 September 1843–13 January 1844, images 311–16, MPLOC.

12. Denominated as "for experiments" in a memo of "monies wanted immediately." Bound vol., 13 September 1843–13 January 1844, image 272, MPLOC. (Undated but included with the January items.)

13. Proposed receipt of Bartlett, December 18, 1843, MPLOC. (Emphasis mine.)

14. Note of the Secretary of the Treasury endorsing new agreement with Bartlett, December 26, 1843, MPLOC.

15. Ibid.

16. Leonard Gale to Samuel Morse, December 29, 1843, MPLOC. In this lengthy letter Gale somewhat undiplomatically lays the ruin of his finances and his health at Morse's door. He then goes on to suggest a course that will clearly further cement this ruin, obviously to demonstrate his continued loyalty to Morse.

17. Ibid.

18. Ezra Cornell to MaryAnn, February 16, 1844, 1/9, ECPKL.

Chapter 28

1. Ibid., prior note, "proposed receipt" of December 18, marked on the back by Morse, "not acceded to by Mr. Smith."

2. December 20, 1843, MPLOC. Letter to be sent to the Secretary of the Treasury which was clearly drafted by Smith for Morse to sign and which was marked (subsequently) "doubtful if sent." The reasons why this letter was never sent are complex and not as simple as Morse's notation that Smith had objected to its contents. The letter is the first official notice of Morse's going to an above-ground pole solution, which was likely to raise all kinds of new questions for which Morse, as yet, had no good answers.

3. Draft of letter dated only "? 1843" to Smith. Bound vol., 13 September 1843–13 January 1844, image 291. "Statement in regard to lead pipe contract of F.O.J. Smith," also undated, MPLOC.

4. Samuel Morse to Secretary Spencer, December 23, 1843, MPLOC.

5. Secretary Spencer to Samuel Morse, December 26, 1843, MPLOC.

6. Samuel Morse to Secretary Spencer, December 27, 1843. This was in part a sop to Smith, as Smith had been promising this position to Cornell for months.

7. F.O.J. Smith to Tatham Bros., December 26, 1843, MPLOC.

8. Ibid.

9. Samuel Morse to James Fisher, December 29, 1843, MPLOC. Marked "Private." (First emphasis in the original; second emphasis mine.)

10. The undated letter drafted by Smith apparently for Morse to sign demands an additional $4,472 to close out the contract. Apparently it was never sent. Bound vol., 13 September 1843–13 January 1844, images 275–6, MPLOC.

11. Henry Ellsworth to Samuel Morse. Bound vol., 13 September 1843–13 January 1844 (first two pages missing), images 293–6, MPLOC.

12. Smith's memorandum of savings on the Tatham contract, November 25, 1843. Bound vol., 13 September 1843–13, January 1844, images 157, 161 and 300, MPLOC. The last image has a notation regarding the savings: "$606.49 being Smith's and $606.49 being mine to be credited to the United States *if payments are completed*" (emphasis mine).

13. Edward Lind Morse, ed., *Samuel F.B. Morse: His Letters and Journals*, vol. 2, 213.

14. Morse now fully understands that Smith had set a snare purposefully to discredit him, putting himself in a position to blackmail him for payment on the Bartlett contract. S.F.B. to Sidney Morse, December 30, 1843, MPLOC.

15. Smith's motives may not have been entirely selfish, as it appears Bartlett was suffering reverses in his farming operations in this period, and that following summer would lose his entire potato crop to the blight. J.E. Teschemacher, Extract of Letter, *The Farmer's Magazine* 13, series 2 (no date): 530.

16. S.F.B. Morse to Sidney Morse, December 30, 1843, MPLOC.

17. Ibid.

18. Alfred Vail, *The New England Magazine* 4 (January 1, 1844): 457.

19. Secretary of the Treasury Spencer to Samuel Morse, December 29, 1843, MPLOC.

20. Ibid.
21. Ibid.
22. Samuel Morse to Secretary Spencer, January 3, 1844, MPLOC.
23. Secretary Spencer to Samuel Morse, January 9, 1844, MPLOC.
24. Smith's rewritten memorial to Congress, dated January 7–15, MPLOC.
25. See Smith's memorial to Congress cited above. Morse's of January 3, 1844, declining to join Smith's appeal to the president (on back). Morse declining to join Smith's appeal to Congress, January 12, 1844 (noted on the back). The embarrassment is evidently too much, as Morse tells Smith he will only have anything further to do with him in writing, MPLOC.
26. Morse's note on the back of copy of Smith's appeal to Congress. January 12, 1844, MPLOC. Appeal signed by Bartlett, dated January 15, 1843, MPLOC.
27. Samuel Morse to the Secretary of the Treasury Spencer, January 7, 1843, and January 17, 1843 (two letters), MPLOC.
28. Says "received of F.O.J. Smith by the hands of Professor Morse" Voucher for $1653.47 marked "For closing the Tatham contract." Dated January 22, 1843, MPLOC. This would not be the end of the matter as far as Smith was concerned.
29. S.F.B. Morse to Sidney Morse, December 30, 1843, MPLOC.
30. S.F.B. Morse to Sidney Morse, January 13, 1844, MPLOC.
31. Silverman, *Lightning Man*, 231.
32. Ibid., 100. Silverman suggests that there was an ongoing romantic attraction between him and Annie. The evidence supports this. Morse, as his other letters indicate, was ill and emotionally vulnerable at the time. On February 7, 1844, the anniversary of his wife Lucretia's death, he writes a letter to Annie Ellsworth, enclosing a gift of a book of etiquette and a Bellini score which he refers to as "my favorite Bellini." It is very likely that the score was to *Romeo and Juliet*, as Morse had attended this performance in Rome. Clearly the timing and content of the gift were evocative of an unspoken romantic attraction.

Chapter 29

1. Fisher had already told him why: because Morse had changed the test conditions. The fact that Morse would not or could not acknowledge this seems indicative of some kind of psychological break.
2. Gale to Samuel Morse, December 29, 1843, MPLOC.
3. James Fisher to Samuel Morse, January 1, 1844, MPLOC. Fisher had warned Morse that it was possible that no amount of insulation in a long underground cable would prove effectual in preventing short circuits. Having written Henry on this same subject, he had apparently shown Morse Henry's reply back in April.
4. James Fisher to Joseph Henry April 15, 1843, and Henry's reply, April 17, POJH, vol. 5, 321–3.
5. James Fisher to S.F.B. Morse January 1, 1844, MPLOC. (Emphasis mine.)
6. Henry's reply, April 17, POJH, vol. 5, 321–3.
7. Fisher's approval, dated November 15, 1843, for ten miles of pipe, MPLOC.
8. The question seemed to continue to obsess Morse, partly due to guilt over Fisher, but also probably because he realized he would at some point have to cross large bodies of water using some kind of conduit.
9. Samuel Morse to Joseph Henry, February 7, 1844, MPLOC. At least Henry seems to have answered the question Gale posed as to whether the burning of the insulation was due to conduction or induction. Unfortunately, his answer was wrong. Events would prove the singeing was due to Tatham's hot annealing manufacturing process and had nothing to do with the current passing through. Morse evidently did not realize this until January. POJH, vol. 6, 21–2. (Here again we see Morse's psychological tic of framing a set of numbered interrogatories when he is in a highly disturbed emotional state.)
10. Leonard Gale to Morse, December 29, 1843, MPLOC.
11. Charles G. Page, "On the Probable Conduction of Galvanic Electricity through Moist Air," *AJS*, vol. 52 (1846): 204–9. This most likely took place on January 22, 1844, and was the reason for Morse's summons to Cornell is preserved in the Cornell papers, 1/8, ECPKL. In the same issue of Silliman's is Page's article on galvanic induction.
12. "On reading your letter on the subject of the Telegraph in the newspapers I was struck with the idea that you had probably met with the very difficulty my researches have led me to anticipate." Joseph Henry to S.F.B. Morse, January 24, 1844." MPLOC.
13. Henry's letter of January 24, 1844, offering to discuss with Morse Henry's new

discoveries regarding insulators. Morse does not bother to reply in writing until he has his own (Cornell and Vail's) version of the pole insulator designs in hand, only then seeking Henry's opinion in person. It seems in this case Henry was using Morse as a sounding board for his own as yet unformulated theories regarding the likelihood of voltaic potentials causing short circuits, a problem that had plagued Henry ever since he had produced the first overwound electromagnet, and which thus far he had failed to answer. But Morse, as usual, probably took Henry's unformed theory as gospel.

14. *Ibid.* This is a long letter describing Henry's work with insulators. He refers to Morse's article in the newspapers regarding the insufficiency of insulation as the cause for the wires' burning up. Morse had admitted to a problem with the insulation in the papers (see note above). Henry hypothesized it was due to a short circuit based on the length of the wire. His theory was that the voltaic potential between wires in proximity increased as the length of the wire increases. Also see Gale to Morse of December 29, 1843, MPLOC. Gale in this letter throws Fisher "under the bus," saying, "I immediately went to Dr. Fisher ... to take charge of my business ... but to my inexpressible surprise and disappointment I found he was already engaged in other business."

15. John Holt, "The Earth as an Electrical Conductor," *The Electrical World* 24, no. 12, 290. Wheatstone was using bare iron wires on an overhead pole with a ground return, the method having been discovered first by Steinhall in 1837.

16. Joseph Henry to S.F.B. Morse, January 24, 1844, MPLOC. This apparently unprompted letter from Henry was probably as a result of Page's informing him of the experiment on the roof of the Patent Office. Ironically, it would confirm exactly what Fisher was saying here, and for exactly the same reasons. Henry's assumptions here turned out to be wrong (as pointed out above), but clearly the result influenced Morse sufficiently that he committed fully to the overhead solution.

17. Ezra Cornell to MaryAnn, February 4, 1844, 1/8, ECPKL.

18. Burbank would supply the chestnut poles for most of the line. Morse had advertised for poles on February 7, but there is a possibility this may have been for the line inside the city. Apparently Morse bought the first batch from someone else. David Burbank to Samuel Morse, March 9, 1844, MPLOC.

19. Samuel Morse to Joseph Henry, February 7, 1844, MPLOC.

20. Samuel Morse to Joseph Henry, February 7, 1844, MPLOC. The tone of this letter is far less deferential than preceding ones. He virtually orders Henry to meet with him on the date he desired.

21. POJH, vol. 5, 22n; *The Daily Madisonian*, February 6, 1844, advertisement for sealed bid. Seven hundred poles (at 200 feet apart) would have been enough for approximately 27 miles, indicating that Morse intended using the underground line from Relay.

Chapter 30

1. Robert Post, *Physics, Patents and Politics* (New York: Science History Editions, 1976), 66. Evidence is Morse had considered the possibility in 1842, but by 1843 had rejected it for reasons that had to do with Henry's erroneous speculations on atmospheric (celestial) magnetism.

2. Anton Huurdeman, *The Worldwide History of Telecommunications* (Hoboken: John Wiley, 2003), 67.

3. Cornell, Ezra. 1844. "Machine for Cutting Trenches and Laying Pipes." U.S. Patent #3456, issued February 28, 1844.

4. Ezra Cornell to MaryAnn, January 19, 1844, 1/8, ECPKL.

5. *Ibid.* That this is the magnetic relay becomes obvious from later letters to Smith when he refers to conversations they had at this time regarding it.

6. Calvert, "The Electromagnetic Telegraph." http://mysite.du.edu/~jcalvert.

7. Taylor, *An Historical Sketch of Joseph Henry's Contribution*, 97. Morse claims there was only one visit to Henry, which contradicts both Cornell and Henry's recollections. Both say that there was a second visit on March 1, and it was on the second visit apparently that Henry expressed his preference for Cornell's design for the insulator.

Chapter 31

1. Samuel. Morse to Ezra Cornell, note dated at the bottom, January 22, 1844, MPLOC.

2. Two letters, Ezra Cornell to MaryAnn Cornell, October 4, 1843, datelined Portland, and October 29 datelined Baltimore, 1/8. ECPKL.

Notes—Chapter 31

3. Letter from Edward to Ezra Cornell starting "Dear Brother," November 30, 1843, 1/8, ECPKL.
4. S.F.B. Morse, voucher dated November 10, 1843, delivered under cover of the letter of appointment, MPLOC.
5. Ezra Cornell to MaryAnn, October 29, 1843, 1/7, ECPKL.
6. Ezra Cornell to MaryAnn, January 1, 1844, 1/8, ECPKL.
7. Ibid.
8. Secretary of the Treasury John C. Spencer.
9. Ibid.
10. Ibid.
11. Ezra Cornell to MaryAnn Cornell, January 19, 1844, 1/8, ECPKL.
12. Ibid.
13. Ezra Cornell to MaryAnn, February 4, 1844, 1/8. ECPKL.
14. Ibid.
15. Ezra Cornell to MaryAnn, February 4, 1844, 1/8. ECPKL. By the 24th he is rooming with someone named "Monroe" (Charles Monroe?) not found on pay vouchers. Monroe will assist Cornell later in his efforts to patent the relay. He was one of the two witnesses to Morse's original patent (1838) and one of the first investors in the Magnetic Telegraph Company. Apparently he was a Maine clockmaker. Monroe, like Cornell, was probably either a Quaker or ex-Quaker, as he begins his letters "Friend Cornell."
16. Ezra Cornell to MaryAnn, February 16, 1844, 1/9, ECPKL.
17. Ezra Cornell to MaryAnn, March 17, 1844, 1/9, ECPKL.
18. This is the first red flag raised with regard to Morse's rather broad interpretation of his employment contract (not found), when it came to intellectual property.
19. Ezra Cornell to MaryAnn, May 13, 1844, 2/1, ECPKL.
20. Ibid. Apparently he would only be successful in selling some interest to Smith. Silverman says he bought a half-interest. *Lightning Man*, 234. He likely paid for it in promises, as he did everything else.
21. See Morse's sketch of the cable layer on the back his letter to James Fisher of December 2, 1843, MPLOC. There is no other good reason for it to appear here.
22. "Annual Report of the Commissioner of Patents for the year 1843," 28th Congress, Senate document 150, 246–7. By way of contrast, Cornell's patent for the magnetic relay would receive a single line in the report of 1845. The mention was entirely at Ellsworth's discretion, as the patent had not been issued by the time of the report.
23. Ibid., 246. See Vail's description (*New England Magazine* 4, 457).
24. Ezra Cornell to MaryAnn, February 16, 1844, 1/9, ECPKL.
25. Cornell had the opportunity to see Daniel Webster argue before the Supreme Court in the Stephen Girard case. There is an interesting citation of Webster's in response to arguments citing "the case of the Jews" when making bequests for religious schools. He cites a case in British Common Law in which the court ruled that the bequest to establish a "Talmud Torah," a Jewish religious school, was not granted charitable status, because it promoted religion. Curiously, the precedent seems 180 degrees opposite to the point Webster is trying to make; nevertheless, he introduces this in context, apparently solely to dispose of "the case of the Jews." Here he cites Sheldon, *On Mortmain and Charitable Uses*, 8vo (London: 1836), 105.
26. Webster's case is *Vidal v. Girard*, 43 U.S. 127 (1844).
27. Ezra Cornell to MaryAnn, February 16, 1844.
28. "Schenectady Cabinet," *Philadelphia Chronicle*, March 7, 1844.
29. It was Alvin's second intended victim, his sister Deborah Cornell, who had appealed for commutation to Governor Bouck on the grounds of insanity. Bouck had handed the case to Beck and Amariah Brigham, Superintendents of the Utica Insane Asylum. On February 22, Beck had concluded that Cornell was guilty, but asked that additional witnesses who had not come to trial be deposed. In light of their testimony, Beck concluded Alvin was insane and the sentence was commuted. Theodoric Romeyn Beck and Amariah Brigham, "Analysis of the testimony on the Trial of Alvin Cornell for Murder, and of the subsequent proof which led to the Commutation of his Punishment," reprinted in *Transactions of the Medical Society of the State of New York* 6 (1846).
30. "Analysis of the Testimony on the Trial of Alvin Cornell," 215.
31. "The Reprieve of Cornell," *Brooklyn Daily Eagle*, April 30, 1844; *Philadelphia Chronicle*, April 30, 1844; *Baltimore Sun*, May 1, 1844; *New York Tribune*, April 29, 1844.
32. Ezra Cornell to MaryAnn, April 14, 1844, 1/10, ECPKL, chiding her for not writing.

33. *New England Magazine* 4; letter of Alfred Vail, March 19, 1844, 457. Vail says they reached Beltsville on the 19th with the poles, but probably he means that they were not set up, only lying on the ground next to the tracks. See Morse's letter of March 18, 1844, instructing Cornell to reach Beltsville by the end of the week.
34. Morse's note April 27, 1844, marked "rainy and cold," MPLOC.
35. Ezra Cornell to MaryAnn, April 14, 1844, 1/10, ECPKL.
36. Ezra Cornell to MaryAnn, May 2, 1844, 1/10, ECPKL.
37. It appears the telegraph was initially set up at the Capitol and later moved to the Supreme Court for the public demonstration on May 24. Samuel Morse to D. Aying(?), May 4, 1844, MPLOC.
38. Samuel Morse to Alfred Vail, May 1, 1844, MPLOC.
39. *Scribner's Magazine* 11 (1892): 654.
40. Ezra Cornell to MaryAnn, May 13, 1844, 2/1, ECPKL.
41. S.F.B. Morse to Sidney Morse, May 7, 1844, MPLOC.
42. Samuel Morse to Alfred Vail, recopied, May 14, 1844, MPLOC.
43. Ezra Cornell to MaryAnn, May 13, 1844, 2/1, ECPKL.
44. Receipt, datelined Annapolis Junction, May 20, 1844, Alex Sumerall, MPLOC.
45. Receipt dated May 23,1844 to John Huxley, witnessed by Cornell, and Mathew Lane dated May 22, 1844, MPLOC.
46. The message sent over the line—"What Hath God Wrought," scripted by Annie Ellsworth—was actually the second transmission.
47. Cornell, *True and Firm*, 91.
48. *The Madisonian*, May 27, 1844.
49. Ezra Cornell to MaryAnn, January 19, 1844, 1/8, ECPKL.
50. Receipt, June 4, 1844, bonus for completion of project, MPLOC.
51. *Journal of the House of Representatives* 28, Issue 1 (June 6, 1844): 1009. Morse's letter was referred to the Commerce Committee.

Chapter 32

1. Silverman, *Lightning Man*, 227.
2. Samuel Morse to Secretary of the Treasury George Bibb, January 28, 1845, MPLOC.
3. Samuel Morse to Henry Rogers, January 28, 1845. Smithsonian, Special Collections (Dibner), MSS 001777 A
4. Prime, *Morse*, 511. *Journal of the House of Representatives* 28, Issue 1, 999. Senate bill dated February 22, 1845, MPLOC, authorizing the extension of the telegraph to New York. It did not pass.
5. Samuel Morse to Cave Johnson, PMG, March 14, 1845, MPLOC. Refers to Rogers as "brother of my assistant." It is clear from other sources they were related.
6. Rogers reply to Morse, January 29 (referencing a "private note"), Morse to Secretary Bibb, January 30, 1845. Bibb to Morse, January 30, 1845. Letter from Secretary of the Treasury approving paying both Vail and Rogers out of the unexpended funds, MPLOC.
7. Samuel Morse to F.O.J. Smith, April 29, 1844, and back note on letter to Vail of May asking Gale for Power of Attorney to act on behalf of the patent owners, MPLOC.
8. Alfred Vail to Samuel Morse, June 3, 1844, MPLOC.
9. Samuel Morse to F.O.J. Smith, June 10, 1844, MPLOC.
10. *Ibid*.
11. Harlow, *New Wires and Old Waves*, 84.
12. Samuel Morse to Leonard Gale, March 11, 1845, MPLOC.
13. *Ibid*.
14. Amos Kendall to Samuel Morse, March 4, 1845, MPLOC.
15. Kendall owned slaves himself.
16. Amos Kendall to Samuel Morse, March 14, 1845, MPLOC.
17. Samuel Morse to Amos Kendall, April 16, 1845, MPLOC. This letter obviously crossed in the mail with Kendall's of the same day, saying he had failed totally.
18. Amos Kendall to Samuel Morse, April 16, 1845, MPLOC.
19. Silverman, *Lightning Man*, 261.
20. Oliver Larkin, *Samuel F.B. Morse and American Democratic Art* (New York: Little, Brown, 1954), 155.
21. This was paid out in installments from his salary. Amos Kendall to Ezra Cornell, July 13, 1845, and Ezra Cornell to MaryAnn, October 14, 1845, 2/11. ECPKL.
22. Prime, *Morse*, 514.
23. *Washington Intelligencer*, February 25, 1869. "Gentlemen I observe in your paper of this morning a letter from Hon. Amos Kendall to W.W. Corcoran Esq. complimenting him on being the first subscriber of $1,000 to form a company in May 1845 to erect a line of

telegraph between the cities of New York and Philadelphia." His collection would become the Corcoran Gallery.

24. *Ibid.* This may have been cash or more likely may have been a buyout of the Bartlett contract.

25. Greenough was an old friend of Morse's and would act as his attorney henceforth on all patent matters.

26. Post, *Physics, Patents and Politics*, 48, 70. Keller had resigned as patent examiner in 1844 to partner with J.J. Greenough. Greenough was the nephew of Horatio Greenough, sculptor, an old friend of Morse's who had done a bust of Morse.

27. Prime, *Morse*, 513.

28. Vail, *Early History of the Electro-Magnetic Telegraph*, 22.

Chapter 33

1. Ezra Cornell to Samuel Morse, July 12, 1849. Defendants Proofs, 22–6. http://babel.hathitrust.org/cgi/pt?id=nyp.33433020656090;view=1up;seq=9, last accessed November 6, 2015, 21–2. In his deposition, Cornell would complain about this treatment he received from Morse after all his loyalty. NYPL.

2. Barnum had coined the phrase "Never give a sucker an even break." He was a strong proponent of the principle that where you have nothing concrete to offer, baffle them with "BS."

3. Pro Forma ad for Boston at Harper's PianoForte rooms and Harding's School, undated, image #149, 2/4, ECPKL. Cornell was now billing himself "Late Superintendent of Construction for Morse's E.M. Telegraph."

4. Samuel Morse to Ezra Cornell, December 3, 1844, ECPKL.

5. Orrin Wood to Ezra Cornell, January 2, 1845, 2/5, ECPKL.

6. Samuel Morse to Ezra Cornell, December 3, 1844, 2/4. ECPKL. Morse's reply to these charges referring to Cornell's letter of November 23 (not found).

7. F.O.J. Smith to Ezra Cornell, December 3, 1844, 24, ECPKL.

8. Ezra Cornell to Orrin Wood, January 11, 1845, 2/5, ECPKL.

9. Jeremiah Beebe to Ezra Cornell, January 27, 1845, 2/5, ECPKL. Beebe had been affluent, but had suffered significant losses in the panic of 1837, when he had been forced to sell off most of his holdings and businesses. By February of the following year, Cornell had still not repaid Beebe, and Beebe started this letter bluntly, "Will you send me the money or must I collect it?" Jeremiah Beebe to Ezra Cornell, February 9, 1846, 3/7, ECPKL.

10. Ezra Cornell to MaryAnn, July 6, 1845, ECPKL. 2/9, describes his attempts to fend off Beebe.

11. Ezra Cornell to Samuel Morse, July 12, 1849. Defendants Proofs, 22–6.

12. F.O.J. Smith to Colonel J.L. Graham, January 9, 1845, 2/5, ECPKL.

13. O.S. Wood to Ezra Cornell, January 2, 1845, 2/5, ECPKL.

14. Undated, image #160, 2/4, ECPKL.

15. O.S. Wood to Ezra Cornell, January 2, 1845, 2/5, ECPKL.

16. David Burbank to Ezra Cornell, January 15, 1845, 2/5, ECPKL.

17. David Burbank to Ezra Cornell, February 24, 1845, 2/6, ECPKL.

18. Orrin Wood to Ezra Cornell, March 1, 1845, 2/6, ECPKL. The letter does not specify the usage, but considering Cornell's avid interest and support for the Ithaca fire department and subsequent events, this was the likely application.

19. Referred to in Morse's reply of the 17th. Not found.

20. Samuel Morse to Ezra Cornell, March 17, 1845, 2/6, ECPKL.

21. *Miscellaneous Documents of the House of Representatives* (Washington, D.C.: GPO, 1891), Under Congressional Edition, vol. 2788, 651. This had been tacked onto the diplomatic appropriations bill and no action was taken regarding the purchase.

22. Not found. See reference in following note of April 10.

23. Morse was in Baltimore making arrangements to move the telegraph from the rail station to the Post Office. Cornell was in New York at the time. Curiously, there is a payment receipt, supposedly signed by Cornell, dated April 8, Washington (two days before the letter denying payment), MPLOC.

24. F.O.J. Smith to Ezra Cornell, April 24, 1845, 2/7, ECPKL.

25. Even though Wood's position as assistant station master had been approved by Cave Johnson, in mid–March, Morse had yet to pay him. In Morse's summary statement from March 22 to June 1, there are three lines of salary, which are obviously Morse, Vail and Rogers, and one more at $0, which is obviously Wood (Tanzinger had been let go). This is the summary. The abstract itself is missing. Possibly Wood

agreed to forego salary in part for the rights to the Albany-Buffalo line, as next we see him in Utica in the employ of Theodore Saxton Faxton, who has that contract, MPLOC.

26. Ezra Cornell to Samuel Morse, April 17, 1845, 2/7, ECPKL.

27. Samuel Morse to Ezra Cornell, May 5, 1845, 2/7, ECPKL. The offer was insulting. To show how little he cared for Cornell's opinion, rather than drafting a new letter, Morse had simply crossed out the first figure of $10.50 per day.

28. Receipt for $15 dated May 19, 1845, MPLOC.

29. J.J. Speed to Ezra Cornell, May 17, 1845, 2/7, ECPKL.

30. See Morse's diagrams of his apparatus, April 27, 1844, MPLOC. On Morse's test-line diagrams, the two wires are clearly marked the "East" and "West," indicating he was still using a separate sending and receiving line.

31. Though he initially (in his later testimony to the Supreme Court) claimed otherwise, Morse at this date was clearly using two separate wires for sending and receiving and continued to do so until at least the end of 1844. Eventually, once the newer relay was installed, the second line was utilized for sending redundant messages, for which there was a slight extra charge (25 cents).

Chapter 34

1. Cornell, *True and Firm*, 91.

2. At some point Cornell's relay had been added to the Baltimore-Washington line. See O.S. Wood to Ezra Cornell, November 19, 1845, 3/1, ECPKL, in which Wood complains about the relays: "The relay magnets get out of adjustment too easily. Much worse than those at Balt. and Wash."

3. Contract of Amos Kendall and F.O.J. Smith with Ezra Cornell on behalf of the Magnetic Telegraph Company, dated May 29, 1845, 2/7, ECPKL.

4. Ibid.

5. The autodidact Cornell had conceived this improvement in February 1844 while at the Patent Office. His letters to his wife suggest he no doubt had realized immediately the implications of the device for commercial application. Obviously he had confided in Smith at the time, as when he later got in a wrangle with Morse over the relay, he adduced Smith's testimony that he had first conceived it while ensconced in the Patent Office. The benefits of what Morse would come to call "marginal circuits" in general (which we discuss in some detail later), when they did become apparent to Morse, would provoke a reaction that provides significant insight into Morse's personality. Morse would desperately seek control from the summer of 1845 until 1846, when at last he managed to substitute a device supposedly of his own design to replace Cornell's.

6. Ezra Cornell to Sylvester Munger, August 18, 1845, 2/10, ECPKL, ordering four magnets "like the one I showed you in June last."

7. "Instructions," Amos Kendall to Ezra Cornell, May 30, 1845, 2/7, ECPKL.

8. Ezra Cornell to MaryAnn, June 16, 1845, 2/8/, ECPKL.

9. Amos Kendall to Ezra Cornell, June 15, 1845, 2/8, ECPKL.

10. Amos Kendall to Ezra Cornell, July 14, 1845, 2/9, ECPKL.

11. Amos Kendall to Samuel Morse, June 7, 1845, MPLOC.

12. *American Telegraph Magazine* 1, no. 1 (1852): Appendix, Deposition of Henry of O'Reilly's Title.

13. Amos Kendall to Samuel Morse, August 5, 1845, MPLOC. Kendall mentions that the New York to Buffalo line had already raised $40,500 in subscriptions.

14. Howell and Tenney, *History of the County of Albany*, vol. 1 (New York: Munsell, 1886), 319–24. Butterfield, Faxton, and Livingston were all associated with the Albany to Buffalo Express. Butterfield would later create the famous western Butterfield Overland Stage, and his company would be merged into American Express. Henry Wells would become the Wells of Wells Fargo. This is not entirely accurate, as Wells and Livingston also had a partial interest in the Albany-Buffalo line. Also Colt, once he (apparently) finished the Coney Island line, went to Boston to help Smith with the Boston line. Samuel Colt to Ezra Cornell, November 27, 1845, 3/2, ECPKL.

15. Burbank had agreed to supply Morse five hundred chestnut poles at ninety-eight cents each, and then disposed of the lead pipe for him at the same price, rendering the switch to overhead poles virtually cost-free This was possibly his reward for that favorable deal.

Chapter 35

1. Unsigned from Smith to Cornell, November 16, 1845, 3/1, ECPKL. Smith mentions

Davis as the contractor for the magnets that Page and Kendall had preferred. Reid mentions Clark & Son of Philadelphia, but these were made in 1846 and were probably on the Breguet model.
 2. http://www.telegraph-history.org/manufacturers mentions J. Burritt and son in Ithaca. Cornell was using Munger for the relays and the registers, and probably also employed the workshop of James Eddy. Cornell had met with Eddy in New York in June (see letter to MaryAnn of June 16, 1845, 2/8, ECPKL, carried by Eddy). Eddy ran a workshop in Ithaca. F.O.J. Smith's letter of June 18, 1845, to Cornell shows he already had charge of approving any equipment purchase. 2/8, ECPKL. Speed was a merchant and a presidential elector, but also an inventor, having invented an astronomical clock in 1847 (mentioned in Silliman's). Kendall's letter of September 29 confirms this. Amos Kendall to Ezra Cornell, September 29, 1845, 2/10, ECPKL.
 3. Ezra Cornell to Charles Grafton Page, October 15, 1845, 2/12, ECPKL.
 4. *Documents of the Senate of the State of New York*, vol. 6 (Albany: Argus, 1872), 115.
 5. James Reid, *The Telegraph in America* (New York: Polhemus, 1886), 302.
 6. John Butterfield to Ezra Cornell, August 6, 1845, 1/10, ECPKL.
 7. *Ibid.*
 8. Ezra Cornell to S. Munger, August 18, 1845, 2/10, ECPKL.
 9. Ezra Cornell to MaryAnn, September 11, 1845, 2/10, ECPKL.
 10. Ezra Cornell to MaryAnn, September 19, 1845, 2/10, ECPKL.
 11. Ezra Cornell to Charles Page, October 15, 1845, 2/11, ECPKL.
 12. *Ibid.*
 13. *Ibid.*
 14. Charles Grafton Page to Ezra Cornell, October 18, 1845, 2/12, ECPKL.
 15. *Ibid.*
 16. Cornell to Page, October 15, 1845, 2/12, ECPKL.
 17. Ezra Cornell to Charles Page, October 20, 1845, 2/12, ECPKL. Appears to be in Kendall's hand.
 18. Ezra Cornell to MaryAnn, October 30, 1845, 2/12, ECPKL. "I sent my papers on to Washington this day for a patent for improvements to the Tel[egraph] ... my arrangement will be to give the partners half to get it patented abroad."
 19. Charles Monroe to Ezra Cornell, November 6, 1845, 2/13, ECPKL. The letter had also contained a call for Kendall's removal as head of the company.
 20. Ezra Cornell to Amos Kendall, November 26, 1845, 3/2, ECPKL.
 21. O.S. Wood to Ezra Cornell, October 19, 1845, 2/12, ECPKL.
 22. Wood to Cornell, October 30, 1845, 2/12, ECPKL.
 23. O.S. Wood to Ezra Cornell November 1, 1845, 2/12, ECPKL.
 24. O.S. Wood to Ezra Cornell, November 14, 1845, 3/1, ECPKL.
 25. Obviously there was some marital strife going on at the time, as Cornell had signed his letter of September 30 and October 1, "Faithfully," which he never had done before. Also, his letters of October have constant references to his definition of a "virtuous woman." ECPKL.
 26. Frederick Hills, *New York State Men* (Albany: Argus, 1910), 54.
 27. *Ibid.*
 28. Conducted jointly by T.R. Beck and Amariah Brigham, the latter of whom was Superintendent of the Utica Asylum.
 29. So, eager as he is to see the telegraph line completed, Ezra Cornell is also relying on Wood in several respects; not only to watch out for his economic interests but to act as proxy for the family name. As the telegraph inches closer and closer to the actual site of the Alvin Cornell murder, this seems ironically to dispel the idea of a familial curse. In a psychological sense, the instrument of Cornell's potential salvation becomes the same as that of his potential destruction.

Chapter 36

 1. S.F.B. Morse to Sidney Morse, March 28, 1846, MPLOC. Long wail to Sidney about how God had asked him to sacrifice his son Isaac (painting) so he could pursue a higher calling temporarily.
 2. Charles Chapman to S.F.B. Morse, May 16, 1845, MPLOC.
 3. Certificate dated June 12, 1845, MPLOC.
 4. Prime, *Morse*, 604.
 5. F.O.J. Smith to Samuel Morse, June 11, 1845, MPLOC.
 6. As Morse was well aware, Wheatstone and Cooke's telegraph was already well established in England, and the French had their own system based on railroad semaphores.

7. Smith's offer was good for one month and dated March 1, 1845, MPLOC.
8. Samuel Morse to F.O.J. Smith, July 16, 1845, MPLOC.
9. Vail, *Early History of the Electro-Magnetic Telegraph*, 20.
10. Alfred Vail to Samuel Morse, August 24, 1845, MPLOC.
11. Grant of rights to the telegraph in Cuba, July 31, 1845, MPLOC.
12. Acknowledging that the business reasons were bogus, Morse's son Edward Lind says the sole purpose of the trip was for Morse to satisfy himself that his telegraph was superior to the European one. While Morse was unquestionably a man of monstrous ego, he was certainly more intelligent than his son gives him credit for. Edward Lind Morse, ed., *Samuel F.B. Morse: His Letters and Journals*, vol. 2, 257.
13. Samuel Morse to Louis Breguet, October 27, 1846, MPLOC. This letter lays out explicitly the chain of events but substitutes himself (Morse) for Cornell as the originator of the relay, which he describes as weighing close to 300 English pounds.
14. Post, *Physics, Patents and Politics*, 165.
15. There is a glaring historical elision here which needs to be corrected. It was Cornell's device that made Henry's discovery practicable, not the other way around. The contention in the years following would center solely around the conflict between Morse and Henry (and peripherally Wheatstone and what weight of credit should be allotted to each of them). Cornell's part was entirely discounted. When Page finally came to challenge Morse's claim in 1860, he would cite only Wheatstone as a prior inventor. As mentioned, through the machinations we lay out in greater detail in the next chapter, the invention of the magnetic relay, like the telegraph itself, would in the end be attributed to Morse and Morse alone.
16. Cornell indicates in his patent that Morse's or any other telegraph system theoretically could be appended to the relay. Cornell, Ezra. 1845. "Improvement in the Mode of Operating Electro-magnetic Telegraphs." U.S. Patent #4318, issued December 20, 1845.
17. William Taylor, *An Historical Sketch of Joseph Henry's Contribution to the Electro-Magnetic Telegraph* (Washington, D.C.: GPO, 1879), 65. (Emphasis mine.)
18. L.J. Davis, *Fleet Fire* (New York: Arcade, 2003). Kindle, https://books.google.com/books?id=Gp_rW9f-Tp4C&pg=PT41&dq.

19. There can be no doubt that this was his intent from the beginning. See Morse's letter to Louis Breguet of October 27, 1846, MPLOC.
20. Articles of Agreement between Samuel Morse and Edward Kent, November 30, 1846, MPLOC.

Chapter 37

1. Henry, in his 1849 deposition, refers to Morse's use of a "secondary circuit" after his return from Europe in 1838, but this is misleading, as a truly independent circuit was not introduced until late in 1845, and this latter device was the one based on Cornell's relay. Morse would later admit that what Henry was referring to in the deposition was his in-line repeaters. Hochfelder says (in the online article) Vail "finessed his explanation ... in order to avoid describing the receiving magnet and local circuit."
2. Taylor, *An Historical Sketch of Henry's Contribution*, 50 (fn).
3. Samuel Morse to Ezra Cornell, December 3, 1844, ECPKL. Even at this late date Morse felt obliged to defend himself to his former employee, Ezra Cornell, and he did so not by just dismissing the charges out of hand as someone confident of his facts might have, but by reviewing the entire fracas in great detail.
4. James C. Fisher to Samuel Morse, May 19, 1845, MPLOC.
5. Joseph Henry to Charles Wheatstone, February 27, 1846, POJH, vol. 6, 385.
6. William Rhees, *Smithsonian Miscellaneous Collections*, vol. 21 (Washington, D.C.: Smithsonian Press, 1881), 382.
7. *AJS*, vol. 31, no. 1 (1837).
8. *Harpers New Monthly Magazine* 55 (1877): 585.
9. A peripheral circuit had been added at some later point, but it had been omitted in any public description of the apparatus for reasons that will become clear.
10. In fact, it had been established by then or shortly thereafter, that plain, uncoated iron wire would serve as well as copper. Faxton would finance the New York-Albany line by selling all the copper left over from the Buffalo line and buying iron wire.
11. Morse would later go to great lengths to attempt to claim he had conceived of the relay back in 1837 but had chosen not to implement it. Gale would swear to his claims in several

hearings and cases. It is evident from subsequent events that this claim was false.

12. It is fair to say Henry had a highly developed working concept of voltaic potentials, but not much theoretical understanding beyond that of Ohm's law.

13. Receipt for $45, Morse to Charles Page, dated July 8, 1845. Marked by Morse on back, "magnet." Page later claimed that he never charged Morse for the use of his magnets, but Vail's later testimony to the Supreme Court, as well as Morse's own invoices, indicate differently. Mabee, *American Leonardo*, 413 (fn). Kendall had informed Cornell of this, but in the interim Page had stopped making his version of the relay. Ellsworth's letter noted below indicates Page was still in New York on July 7.

14. Charles Page, "Notice of a Spiral Magnet" and "Description of a new plate, or quantity, Helix for Electro Magnetic Apparatus," *Journal of the Franklin Institute*, third Series, vol. 3 (1842): 166–7 and 167–8 (cited Post, *Physics, Patents and Politics*, 194–5). (This journal was the unofficial organ of the Patent Office.)

15. Morse apparently had been laboring under the same misapprehension until Page corrected it.

16. If made from hard iron, the magnet would become a permanent magnet, thus voiding its utility as an actuator. There is evidence that Cornell was aware of this fact. Ezra Cornell to Sylvester Munger (silversmith), August 18, 1845, 2/10, ECPKL. He says, "The magnet must be made from the finest and best iron ... [and must] not be hardened." The later problems Cornell had indicate this instruction was ignored.

17. This (three days) is an assumption, but based on the receipt above, Morse's going rate for contract work at the time was apparently $15 per day.

18. Once Cornell started selling his relay, Page would accuse him of stealing it to create his own. Page's outrage, under the circumstances, seems manufactured.

19. Charles Page to Samuel Morse, July 28, 1845, MPLOC.

20. Post, *Physics, Patents and Politics*, 70. See Page's of July 24th to Morse marked "Private Business," cited.

21. *Ibid.* Page suffered from a common malady of brilliant men in that he had an external locus for his moral center, entrusting it entirely to equally or perhaps more brilliant but more unscrupulous men than himself, like Morse.

22. Charles Monroe to Ezra Cornell, November 6, 1845, 2/13, ECPKL. Monroe says Page is now insisting on a working model of the relay before he will grant the patent. Page had passed on Cornell's patent the day after Morse's return to Washington from Europe. Morse obviously no longer even considered it a threat.

23. POJH, vol. 6, 569 (fn). Despite his enthusiasm, the duplicity obviously did not sit well with Page, as it was not long after this he began exploring employment options outside the Patent Office. By 1860 he had done a complete 180-degree turnabout, opposing Morse's renewal for the relay patent.

24. Henry L. Ellsworth to Samuel Morse, July 7, 1845, MPLOC. (Emphasis mine.) The "marplot" Ellsworth was referring to in this case was an amalgam of Cornell and Smith.

25. Samuel Morse to Thomas Benton (Senator), January 12, 1846, MPLOC.

26. Not found but referred to in Morse's of the 24th (see next note).

27. Charles Page to Samuel Morse, July 24, 1845, MPLOC. (Emphasis original.)

28. This is to some degree speculation, but with a great deal of explanatory power. Morse certainly would have anticipated that Cornell might learn Page had worked to improve his invention while he was away in Europe. It certainly explains Cornell's apparent shock at Morse's actions on his return. He still thought Kendall's contract protected him. The key was that Morse had the right, according to the contract, to "improve" any designs Cornell might claim as his invention.

29. Ezra Cornell to MaryAnn, July 11, 1845, 2/9, ECPKL. The question naturally arises why did not Cornell protest Page's interference immediately. Cornell had been struggling with his version of the device for a year and a half and may have at this point welcomed Page's assistance, but he was probably away in New Jersey on Kendall's orders at the time of Page's visit to New York. The letters are datelined New York, but Cornell confesses he was actually spending very little time there.

30. Ezra Cornell to Charles Grafton Page, October 15, 1845, 2/12, ECPKL.

31. Roger Sherman, ed., *Journal of the American Scientific Instrument Enterprise* 2, no. 6 (1988): 34–47. It is unlikely Page was fabricating them himself. They were probably being fabricated for him by Daniel Davis.

32. Charles Page to Samuel Morse, July 28, 1845, MPLOC. Page observes that "the shaft is undoubtedly too long in the one you carried

with you but it won't matter much one way or the other." (Emphasis mine.) This would eventually be the one that he brought with him to Europe as evidenced by Morse's letter to Louis Breguet dated October 27, 1846, MPLOC.
33. *Ibid.*, Page to Morse.
34. Prime, *Morse*, 559. Page to Amos Kendall, February 22, 1848.
35. Charles Grafton Page to Alfred Vail, July 25, 1845, MPLOC.
36. Alfred Vail to O.S. Wood, August 29, 1845, 2/10, ECPKL.
37. Amos Kendall to Ezra Cornell, September 29, 1845, 2/10, ECPKL.

Chapter 38

1. O.L. Holly, ed., *The New York State Register for 1845* (New York: Dirstable, 1845), 255.
2. The fact that it was Page's magnet that he carried with him is explicitly laid out in Morse's letters to Louis Breguet of October 27, 1846, and to François Arago of October 30, 1846, MPLOC.
3. Samuel Morse to Consul Robert Walsh, November 11, 1845, MPLOC. The sealed box contained Page's relay as stated in the letter to Arago cited above.
4. Secretary James Buchanan to Samuel Morse, July 18, 1845. Morse, *Samuel F.B. Morse: His Letters and Journals*, vol. 2, 248.
5. S.F.B. Morse, letter to Sidney Morse, November 1, 1845, MPLOC. Apparently he brought a working model of the telegraph as well as the relay. There would have been no reason to secret the telegraph itself in the diplomatic pouch, as it was patented.
6. Ezra Cornell to Mary Ann Cornell, July 11, 1845, 2/9, ECPKL.
7. When Kendall wrote from Boston, he abandoned this salutation. When he returned to Washington. it resumed.
8. Amos Kendall to Ezra Cornell, July 13, 1845, 2/9, ECPKL. (See letter cover.)
9. Ezra Cornell to MaryAnn, July 25, 1845, 2/9, ECPKL.
10. Ezra Cornell to MaryAnn Cornell, July 6, 1845, 2/9, ECPKL.

Chapter 39

1. *New York State Register, ibid.* There were premium cabins available for $150. Most likely Morse occupied the latter.
2. Son of Henry L. Ellsworth and Annie's brother.
3. Letter of Ellsworth as *charge d'affaires* to the Court of Sweden to Morse, August 4, 1845. It seems he parted company with Ellsworth at Hamburg, according to a second letter of December 22, 1845. Both are in MPLOC.
4. Henry W. Ellsworth to Samuel Morse, August 4, 1845, MPLOC.
5. Samuel Morse to B. Brompton, September 16, 1845, MPLOC.
6. Alfred Vail to Samuel Morse, August 24, 1845, MPLOC. The letter was intended to announce Vail's intent to proceed with publishing his book even without Morse's approval. The news about Bodisco must have taken Morse by surprise as he changed his entire itinerary, eliminating the excursion to Russia.
7. S.F.B. Morse to Sidney Morse, November 1, 1845, MPLOC.
8. *Journal des Chemins de Fers* 4 (1845): 703. Advertisement taken out by M. Banner (Venzat and Banner) to demonstrate his "metalized" (creosoted) railroad ties at the Hotel Wagram.
9. Vail, *The American Electro-Magnetic Telegraph*, 90.
10. S.F.B. Morse to Sidney Morse, November 1, 1845, MPLOC.
11. B. March 21, 1768, d. May 16, 1830.
12. S.F.B. Morse notes, October 31, 1845, MPLOC.
13. *The Athenaeum* (July to December 1883): 572. (Announcing Breguet's death.)
14. S.F.B. Morse to Sidney Morse, November 1, 1845, and December 2, 1845. Both make reference to "Edward," making it clear this is a family member. It could not have been Edward Lind, so it must have been James Edward.
15. Bryant and Voss, ed., *Letters of William Cullen Bryant*, vol. 2 (New York: Fordham, 1977), 292.
16. Eliakim and Robert S. Littel, eds., *Littel's Living Age*, vol. 7 (Boston: Waite, 1845).
17. Bound vol., 4 March–29 December 1845, images 306–7, MPLOC. When Foy tested it, it was actually five times faster, which he was gracious enough to admit.
18. S.F.B. Morse to Sidney Morse, November 1, 1845, MPLOC.
19. Louis Breguet's invitation to Samuel Morse, October 29, 1845, MPLOC. Refers to Arago informing him of Morse's arrival. It is addressed to Morse at the Hotel Wagram, Rue de Richelieu. This is a mistake. The hotel was on the Rue de Rivoli. (The mistake was possibly

intentional, a way to let Morse know that he was by no means highly anticipated.)

20. James Lequeux, *Francois Arago* (Springer: New York, 2008), 141.

21. S.F.B. Morse to Sidney Morse, November 1, 1845, MPLOC.

22. Samuel Morse to François Arago (draft), November 1, 1845, MPLOC. One may argue that the relay had become Morse's main reason for the European trip. Even had he been successful in obtaining a European patent, which was unlikely, Morse had no legal standing in any European court of law or the means and recourse to enforce it. Economically speaking, it was a fool's mission from beginning to end, and Morse was nobody's fool. The reasons proffered were probably mostly smoke and mirrors to get the patent partnership to underwrite the trip. He did, however, have a profound interest in assuring that his posterity would be fairly represented in European accounts and histories, and in proving to himself the superiority of his system over the European ones, hence the nineteen-page letter to François Arago.

23. Transcribed from *Galignani's Newspaper* of November 6, 1845, MPLOC. I say "for the time being" because decades later Morse's system would be adopted.

24. Bill from Breguet of November 8, 1845, for "Boussoles," (compasses) MPLOC. On the back it is marked "Boussoles and Galvanometer." It becomes evident from later correspondence that this was actually the bill for the two relays.

25. Samuel Morse to Consul Robert Walsh, November 11, 1845, MPLOC. The sealed box contained Page's relay.

26. *New York Observer*, April 20, 1839.

27. This is what actually happened: as the French lines got longer, they ran into the same problem as had Morse. Louis Breguet to Samuel Morse, September 13, 1846, and Samuel Morse to Louis Breguet, October 27, 1846, both MPLOC. Morse had evidently also inquired of Walsh if the box was still sealed. See Walsh's reply to Morse's letter (not found): "Your deposit of the Telegraph is safe." Robert Walsh to Samuel Morse, October 29, 1846, MPLOC.

28. *American Railroad Journal* 18 (December 11, 1845); S.F.B. Morse to Sidney Morse, December 2, 1845, MPLOC.

29. Samuel Morse to Charles Page, April 3, 1860, MPLOC.

30. Mabee, *American Leonardo*, 292. This is generally assumed to have happened sometime during Christmas, but likely it was when Morse met with Page on the 19th or 20th to discuss the relay patent.

31. Charles Monroe to Ezra Cornell, November 6, 1845, 2/13, ECPKL.

32. It is clear that Cornell in fact first filed an application for the entire relay patent, not only from the drawings furnished to O'Reilly in the autumn of 1845, but from other statements Morse made to Vail in February of 1846. Morse to Alfred Vail, February 16, 1846, MPLOC. Page evidently had counseled him to simplify his claim, as what passed on December 20 was an incomplete version of what was shown to O'Reilly, and was faulty in its claims. See diagram accompanying Ezra Cornell to Henry O'Reilly, December 5, 1845, MPLOC.

33. Cornell, Ezra. 1845. "Improvement in the Mode of Operating Electro-magnetic Telegraphs." U.S. Patent #4318, issued December 20, 1845.

34. Ezra Cornell to MaryAnn, December 26, 1845, 3/5, ECPKL. Postscript probably of the 28th: "I hear from Mr. Kendall to day that my patent is granted."

35. Morse, S.F.B. 1846. "Improvement in electro-magnetic telegraphs." U.S. Patent #4453, issued April 11, 1846.

36. This occurred when Morse tried to renew the patent in 1860. Page, however, based his objections entirely on prior invention by Wheatstone (not Cornell). He seemed to loathe by then to even mention Cornell's name, possibly because of the latter's financial success. As Kendall pointed out, Page's objection in 1860 was surely without precedent, since Page himself had been the one to grant the patent in the first place.

Chapter 40

1. Theodore Faxton Saxton to Ezra Cornell, October 3, 1845, 2/11, ECPKL.

2. Vail, *Early History of the Telegraph*, 20.

3. Vail tried to get Wood to act as his sales agent, selling what was essentially a 12-page pamphlet for a dollar. When Wood balked, Vail reissued it at twelve cents.

4. Morse later claimed to Henry he had never seen the book. This is unlikely. Morse would not have gone to the length of contacting George Vail were he not aware of the contents.

5. Vail, *Early History of the Electro-Magnetic Telegraph*, 20.

6. The publishing of the major volume was worked out with Lea and Blanchard of Philadelphia, who, that year, were also publishing the five-volume account of the Wilkes expedition. They obviously had plenty of skilled engravers on hand of whom Vail availed himself.

7. Vail, *Early History of the Electro-Magnetic Telegraph*, 20.

8. Ibid., 21, George Vail to Alfred Vail, August 7, 1845. Telegraph, 20-1.

9. Alfred Vail to Samuel Morse, July 20, 1845, MPLOC.

10. Alfred Vail to Samuel Morse, August 24, 1845, MPLOC. (Emphasis in original.)

11. Ibid.

12. The underlines in Vail's letter of August 24 would indicate that he had figured out that Cornell is the actual target of Morse's plan.

13. Edward Howard, "The States of Vail's Description," *Maryland Historical Magazine* 62, no. 3 (1967): 355.

14. Alfred Vail to O.S. Wood, August 29, 1845, 2/10, ECPKL.

15. Daniel Davis, *Manual of the Telegraph* (self-published, 1851), 44 pp.

16. Orrin Wood to Ezra Cornell, October 29, 1845, 2/12, ECPKL. So evidently Davis's manual was already available.

17. It appears the larger work came out in late October or early November as per Henry's letter to Alexander Dallas Bache of November 12. It is not clear which one Vail is trying to get Wood to sell for him.

18. O.S. Wood to Ezra Cornell, November 1, 1845, 2/13, ECPKL.

19. O.S. Wood to Ezra Cornell, October 30, 1845, 2/12, ECPKL. Henry Rogers had come up with a pamphlet of his own, "The Telegraph Dictionary and Seaman's Signal Book," that he tried to get Wood to sell as well. Ironically this volume would come to have far greater significance than Vail's, by establishing the marine semaphore system based on Morse code. The pamphlet was reworked in 1847 as Rogers and Black's *American Semaphoric Signal Book for the Use of Vessels Employed in the United States Naval, Revenue and Merchant Service*, and was officially adopted by the government for all naval and merchant vessels shortly after.

20. O.S. Wood to Ezra Cornell, November 19, 1845, 3/1, ECPKL.

21. Canfield and Clark, *Things Worth Knowing About Oneida County* (Utica: Griffiths, 1909), 80.

22. Alfred Vail to O.S. Wood, August 29, 1845, 2/10, ECPKL. Obviously Wood was already employed by Saxton at this time, but Vail said he would recommend him for the position.

23. Not much is known about Reid's early life except that, after emigrating from Scotland, he had entered O'Reilly's employ in 1837 at the age of 18. Aside from his part in running the Pittsburgh line, he was later known as the author of the most comprehensive history of the telegraph. Obituary in *Western Electrician* 28 (1901): 300. Though he had no formal scientific training, Reid would prove to have a facile mind and a quick grasp of things technical.

24. Robert Sabine, *The History and Progress of the Electric Telegraph* (New York: Van Nostrand, 1869), 113-14 and 124-5. The magnetic relay would not become just the basis for solving these practical problems but, through the further application of Ohm's law, would be the foundation of future improvements to telegraphy such as "duplexing," sending messages in opposite directions at the same time over the same wire (though Morse had noted this before).

25. Theodore Saxton Faxton to Ezra Cornell, November 26, 1845, 3/2, ECPKL. Asks his opinion on "the practicability of naked wire."

26. The concerns about "celestial magnetism" had sprung from an offhand comment Henry made regarding a theory Faraday had proposed but not yet proved.

27. Alfred Vail, *Description of the American Electro-Magnetic Telegraph*, 134.

28. Bache had succeeded Hassler in this position. Ferdinand Hassler was a noted scientist who headed the U.S. Coastal Survey. He would later become the first president of Girard College.

29. Joseph Henry to Alexander Bache, November 12, 1845, POJH, vol. 6, 325-7. Henry thereafter could not tolerate even the mention of Vail's name without flying into a rage. Alfred Vail to S.F.B. Morse, May 17, 1847, MPLOC.

30. Alexander Bache to Joseph Henry, November 17, 1845, POJH, vol. 6, 331. It is doubtful that Bache actually read it, as he refers to "letters of Prof. Henry and other unworthies." No letters of Henry appear in the book.

31. There is evidence that Henry routinely denigrated Morse's claims in his lectures at Princeton and claimed priority in the invention of the telegraph for himself. Moyer, *Henry*, 108 and this quote, 243.

32. Joseph Henry to Charles Wheatstone, February 27, 1846, POJH, vol. 6, 382-5.

33. *Ibid.*
34. *Ibid.*
35. David Hochfelder, *Joseph Henry: Inventor of the Telegraph?* Accessed February 20, 2015, http://siarchives.si.edu/oldsite/siarchives-old/history/jhp/joseph20.htm.
36. Alfred Vail to Samuel Morse, July 30, 1845, MPLOC. See Morse's note at the bottom, which was obviously added at a later date.
37. Alfred Vail to Joseph Henry, July 17, 1846. POJH, vol. 7, 450. Moyer agrees this belated "mea culpa" was probably ghostwritten by Morse. In it Vail promises to rectify the errors if he finds them. From the subsequent letter to Bache it is clear Henry does not believe him in the least. Moyer, Henry, 243.
38. *Ibid.*, 243–4.
39. *Ibid.*, 245.
40. This breach, unlike the one between Morse and Henry, would heal, and by the 1840s the Henrys would be regular guests at the Beck residence in New Brunswick. In one of his rare forays into patronage politics, Henry would help secure Lewis Beck a commission from the Patent Office to investigate the adulteration of breadstuffs in the U.S. The results were published by Beck under the title "Breadstuffiana," 1848.
41. Joseph Henry to Charles Wheatstone, February 27, 1846, POJH, vol. 6, 385.
42. POJH, vol. 6, 327 (fn). Rothenberg points out that Henry in 1850 wrote an analysis of what was perhaps an overreaction on his part. This analysis is presently in the Smithsonian, but it does not add much to the understanding of events.
43. Joseph Henry to Alexander Bache, November 12, 1845, POJH, vol. 6, 327 (fn). Rothenberg points out this allegation of Vail's is prima facie ridiculous.
44. "Cross-Interrogatories Propounded by Complainant to Prof. Joseph Henry," May 21, 1850, Vail Telegraph Collection, Box 9A, Smithsonian Institution Archives.

Chapter 41

1. To be fair, it appears it was Faxton and Wood who first came up with the idea of using uninsulated wire.
2. *Rochester History* 7, no. 1, Dexter Perkins, Henry O'Reilly, 19. www.rochester.lib.ny.us/~rochhist/v7_1945/v7i1.pdf, last accessed, December 12, 2015, 13–15.
3. Ronald Shaw, *Erie Water West* (Lexington: University of Kentucky Press, 1966), 308,

319 and 421. He also had been a strong advocate for conversion to private ownership.
4. Henry O'Reilly to Ezra Cornell, December 5, 1845, 3/3, ECPKL. Refers to Cornell's previous letter regarding that fact (probably of the 17th), not found.
5. T.S. Faxton to Samuel Morse, January 14, 1846, MPLOC.
6. Reid, *Telegraph in America*, 117. Reid may be mistaken about the size of the relay. Vail was probably using Page's relay, but he kept all the equipment, including the registers, locked up.
7. *Rochester History*, Henry O'Reilly, 19.
8. Ezra Cornell to Edward Cornell, November 5, 1845, 2/13, ECPKL.
9. Henry O'Reilly to Ezra Cornell, November 29, 1845, 3/2, ECPKL.
10. O'Reilly to Cornell, November 30, 1845, 3/2, ECPKL.
11. Henry O'Reilly to Ezra Cornell, November 15, 1845, 3/1, ECPKL.
12. Samuel Colt to Ezra Cornell, November 27, 1845, 3/2, ECPKL. Requesting two magnets immediately for Boston. (Note on back refers to Kendall drawing the agreement.)
13. O'Reilly to Cornell, December 5, 1845, 3/3, ECPKL. (Emphasis in original.)
14. Ezra Cornell to T.S. Faxton December 28, 1845, 3/5, ECPKL. "Were it not that I take this course in the matter" (asking Faxton to sign the paper) "they would decide at the Patent Office when I apply for a patent that I have abandoned my claim."
15. James Reid to Henry O'Reilly, December 22, 1845, 3/4, ECPKL. Reid evidently did not comprehend that the circuit was controlled by the key at the remote station. The fact that Cornell's magnets were poorly made just compounded the confusion.
16. Henry O'Reilly to Ezra Cornell, December 30, 1845, 3/5, ECPKL.
17. Henry O'Reilly to Ezra Cornell, December 22, 1845, 3/4, ECPKL.
18. *Journal of the Franklin Institute* 3, series 3, ed. Mapes and Jones (1842), 341.
19. Amos Kendall to Ezra Cornell, October 27, and 29, 1845, 2/12, ECPKL.
20. Amos Kendall to Ezra Cornell, September 29, 1845, 2/10, ECPKL enclosing Henry Rogers's to Kendall of September 14 (MPLOC). Obviously this was Rogers's letter containing Page's suggestions.
21. A.C. Goell to Ezra Cornell, December 12 and 16, 1845, 3/4, ECPKL. Complaining about lack of materials. Amos Kendall to Ezra Cornell, December 9, 1845, 3/3, ECPKL.

22. Theodore Saxton Faxton to Ezra Cornell, December 17, 1845, 3\4, ECPKL. Receipt of glass caps. A.C. Goell to Ezra Cornell, December 16, 1845, 3\4, ECPKL. "I am much surprised to learn that you have not a sufficient supply of glass caps and wires for my end of the line."
23. Henry O'Reilly to Samuel Morse, December 23, 1845, MPLOC. (Emphasis in original.)
24. *Ibid.*
25. Henry O'Reilly to Ezra Cornell, December 30, 1845, 3/5, ECPKL.
26. Henry O'Reilly to Ezra Cornell, December 30, 1845, 3/5, ECPKL. Once again in Reid's hand but under O'Reilly's signature.
27. Henry O'Reilly to Samuel Morse, February 9, 1846, MPLOC. Payment for two relays.
28. James Reid to Henry O'Reilly, December 22, 1845, 3\4, ECPKL.
29. F.O.J. Smith to Ezra Cornell, December 13, 1845, 3\4, ECPKL. Containing Colt's low opinion of Cornell's relay magnet.
30. Daniel Davis, *Manual of Magnetism* (Boston: self-published, 1842). The pamphlet that he had prepared for the subcontractors, and that Cornell had distributed along with Vail's, was in fact more detailed than Vail's.
31. Ezra Cornell to Amos Kendall, November 26, 1845, 3/2, ECPKL. "I judge that the principal difficulty they have had to contend with has been a poor battery, caused by the constant action of their dependent circuit and a want of experience in their management of it." (Emphasis in original.)
32. O'Reilly to Cornell, December 26, 1845, 3/5, ECPKL. Reid, unlike O'Reilly, would become a firm supporter of Morse. He would succeed Amos Kendall and also end up in charge of the Harrisburg-Lancaster line, and eventually Faxton's New York, Albany and Buffalo line. (Emphasis in original.)
33. Ezra Cornell to O'Reilly, penciled in "December 2." Clearly incorrect. This should be January 2, 1845, 3/3, ECPKL.
34. *Ibid.*

Chapter 42

1. Ezra Cornell to Orrin Wood, December 15, 1845, 3/4, ECPKL.
2. This was stock in lieu of pay. *Ibid.*, Smith, Defendant's Proofs, 24.
3. *Ibid.*

4. Ezra Cornell to MaryAnn, December 26, 1845, 3/5, ECPKL. Evidently Smith had not made good on his offer to have his tailor make him up one.
5. James Eddy to Ezra Cornell, December 19, 1845, 3/5, ECPKL. The scheme foundered when Eddy could not come up with enough investors.
6. Bound vol., 30 December 1845–8 December 1846, image 35, top extract, MPLOC. (Probably January 12.)
7. Ezra Cornell to Theodore Saxton Faxton, January 8, 1846 (with enclosure), 3/6, ECPKL.
8. Theodore Saxton Faxton to Samuel Morse, January 8, 1846, MPLOC.
9. Samuel Morse to Theodore Saxton Faxton, January 10, 1846, MPLOC.
10. T.S. Faxton to Ezra Cornell, January 16, 1846, 3/6, ECPKL. (Agreeing to sign Cornell's waiver.)
11. See Quimby's interesting letter to Morse suggesting a figurative monument to Morse. A.B. Quimby to Samuel Morse, May 23, 1846. This is the same Quimby who was an instructor at the Albany Academy.
12. This raises the question why Morse would insist on relying on Cornell to devise a workable submarine cabling system when Colt, already an acknowledged expert in the field, was then laying one in Long Island. The link across the Hudson was the nerve center for the entire Northeast grid and key to the eventual Atlantic telegraph, and Morse evidently did not want to hazard someone with a solid scientific reputation like Colt's laying claim to it. Yet his continued reliance on Cornell (who was by now detested) over Colt, is still somewhat mystifying. It is as if the two parts of his psyche were struggling over what to do about it.
13. Amos Kendall to Ezra Cornell, January 2, 1846, 3/6, ECPKL.
14. Alfred Vail to Samuel Morse, January 11, 1846, MPLOC. (Not found but referred to in Morse's response the next day.)
15. Samuel Morse to Alfred Vail, January 12, 1846, MPLOC.
16. Samuel Morse to Alfred Vail, October 7, 1846, MPLOC.

Chapter 43

1. The Power of Attorney is in Kendall's hand and is unsigned and undated, suggesting Morse may have balked at signing this. Penciled, "August 1845," MPLOC.

2. Donald Cole, *A Jackson Man* (Baton Rouge: Louisiana State University Press, 2004), 4.

3. *Ibid.*, 247.

4. The formation of the Magnetic Telegraph Company by Kendall (in May) had been ostensibly for the purpose of creating only this line. While the overall "master plan" must have been formed well prior to this, the details don't seem to jell until Kendall's letter to Morse of June 7, 1845, MPLOC.

5. Amos Kendall to Ezra Cornell, August 22, 1845, 2/10, ECPKL.

6. Amos Kendall to Ezra Cornell, December 11, 1845, 3/3, ECPKL. "I open my letter at the request of Mr. Vail."

7. Scharf and Westcott, *History of Philadelphia*, vol. 3 (Philadelphia: Everts, 1884), 2130.

8. John T. Scharf, *History of Baltimore City and County* (Philadelphia: Everts, 1881), 506.

9. *The National Magazine* 11 (1889): 92.

10. Theodore Faxton Saxton to S.F.B. Morse, January 14, 1846, MPLOC. Faxton would soon become a staunch ally and friend of Cornell, not only contracting him to erect the New York City–Albany line, but later as a partner in his oil and coal business.

11. F.O.J. Smith to Ezra Cornell, December 13, 1845, 3/3, ECPKL.

12. Not found but referenced in the following from Smith of the 17th.

13. F.O.J. Smith to Ezra Cornell, December 17, 1845, 3/3, ECPKL.

14. O.S. Wood to Ezra Cornell, December 26, 1845, 3/5, ECPKL. Wood had echoed Smith's sentiment, saying, "Don't send your relay-magnets til they are right," and suggesting Cornell get the contractors to pay for them.

15. Theodore Saxton Faxton to Samuel Morse, January 8, 1846, MPLOC. There is another copy of essentially the same letter under the same dateline, but addressed to Morse, Kendall and Smith. Apparently Faxton thought better of sending that. How Morse obtained it is open to question, but it is likely he had a spy in Faxton's office. See Faxton to Morse, Kendall and Smith, January 8, 1846 (misfiled), Bound vol., 9 December 1846–17 May 1847, image 36.

16. T.S. Faxton's reply to Morse, January 14, 1846, MPLOC.

17. Samuel. Morse to Theodore Faxton, January 10, 1846, MPLOC.

18. Based on his letter to Faxton of January 10, 1846, it is clear that Morse had little faith in Page's alternate solution this time.

19. Samuel Morse to Alfred Vail, January 3, 1846, MPLOC. [*sic*] Cutting edge.

20. *Ibid.* Apparently one of Breguet's magnets had been broken by accident and Morse was attempting to fix it. Apparently it had not been fixed by the time of the test.

21. Amos Kendall to Ezra Cornell, January 12, 1846, 3/6, ECPKL.

22. Alfred Vail to Samuel Morse, January 24 and 26, 1845, MPLOC.

23. *Ibid.*, January 26.

24. Samuel Morse to Louis Breguet December 29, 1845 (copied to Green Co., Paris), requesting Breguet deliver to Green thirty more of the small electromagnets of the type Morse had purchased while in Paris.

25. See invoice from Breguet et Fils, March 9, 1846, MPLOC.

26. Morse would eventually settle Smith's claims by slyly offsetting them against his own purported expenses for the 1838 European trip. He asserts in a letter to Smith (without letting on that the government had already picked up the tab), that having made the trip to secure the English and French versions of the patent on behalf of all the patentees (not for settling the relay question), that the patent partnership, which included Smith, was entirely responsible for his travel expenses in this case as well. Morse lets this plan slip with evident glee to brother Sidney in a letter after his return. In Morse's moral calculus, if Smith could double dip, he could as well. See Morse to F.O.J. Smith, February 29, 1844, and letter to his brother Sidney, March 4, 1844, saying (with regard to Smith's dunning him for payments to Bartlett), "I think I have got him in check" by a counterdemand for four months of his (Morse's) services in Paris (over what they had agreed). Both MPLOC.

27. S.F.B. Morse to Alfred Vail, February 16, 1846, MPLOC.

28. Ezra Cornell to T.S. Faxton, January 15, 1846, 3/6, ECPKL. (Emphasis added.) Cornell offered to supply Page's magnets instead of his own at $75 apiece.

29. F.O.J. Smith to H.W. Paine, the telegraph agent at Springfield (hand-delivered by Cornell), March 16, 1846, and letter from T.S. Faxton to Cornell, March 10, 1846 (also see letter of May 12, 1846, from Damon requesting relays), 3/11, ECPKL.

30. Samuel Morse to T.S. Faxton, April 23, 1846, MPLOC.

Chapter 44

1. Samuel Morse to François Arago (draft), November 1, 1845, MPLOC. Discussing electrifying the entire body of water to pass the signal instead of a cable.
2. Amos Kendall to Ezra Cornell, November 10, 1845, 3/3, ECPKL.
3. F.O.J. Smith to Ezra Cornell, July 9, 1845, 2/9, ECPKL. From this letter it is clear Smith first contemplated crossing the Hudson at or near the base of Chambers Street.
4. Amos Kendall to Ezra Cornell, July 19, 1845, 2/9, ECPKL.
5. Also called "Audubon Park." Morse was a friend of Audubon's son, John Woodhouse Audubon, since his days as president of the National Academy. Audubon and his brother were associate members of the Academy. Thomas Cummings, *Historic Annals of the National Academy of Design*, vol. 3 (New York: Sackett & Cobb, 1861), 305.
6. Letters to MaryAnn, August 10, 13 and 17, 1845, and Alonzo, August 17 and 31, 1845. He asks Alonzo to go see how Munger is getting along with the relays. The attempts failed, as evidenced by the fact that MaryAnn did not show up at the Utica Fair. Letter to Sylvester Munger of August 17, 1845. All 2/10, ECPKL.
7. Amos Kendall to Ezra Cornell, August 22, 1845, 2/10, ECPKL.
8. Henry Rogers to Amos Kendall, August 29, 1845, 2/10, ECPKL. Refers to Kendall's inquiry of the 28th.
9. Ibid.
10. Henry Rogers to Amos Kendall, September 14, 1845, 2/10, ECPKL.
11. Amos Kendall to Ezra Cornell, September 29, 1845, 2/10, ECPKL.
12. Ibid.
13. Amos Kendall to Ezra Cornell, September 29, 1845, 2/10, ECPKL. Enclosing Henry Rogers's to Kendall of September 14 (MPLOC). Rogers's letter reiterated Page's suggestions that Rogers had included in his letter of August 29.
14. Ibid.
15. Taylor and West's was essentially a copper wire encased in India rubber, wrapped in hemp and then encased in lead. "History of the Atlantic Cable and Undersea Communications," Steven Roberts. Online. atlantic-cable.com/Cables/Domestic/. Last accessed January 2, 2015. Morse, while in England at this time, had no doubt become aware of these efforts by the Admiralty, and with Wheatstone by this time having shut him out, the prospect of showing up the English, and particularly Wheatstone, no doubt would have set him salivating.
16. Amos Kendall to Ezra Cornell, October 9 and 29, 1845, 2/11, ECPKL.
17. Amos Kendall to Ezra Cornell, October 11, 1845, 2/11, ECPKL.
18. Ezra Cornell to MaryAnn, October 26, 1845, 2/12, ECPKL.
19. *Telegraph Age* 24, no. 1 (1909): 409. Notes the use of masts at Fort Lee. to Audubon's mansion on the upper west side.
20. Amos Kendall to Ezra Cornell, November 24, 3/2, 1845, 3/2, ECPKL.
21. Amos Kendall to Ezra Cornell, November 12, 1845, 3/1, ECPKL. "Mr. Colt has written to Morse and Vail that his line to Coney Island is done." O.S. Wood to Ezra Cornell, November 14, 1845, 3/1, ECPKL. "I am sorry to hear of Colt's ill success."
22. Amos Kendall to Ezra Cornell, November 26, 1845, 3/2, ECPKL. Eddy and Cornell were looking into raising a subscription for a line from Auburn to Ithaca. They would fail to raise enough capital for the project and it was dropped.
23. Ibid.
24. George Beecher to Ezra Cornell, February 9, 1846, referencing Goodyear's cost estimates. 3/7, ECPKL. It was Goodyear's proposed charges that had set Kendall off. Beecher's shop was around the corner from #10 Wall Street at 100 Broadway (across from Trinity Church). Beecher was Goodyear's brother-in-law and Goodyear's exclusive agent in New York.
25. Amos Kendall to Ezra Cornell, December 19, 1845, 3/4, ECPKL.
26. Amos Kendall to Ezra Cornell, December 11, 1845, 3/3, ECPKL.
27. Amos Kendall to Ezra Cornell, December 19, 1845, 3/4, ECPKL.
28. Ibid.
29. Livingston was likely already suffering from tuberculosis (consumption). *New York Tribune*, April 29, 1849, under testimonials to a patent medicine, he was said to be suffering from "dyspepsia" of several years' standing.
30. Ezra Cornell to MaryAnn, November 3, 1844, 2/3, ECPKL.
31. Ezra Cornell to Amos Kendall, December 28, 1845, 3/5, ECPKL; *Electrical Review* 53 (1908): 569.
32. The source for this was lost but the author specifically recalls the gist and decided to include it.

33. O.S. Wood to Ezra Cornell, December 31, 1845, 3/5, ECPKL.
34. Ezra Cornell to Amos Kendall, December 21, 1845, 3/4, ECPKL.
35. *New York Tribune*, December 22, 1845. chroniclingamerica.loc.gov. Obviously the reporter was personally on hand at Audubon's mansion, as Cornell had written just that morning to Kendall that he hoped to get the line in the river by nightfall.

Chapter 45

1. *New England Magazine* 4 (1891): 459.
2. Samuel Morse to Ezra Cornell, January 10, 1846, 3/6, ECPKL.
3. Undated. Marked experiments. Penciled in 1844. Bound vol., 22 October 1844–1 March 1845 images 206–8.
4. Cornell, *True and Firm*, 97.
5. Bound vol., 30 December 1845–8 December 1846, image 35, top extract, MPLOC. "Take care specially to not charge him [Cornell] wrongfully which you will be apt to do in the present circumstances." Samuel Morse to Alfred Vail (presumably), undated extract (presumably January 12).
6. Samuel Morse to Alfred Vail (presumably), undated extract. Bound vol., 30 December 1845–8 December 1846, image 34, top extract, MPLOC. (Probably January 12.) In these letters to Vail between the 12th and the 16th, Morse is very obviously and consciously writing here for posterity, showing a "kinder, gentler face" than he would under usual circumstances.
7. *Ibid.*, image 35. Extract. Presumably Morse to Vail, dated January 13, MPLOC.
8. Amos Kendall to Ezra Cornell, January 16, 1846, 3/6, ECPKL.
9. Samuel Morse to Vail, dated January 13, *Ibid.*, image 34. "I am anxious to have the line complete before the arrival of the steamer."
10. *New York Tribune*, January 16, 1846. Correspondence from Philadelphia, January 15. chroniclingamerica.loc.gov. Reminiscent of some more recent scandals regarding bulk trading.
11. Samuel Morse to Alfred Vail, January 17, 1846, MPLOC. "We have such opposition all around, the government the press and certain speculators."
12. Samuel Morse to Alfred Vail, January 15, 1846, MPLOC.
13. *Ibid.*
14. Samuel Morse to Alfred Vail, January 16, 1846, MPLOC. "It is now ½ past 12 and I have nothing from Fort Lee."
15. *Ibid.*
16. S.F.B. Morse to Alfred Vail, February 16, 1846, MPLOC.
17. T.S. Faxton to Samuel Morse, January 14, 1846, MPLOC. Marked on back "Ans. 21 Jan. w/ Diagram." Probably Morse had Cornell send the answer to Faxton. It was Cornell's habit, not Morse's, to put the word "Answered" plus the date on the back of correspondence.
18. Samuel Morse to Alfred Vail, January 16, 1846, MPLOC.
19. January 23, 1846. No recipient. "The mode of connecting an intermediate station with a station on the right or left in case of the breaking of the wire or the like you will be pleased with. *It is after your plan of making the changes in the connections.*" In Morse's hand, obviously intended for Cornell. Probably never sent. (Emphasis in original.)
20. *Ibid.*
21. Intermediate sub-station configuration. Witnessed by A. Vail, February 25, 1846, Bound vol., 30 December 1845–8 December 1846, image 358, MPLOC.
22. When Cornell's line was again broken by ice floes three years later, it was replaced with a cable encased with gutta-percha and rubber. Osswald and Menges, *Materials Science of Polymers for Engineers*, 2nd edition (Cincinnati: Hanser Gardner, 2003), 18.
23. *Ibid.*
24. Charles Grafton Page to Samuel Morse, February 10, 1846, MPLOC. Goodyear had patented a waterproof, flexible material that seemingly would have been ideal. Goodyear, Charles. U.S. Patent #4099. Issued July 5, 1845. Vulcanized rubber had been patented the year before. The "little gentleman" Page is sarcastically referring to in the letter is no doubt Cornell. The twisted strands are actually quite innovative and provided greater strength and guarantee against breakage. "The pipe must be dispensed with."
25. Samuel Morse to Alfred Vail, January 15, 1846, MPLOC. Clearly Morse is a being disingenuous here, as he knows the purpose.
26. *New England Magazine* 4, 459. This is an assumption, but Kendall's note of the 18th indicates the line was working through to New York. There was no river traffic this time of year so no danger from anchors, and the only logical explanation is that it was ice floes that severed the line. Apparently the boatmen came up with

their own scheme of imposing a twenty-five cent "surcharge" to the messages with a special franking stamp, pocketing the earnings.

27. Alfred Vail to Samuel Morse, February 18, 184 (copy), MPLOC. Also informing Morse that Clark had ready six receiving magnets on Breguet's design.

28. Alonzo Cornell to Ezra Cornell, March 1, 1846, 3/8, ECPKL.

Chapter 46

1. Samuel Colt to S.F.B. Morse, February 3, 1846, MPLOC.
2. Cornell was by now employed by Smith's New York and Boston Company as a contractor. In October he would become an employee of the New York, Albany and Buffalo Telegraph Company.
3. *Shaffner's Telegraph Companion*, vol. 1, 129–30, 219–20.
4. "I think in regard to any league between Marshall—Livingston and Cornell—co., I don't think it." Bound vol., 30 December 1845– 8 December 1846, image 35, top extract, MPLOC. (Probably January 12.)
5. *Shaffner's Telegraph Companion*, vol. 2, 408–9. Morse claimed he knew and understood the principles of the relay in 1837 but had chosen to omit them from the telegraph at that time.
6. Cornell's use of the solenoid in his receiving magnet had undoubtedly been influenced by Page's improvement and this had been at the root of Page's umbrage at Cornell in October and at the root of Munger's problems with hysteresis.
7. Boston, being the first port of call for ships from the continent, became an important news conduit with most of the large news organizations like AP, eventually building their own dedicated lines for fear of poaching.
8. Henry O'Reilly to Samuel Morse, January 20, 1846. Trying to get Kendall to assume the presidency of the New York-Washington Telegraph Company. This was obviously an attempt to drive a wedge between Morse and Kendall, but though Kendall would accept, the attempt would fail.
9. Livingston and Crawford would abandon this idea, but O'Reilly would proceed using House and Bain's telegraphs leading to the lawsuits of the 1850s.
10. O'Reilly became an investor in the House Telegraph and will for a time be in head-to-head competition with Morse, with each stringing a parallel set of lines.
11. The validity of Morse's claim would eventually be tested in a case before the Supreme Court by 1854. O'Reilly would lose the case, but the court would also deny Morse's last claim to ownership of any device employing electromagnetism.
12. "I think in regard to any league between Marshall—Livingston and Cornell—co., I don't think it." Bound vol., 30 December 1845– 8 December 1846, image 35, top extract, MPLOC. (Probably January 12.)
13. This was stock in lieu of pay. *Ibid.*, Smith, Defendant's Proofs, 24.
14. F.O.J. Smith to Ezra Cornell, December 27, 1845, containing extract of Kendall's letter to Smith, 3/5, ECPKL.
15. *Ibid.*
16. F.O.J. Smith to Ezra Cornell, February 12, 1846, 3/7, ECPKL. Cornell started referring to his device, after Morse's patent was granted, as an "independent circuit" rather than a "relay." Cornell to Faxton, April 18, 1846, 3/10, ECPKL. Cornell had learned from Smith of Morse's patent in Smith's letter to Cornell, April 7, 1846, 3/10, ECPKL.
17. Ezra Cornell to the editor of the *New York Herald*, undated, 3/7 image 298–9, ECPKL. Referring to letter of February 9, 1846, in the *Herald*.

Chapter 47

1. S.F.B. Morse to Alfred Vail, February 16, 1846, MPLOC.
2. Amos Kendall to Ezra Cornell, January 24, 1846, 3/6, ECPKL. This is obviously a compromise Kendall wanted at with Morse. Kendall clearly wants him fired, and this is a way to keep Cornell and allow Kendall to save face.
3. Amos Kendall's letter to the *Washington Union* of January 26, 1846 (reprinted in the *New York Tribune*, February 5).
4. Vail and Avery would both later make anemic attempts to correct the record, but by then no one cared.
5. Samuel Morse to Alfred Vail (presumably), January 13. Bound vol., 30 December 1845–8 December 1846, image 35, MPLOC.
6. Samuel Morse to Alfred Vail, January 12, 1846, MPLOC.
7. *Ibid.*
8. Samuel. Morse to Alfred Vail, February 16, 1846, MPLOC.

9. Morse to Vail, February 20, 1846, MPLOC. Apparently Cornell was working on a new patent in addition to, or to correct, the December 20 one.
10. Ibid.
11. Alfred Vail to Samuel Morse, January 24, 1845, MPLOC.
12. Amos Kendall's letter to the *Washington Union* of January 26, 1846, reprinted in the *New York Tribune*, February 5.
13. J.D. Parks to Ezra Cornell, March 15, 1846, 3/8, ECPKL.
14. S.F.B. Morse to Alfred Vail, February 16, 1846, MPLOC.
15. Ibid.
16. Samuel Morse to Alfred Vail, February 20, 1846, MPLOC.
17. The cost to Faxton eventually will be eighteen dollars. Orrin Wood to Ezra Cornell, April 16, 1846, 3/10, ECPKL.
18. See articles of agreement between Morse and Edward Kent. November 30, 1846, MPLOC. Kent is listed as a practical chemist, assayer and machinist with offices at 116 John St., New York (*New York Tribune*).
19. By February 1846, both Clark in Philadelphia and Davis in Boston were adept enough at turning these out at a reasonable speed and cost that Morse no longer had the problem of relying on Breguet. Kent seems to be a politically motivated choice.
20. Amos Kendall to Ezra Cornell, January 12, 1846, 3/6, ECPKL.
21. Ezra Cornell Pocket Diary, October 5, 1846, http://rmc.library.cornell.edu/ezra/exhibition/telegraph, last accessed December 2015.
22. Mabee, *American Leonardo*, 299; Samuel Morse to T.S. Faxton, March 15, 1848.

Chapter 48

1. Jonathan Lyons, *The Society for Useful Knowledge* (New York: Bloomsbury, 2013), 164.
2. Allied Textile Printers, HAER No. NJ-17, National Park Service.
3. David Hounshell, *From the American System to Mass Production* (Baltimore: Johns Hopkins University Press, 1984), 47.
4. Colt commissioned a painting of the subsequent test of the submarine battery test by a French painter, Antoine-Placide Gibert. (See Lundeberg.)
5. Although after the demonstration he claimed to Smith that he needed power of attorney for the patentees for negotiating with private investors, clearly Colt had put a bug in his ear about the government as the "ideal end consumer."
6. Richard John, *Network Nation* (Cambridge: Harvard University Press, 2010), 79.
7. John, *Network Nation*, 79.
8. Unlike "The Lightning Doctor," Colt would eventually prove correct in this assumption. During the Mexican War, Colt was contracted to supply the army with revolvers. Ibid.
9. *New York Morning News*, October 24, 1845, reprinted in the *Washington Daily Union*, October 25, 1845, under the title "The Question Settled."
10. Three days after Kendall's letter to Cornell, the failure of Colt's line was reported. *Brooklyn Eagle*, November 15, 1845.
11. *New York Tribune*, April 9, 1846.
12. Morse's efforts in this regard were so vigorous it almost seems like he had intentionally set out to humiliate Colt.
13. See the bizarrely grandiose document outlining the plan for a trans-Atlantic telegraph company with the crowned heads of Europe proposed as the major stockholders. Page 153 of Bound vol., 4 March–29 December 1845, image 153. Note: this document is misplaced in the archive as it is much later than 1845, referring to 40,000 miles of telegraph already erected in the U.S.
14. *Shaffner's Telegraph Companion*, vol. 1, 73. Gale's deposition.
15. Ezra to Alonzo Cornell, February 22, 1846, 3/7, ECPKL.
16. Coincidentally, Alonzo's first job would be with the Mutual Life Insurance Company. This was the company from whom Cornell had purchased insurance before his plow trip in 1843. Cornell's uncle was on the board and Ezra had obtained the position for his son.

Chapter 49

1. Bound vol. 31 March 1859–28 March 1860, image 251, MPLOC. Labeled by Morse on the back "Chas. G Page's extraordinary missive sent to those whom it might influence in the matter of my Petition for Extension of Patent of 1846." Dated March 26, 1860.
2. Ibid.
3. Amos Kendall to Samuel Morse, March 24, 1860, MPLOC.
4. Ibid.

5. Charles Mason to Amos Kendall, March 23, 1860, MPLOC.
6. "Argument of Henry O'Reilly Against the Extension of Professor Morse's 1860 patent," Rochester Historical Society, Henry O'Reilly papers.
7. Ibid.
8. Amos Kendall to Charles Page, April 5, 1860, MPLOC.
9. Ibid. (Emphasis mine).
10. Post, *Physics, Patents and Politics*, 165.
11. Ibid., 167, Page to Kendall, April 7, 1860, MPLOC.
12. Norvin Green to Samuel Morse, March 19, 1860, MPLOC.
13. AP, March 12, 1860, appearing in the *New York Times*, March 13. The Pacific Telegraph Act of 1860 (June 16) will provide for bidding out the lines west of Missouri. There will be only three bidders, two of whom will drop out leaving Sibley's Western Union as the winner by default.
14. They would eventually prove successful, and its passage would lead to a successful bid by this same consortium for control of all the telegraph lines west of the Mississippi. The eventual dominance of Western Union in time will make Cornell and his partners rich beyond their wildest dreams.
15. Norvin Green to Samuel Morse, March 19, 1860, MPLOC.
16. Ezra Cornell to MaryAnn, March 11, 1860, 20/8, ECPKL. This is written on the lined notepaper provided by the Willard Hotel (as is Green's).
17. Ibid.
18. Ibid.
19. Bound vol., 31 March 1859–28 March 1860, image 253, MPLOC. Notes on Page's manifesto.
20. Samuel Morse to Charles Page, April 3, 1860, MPLOC.
21. Charles Page to Amos Kendall, April 4, 1860, MPLOC.
22. Amos Kendall to Samuel Morse, April 9, 1860, MPLOC.
23. Amos Kendall to Morse, April 9, 1860, MPLOC. (Postscript.)
24. Ibid.
25. Morse's affidavit of the value of his inventions. Bound vol., 31 March 1859–28 March 1860, images 223–8, MPLOC.
26. Post, *Physics, Patents and Politics*, 165.
27. Morse to Page, April 3, 1860. This is a detailed and highly hostile point-by-point rebuttal of the facts as Page presented them in his broadside. Page's dates were clearly wrong. See Page's receipt to Morse dated July 8, 1845 (well before Morse leaves for Europe). Morse puts Page's receipt and the payment to Breguet for 210 francs made while in Paris on the same expense report dated February 1, 1846 (abstract A, materials), probably to remind himself they were for the same device.
28. Morse, Samuel. 1840. "Improvement in the mode of communicating information by signals by the application of electro-magnetism." U.S. Patent #1647, issued June 20, 1840.
29. "Decision of Hon. Philip F. Thomas, Commissioner of Patents on the Application of Samuel F. B. Morse" (Washington, D.C.: Polkinhorn, 1860), 13. "A willful failure therefore to set forth and describe in the specification whole of the invention as known and understood by patentee or a reservation of any essential part of it or concealment of the most beneficial way of constructing it operates a fraud upon the public and vitiates the patent. Nor can a valid patent be subsequently granted for that of the invention which the patentee has concealed and to describe and claim in his specification for that enable him to take advantage of his own fraud and virtually to secure a monopoly of twenty eight instead of fourteen years for one invention."
30. Ibid., 14, 16–17. It would seem that Gale, as a former sitting Patent Examiner himself, should have been aware of this possibility.
31. Leonard Gale to Samuel Morse, April 12, 1860, MPLOC.
32. Gale to Morse, April 12, 1860, MPLOC. Bill for $1,056 for services rendered and $75 for printing for Polkinhorn.
33. Ibid.
34. Gale to Morse, April 16, 1860, MPLOC.
35. In the context of his subsequent attack on Henry (issued after the Supreme Court decision of 1854), Morse would allude to the fact that he had first gone from a two-wire system to a one-wire system sometime during the erection of the Baltimore-Washington test line, but he failed to give the reasons why he might have neglected to even mention let alone patent it at the time. Later he corrected this, asserting the idea came to him some time in 1845.
36. Morse changed his version of these events several times to suit the circumstances. In his attack on Henry in Shaffner's magazine, he claimed that the single wire was initiated sometime late in the course of the erection of the test line. In later versions he said the idea occurred to him during the erection of the test line but was not implemented until 1846. In his

diatribe against Henry in 1855 in Shaffner's, Morse would eventually back away from his original claim, admitting that the new relay had in fact first been used late in 1844.

37. Meaning that the actuating of the receiving apparatus was a direct result of current in the line.

38. Diagrams of the test line, as noted above, clearly show two lines, one marked "East" and one "West," also with two separate in-line relays. In the original configuration there were two complete circuits (4 lines) and one ground wire. April 27, 1844, MPLOC

39. As mentioned, this assertion by Gale not only almost cost Morse the renewal, but nearly cost him the original telegraph patent as well (Washington: GPO, 1875).

40. Memorial of Samuel Finley Breese Morse, 19.

41. David Hochfelder, "Flash of Genius," paper delivered to SHEAR conference, July 1998, Case Western Reserve, 8. academia.edu/1682289/_Flash_of_Genius_Samuel_F.B._Morse_s_Telegraph_Patents_and_the_Legal_Construction_of_Creativity_1832–1854. Accessed January 2, 2015.

42. *New York Times*, April 6, 1872.

43. Paraphrased slightly for readability.

44. *New York Times*, September 5, 1872.

Coda

1. Mary Ann to Ezra Cornell, November 18, 1855, 17/12, ECPKL.

2. Joseph Henry to Ezra Cornell, December 29, 1860, 20/13, ECPKL.

3. Edward Cornell to Ezra Cornell, December 10, 1860, 20/13, ECPKL.

4. Contrary to the "I'm OK, You're OK" pop psychology that is generously applied in assessing the mental health of many more modern billionaires.

5. According to the letter from D.B. Cornell to Ezra, February 20, 1861, "I had a letter last week stating Edward was insane again." 21/2, ECPKL

6. Referred to above in the chapter, "A Ghost Story."

7. Edward Cornell to Ezra Cornell, May 7, 1861, 21/4, ECPKL.

8. Hiram Robertson to Ezra Cornell, June 9, 1861, 21/5, ECPKL.

9. Edward Cornell to Ezra June 9, 1861, 21/5, ECPKL. Edward Cornell to Ezra June 9, 1861, 21/5, ECPKL. Edward, after not hearing back from Ezra after the letter of May 7, had written both Hiram his brother-in-law and Elijah his father, begging to be released. The letter from Hiram to Ezra is dated June 9, 1861, and the letter to Elijah from Edward, June 4. In this latter letter Edward refers to himself in the third person and remarks fearing for his marriage that several other of the inmates' wives had already left them and remarried. A letter dated May 30 from Edward to Ezra is clearly more rational, perhaps indicating Edward had rational episodes and irrational episodes.

10. Three dollars and fifty cents per week.

11. Edward Cornell to Ezra Cornell, May 30, 1861, 21/4, ECPKL.

12. Renamed Ithaca, from Gratiot Center in 1857. Pine River Township (not to be confused with Pine River, Michigan). Ithaca is in Arcadia Township adjacent to Pine River near Edward's farm. Walter Romig, *Michigan Place Names* (Detroit: Wayne State University Press, 1986), 289.

13. Edward Cornell to Elijah Cornell, June 4, 1861, 36/11 ECPKL. It begins rather clinically, "Dear Parent."

14. Hiram Robertson to Ezra Cornell, June 9, 1861, 21/5, ECPKL.

15. Crane, *Killed Strangely*. Crane draws the parallels between Alvin and Thomas Cornell quite clearly but does not go much into the case of Edward, which, though it does not result in murder, bears some of the same psychological signature of obsessive jealousy.

16. Joseph Henry, Sylvester David Willard and Edward van Deussen, all disciples of T.R. Beck, all would exert a profound influence on Ezra's life at these two key periods.

17. Alvin was the fourth generation offspring of that marriage between Steven and Hannah, as were Ezra and brother Edward.

18. Mary Robertson to Ezra Cornell, July 17, 1861, 21/5, ECPKL.

19. Edward Cornell to Ezra Cornell, July 14, 1861, 21/5, ECPKL.

20. Angeline Mosher to Ezra Cornell, July 16, 1861, 21/5, ECPKL.

21. Mary Robertson to Ezra Cornell, July 17, 1861, 21/5, ECPKL.

Appendix B

1. George Iles, *Leading American Inventors*, 146.

Bibliography

Allen, A.B., ed. *The American Agriculturalist*, vol. 2. New York: Saxton & Miles, 1843.
American Railroad Journal 18, December 11, 1845.
American Telegraph Magazine 1, no. 1, 1852.
Anbinder, Paul, ed. *Albany Institute of History and Art*. New York: Hudson Hills, 1908.
"Annual Report of the Board of Regents of the Smithsonian Institution." Washington: Harris, 1858.
"Annual Report of the Commissioner of Patents for the year, 1843," 28th Congress, Senate document 150.
The Athenaeum, July–December 1883.
Beck, Lewis C. *A Manual of Chemistry*. Albany, NY: Webster & Skinners, 1831.
Beck, T.R. *Eulogium on the Life of Simeon De Witt*. Albany, NY: Skinner, 1835.
Beck, Theodoric Romeyn, and Amariah Brigham. "Analysis of the Testimony on the Trial of Alvin Cornell for Murder and of the Subsequent Proof which led to the Commutation of His Punishment." Reprinted in *Transactions of the Medical Society of the State of New York* 6, 1846.
Bishop, Morris. *A History of Cornell*. Ithaca: Cornell University Press, 1962.
Bowers, Brian. *Sir Charles Wheatstone, FRS*. London: Institution of Electrical Engineers, 2001.
Brigham, Amariah. *Observations on the Effects of Religion on Health and Physical Welfare*. Boston: Marsh, 1835.
Bruce, Robert. *The Launching of Modern American Science*. New York: Knopf, 1987.
Bryant, William Cullen, and Thomas G. Voss, ed. *Letters of William Cullen Bryant*, vol. 2. New York: Fordham, 1977.
Calvert, Dr. James. "The Electromagnetic Telegraph," accessed January 2, 2015. http://mysite.du.edu/~jcalvert.
Canfield, William, and J.E. Clark. *Things Worth Knowing About Oneida County*. Utica: Griffiths, 1909.
"Charters of American Life Insurance Companies." New York: Spectator, 1906.
Chernow, Ron. *Alexander Hamilton*. New York: Penguin, 2004.
Clinton, George W. *Journal of a Tour from Albany to Lake Erie by the Erie Canal in 1826*. Buffalo: Buffalo Historical Society, 1910.
Coe, Lewis. *The Telegraph*. Jefferson, NC: McFarland, 2003.
Cole, Donald. *A Jackson Man*. Baton Rouge: Louisiana State University Press, 2004.
Cornell, Alonzo. *True and Firm*. New York: A.S. Barnes, 1884.
The Cornell Era 39, no. 1, 1909 (monthly).
Craig, William O. *Around the World with the Smithsonian*. Coral Springs, FL: Llumina Press, 2004.
Crane, Elaine. *Killed Strangely*. Ithaca: Cornell University Press, 2002.

Bibliography

The Cultivator 10, no. 3, Albany.
Cummings, Thomas. *Historic Annals of the National Academy of Design*, vol. 3. New York: Sackett & Cobb, 1861.
Curtis, George. *Life of Daniel Webster*, vol. 2. New York: Appleton, 1870.
Dangerfield, George. *Chancellor Robert R. Livingston of New York*. New York: Harcourt Brace, 1960.
Daniels, George. *American Science in the Age of Jackson*. Tuscaloosa: University of Alabama Press, 1968.
Davis, Daniel. *Journal of the American Scientific Instrument Enterprise* 2, no. 6, 1988.
———. *Manual of the Telegraph*. Boston: self-published, 1851.
Davis, L.J. *Fleet Fire*. New York: Arcade, 2003.
DeKay, J.E. *Natural History of New York*. Albany: Thurlow Weed, 1842.
Demarest, William. *A History of Rutgers College*. New Brunswick: Rutgers University Press, 1924.
Documents of the Senate of the State of New York, vol. 6. Albany: Argus, 1872.
Eaton, Amos. *A Geological and Agricultural Survey of the District Adjoining the Erie Canal*. Albany: Benthuysen, 1824.
Faraday, Michael. *Experimental Researches in Electricity*, vol. 1. London: Quaritch, 1839.
Fink, William B. "Stephen Van Rensselaer, the Last Patroon." Ph.D. diss., Columbia Teacher's College, 1950.
Fisher, George. *Life of Benjamin Silliman*. New York: Scribner, 1866.
Geddings, E., ed. *Medical and Surgical Journal and Review*, vol. 1. Baltimore: Lucas, 1833.
Greene, H.W. *Letters Addressed to F.O.J. Smith*. Privately published, 1839.
Groft, Tammis, ed. *Hudson River Panorama*. Albany: Hudson Hills, 1998.
Grondahl, Paul. *Mayor Erastus Corning*. Albany: SUNY Press, 2007.
Gross, Samuel, ed. *Lives of Eminent American Physicians and Surgeons of the Nineteenth Century*. Philadelphia: Lindsay & Blakiston, 1861.
Hackett, David. *The Rude Hand of Innovation: Religion and Social Order in Albany, New York*. New York: Oxford University Press, 1991.
Haines, Selden. *Biographical Sketch of the Class of 1826, Yale College*. Utica: Roberts, 1866.
Hamilton Literary Monthly 26, no. 8 (Hamilton College), April 1892.
Harlow, Alvin. *New Wires and Old Waves*. New York: Appleton, 1936.
Harpers New Monthly Magazine 55, 1877.
Harriman, Walter. *History of Warner*. Concord: Republican Press, 1879.
Harris, Luther. *Around Washington Square*. Baltimore: Johns Hopkins University Press, 2003.
Hazeltine, Gilbert. *Early History of the Town of Ellicott*. Jamestown: Journal Printing, 1887.
Henry, Joseph. "Contributions to Electricity and Magnetism." *Transactions of the American Philosophical Society* 6, no. 2, 1849.
———. "On some Modifications of the Electro Magnetic Apparatus." *Transactions of the Albany Institute* 1, 1830.
———. *Scientific Writings of Joseph Henry*, vol. 2. Washington: Smithsonian Press, 1887.
Hills, Frederick. *New York State Men*. Albany: Argus, 1910.
Hochfelder, David. *Joseph Henry: Inventor of the Telegraph?* Accessed February 20, 2015. http://siarchives.si.edu/oldsite/siarchives-old/history/jhp/joseph20.htm.
Homans, Benjamin. *Army and Navy Chronicle*, vols. 6–7. Washington: Homans, 1838.
Horton, John. *James Kent: A Study in Conservatism*. New York: Appleton, 1939.
Hounshell, David. *From the American System to Mass Production*. Baltimore: Johns Hopkins University Press, 1984.
Howard, Edward. "The States of Vail's Description." *Maryland Historical Magazine* 62, no. 3, 1967.
Howell, George, and Jonathan Tenney. *History of the County of Albany*, vol. 1. New York: Munsell, 1886.

Huurdeman, Anton. *The Worldwide History of Telecommunications*. Hoboken: John Wiley, 2003.
Iles, George. *Leading American Inventors*. New York: Holt, 1912.
Irwin, Douglas. "The Aftermath of Hamilton's 'Report on Manufactures.'" *Journal of Economic History* 64, no. 3 (2004): 800–821.
John, Richard. *Network Nation*. Cambridge: Harvard University Press, 2010.
Jones, Robert. *King of the Alley*. Philadelphia: American Philosophical Society, 1992.
Jordan, John. *Colonial and Revolutionary Families of Pennsylvania*. New York: Clearfield, 1911.
Journal des Chemins de Fers 4, 1845.
Kent, William. *Memoirs and Letters of James Kent*. New York: Little, Brown, 1898.
Kirk, John W. "Historic Moments: The First News Message By Telegraph." *Scribner's Magazine* 11, 1892.
Kroessler, Jeffrey. *New York, Year by Year*. New York: NYU Press, 2001.
Lanier, Henry. *A Century of Banking in New York*. New York: Doran, 1922.
Larkin, Oliver. *Samuel F.B. Morse and American Democratic Art*. New York: Little, Brown, 1954.
Larson, John. *The Market Revolution in America*. Cambridge: Cambridge University Press, 2010.
Lequeux, James. *Francois Arago*. New York: Springer, 2008.
Littel, Eliakim, and Robert S. Littel, eds. *Littel's Living Age*, vol. 7. Boston: Waite, 1845.
Lundeberg, Philip. "Samuel Colt's Submarine Battery." *Smithsonian Studies in History and Technology* 29, 1974.
Lyons, Jonathan. *The Society for Useful Knowledge*. New York: Bloomsbury, 2013.
Mabee, Carleton. *The American Leonardo*. New York: Knopf, 1943.
McAllister, Ethel. *Amos Eaton: Scientist and Educator*. Philadelphia: University of Pennsylvania Press, 1941.
McKay, Richard. *South Street*. New York: Haskell, 1934.
Meacham, Jon. *Thomas Jefferson: The Art of Power*. New York: Random House, 2012.
Morse, Edward L., ed. *Samuel F.B. Morse: His Letters and Journals*, vol. 2. New York: Houghton Mifflin, 1914.
Moyer, Albert E. *Joseph Henry: The Rise of an American Scientist*. Washington, D.C.: Smithsonian Institution Press, 1997.
Munsell, Joel. *Annals of Albany*, vol. 6. Albany: Munsell, 1855.
Murphy, Brian. *Building the Empire State*. Philadelphia: University of Pennsylvania Press, 2015.
The National Magazine 11, 1889.
New England Magazine 4. Letter of Alfred Vail, dated January 6, 1844.
New York Observer (collected) 19, no. 42, 1841.
Oleson, Alexandra. *The Pursuit of Knowledge in the Early American Republic*. Baltimore: Johns Hopkins Press, 1976.
O'Reilly, Henry. "Argument of Henry O'Reilly Against the Extension of Professor Morse's 1860 patent." Rochester Historical Society, Henry O'Reilly papers.
Osswald, Tim, and Georg Menges. *Materials Science of Polymers for Engineers*, 2nd ed. Cincinnati: Hanser Gardner, 2003.
Page, Charles. "Notice of a Spiral Magnet" and "Description of a new plate, or quantity, Helix for Electro Magnetic Apparatus." *Journal of the Franklin Institute*, third Series, vol. 3, 1842.
Perkins, Dexter. "Henry O'Reilly." *Rochester History* 7, no. 1. www.rochester.lib.ny.us/~rochhist/v7_1945/v7i1.pdf.
Philbrick, Nathaniel. *Sea of Glory*. New York: Penguin, 2003.
Prime, Samuel Iranaeus. *The Life of Samuel F.B. Morse, LL.D*. New York: Appleton, 1875.
Rathbun, Richard. *The Columbian Institute*. Washington: GPO, 1917.
Reid, James. *The Telegraph in America*. New York: Polhemus, 1886.
Reingold, Nathan, and Marc Rothenberg, et al., eds. *Papers of Joseph Henry*, vols. 1, 4, 5, 6 and 7. Washington, D.C.: Smithsonian Institution Press, 1972–1996.

Report of the American Institute, September 12, 1842: The American Electro-Magnetic Telegraph. Philadelphia: Lea and Blanchard, 1845.
Report Upon the Condition and Progress of the U.S. National Museum. Washington, D.C.: GPO, 1897.
Reynolds, Cuyler, ed. *Albany Chronicles.* Albany: Lyon, 1906.
Rhees, William. *Smithsonian Miscellaneous Collections*, vol. 21. Washington, D.C.: Smithsonian Press, 1881.
Ricketts, Palmer. *History of the Rensselaer Polytechnic Institute.* On demand: Forgotten Books, 2015.
Robbins, Allen. *A History of Physics and Astronomy at Rutgers.* Louisville: Gateway, 2001.
Roberts, Steven. "History of the Atlantic Cable and Undersea Communications." Online: atlanticcable.com/Cables/Domestic.
Romig, Walter. *Michigan Place Names.* Detroit: Wayne State University Press, 1986.
Rutledge, Anna. *Artists in the Life of Charleston: Through Colony and State.* Philadelphia: American Philosophical Society, 1949.
Sabine, Robert. *The History and Progress of the Electric Telegraph.* New York: Van Nostrand, 1869.
Scharf, John T. *History of Baltimore City and County.* Philadelphia: Everts, 1881.
Scharf, J. Thomas, and Thompson Westcott. *History of Philadelphia*, vol. 3. Philadelphia: Everts, 1884.
Schiffer, Michael. *Power Struggles: Scientific Authority and the Creation of Practical Electricity before Edison.* Boston: MIT Press, 2008.
Scientific American 2, no. 8, 1860.
Seagar, Robert. *Papers of Henry Clay*, vol. 3. Lexington: University of Kentucky Press, 1963.
Sebring, I.B. *Life of Lewis C. Beck, MD.* Schenectady, NY: privately printed, 1934.
Shaffner, Tal. *Shaffner's Telegraph Companion*, vols. 1 and 2. New York: Pudney & Russell, 1954.
Shaw, Ronald. *Erie Water West.* Lexington: University of Kentucky Press, 1966.
Silverman, Kenneth. *Lightning Man: The Accursed Life of Samuel F.B. Morse.* New York: Knopf, 2003.
Smith, F.O.J. Defendants Proofs, New York Public Library. http://babel.hathitrust.org/cgi/pt?id=nyp.33433020656090;view=1up;seq=9
Sweet, S.H. *Documentary Sketch of New York State Canals.* Albany: Benthuysen, 1863.
Taylor, William. *An Historical Sketch of Henry's Contribution to the Electro-magnetic Telegraph.* Washington, D.C.: GPO, 1879.
Thomas, Philip F. *Decision of Hon. Philip F. Thomas, Commissioner of Patents on the Application of Samuel F.B. Morse.* Washington: Polkinhorn, 1860.
Thompson, Silvanus. *Lectures on the Electromagnet.* New York: Johnston, 1891.
Thue, William. *Electrical Power Cable Engineering.* New York: Marcel Decker, 2005.
Transactions of the Society of Arts, Manufactures and Commerce 43. London: Flindell, 1825.
Vail, Alfred. *Early History of the Electro-Magnetic Telegraph.* New York: Hine, 1914.
Vail, Stephen. "Early Days of the First Telegraph Line." *The New England Magazine* 4, 1892.
Wilson, James. *The Memorial History of the City of New York.* New York: New York History Company, 1893.
Wood, Gordon. *Empire of Liberty.* Oxford: Oxford University Press, 2009.
Zelizer, Viviana. *Morals and Markets.* New York: Columbia University Press, 1979.

Index

Academy of Science (French) 199–200; see also Arago
Adams, John Quincy 76–9
Agriculture 22–3, 28, 30, 34, 36–8, 92, 103, 106, 114, 120, 182; county agricultural societies 34, 38, 40–1; New York board of 41–4, 210
air bubbles (in lead pipe) 119–20
Albany 30, 37–65, 66–7, 64–5, 68–9, 71–80, 183; Albany Institute 29, 43, 45–6, 60; see also Albany Academy; Albany Lyceum
Albany Academy 35, 39–40, 42, 44, 46, 57–8, 60–1, 62, 64–5, 69, 73–4, 80, 83, 87–8, 108
Albany Lyceum 45–6, 60, 62, 75
Albion 68
alcoholism (Henry's father) 246
Allied Textile 244
Allston, Washington 49, 96
American Academy of Fine Arts 27, 49
American Academy of Science 27
American Exploring Expedition (Wilkes Expedition) 140–43, 145
American Institute Fair 244, 254–5; plowing contest 120; see also Castle Gardens
American Journal of Science 29, 36, 38, 69–71, 83–4, 143
American Mineralogical Journal 38
American Philosophical Society 20, 25, 28, 33, 35
Ames, Ezra 26, 39–41, 51, 75
amperage 87; see also Quantity/Intensity circuit
appropriations, New York 182
Arago, François 199–200, 204, 225
Army Topographical Engineers 58
Ashburton 196
asphaltum 82, 103–4, 119, 134–5, 157, 227
Atlantic cable 225, 236, 246, 263

Audubon, James 27; Minniesland 217, 219, 224, 226, 229, 234–61
Avery, Thomas 124, 140–1, 159, 174, 179, 191

Babbage, Charles 89
Bain, Alexander 238–9, 251
Ballston Spa 52, 66, 68
Baltimore 93, 112, 120–1, 123, 129–31, 135, 137, 144, 158, 163–5, 169, 173, 225; Pratt Street 121, 163; see also Burbank, David; Mount Clare Rail Station; Whig Convention
Baltimore and Ohio Railroad 221; see also McLane, Louis
bank drafts, missing 135–6
Barlow, Peter 72, 86
Barnaby Mooer plough 92–5, 108–110, 118, 120, 127, 158–9
Barnard, Frederick 254
Barnum, P.T. 173
Bartlett, Levi 106–7; contract dispute 136, 146–52, 164, 187, 191; trenching contract 106–8, 110, 118, 119–20, 125, 147–8, 170
Bartram, John 27
Batteries 72, 87–8, 90, 124, 134; Fisher's tests on 135, 137, 154, 163, 230, 234, 267; Grove Battery 154–163; see also Cruikshank Battery; galvanic cell; submarine battery
Bayard, Harriet 66–7
Beck, Harriet 61
Beck, Lewis 19, 37, 40, 42–6, 57–60, 73–5, 88–9, 143, 208
Beck, T. Romeyn 19, 25, 30, 36–40, 43, 58, 61, 66, 162, 164, 186, 208, 256, 259, 261; breach with Henry 208
Beebe, Col. Jeremiah 92, 96, 110, 174, 217, 264

325

beeswax *see* wire (insulation)
Belgian Academy 187
Bibb, George 169
Blatchford, Samuel 57–8
Bloomingdale rope walk 134–5; *see also* Chase, Ebenezer
Bodisco, Alexander de 198
Boston 49–50, 83, 106, 122, 134, 173, 181, 201, 214, 222, 229, 239
Botany Tour 33
Bouck, William 112, 162
breach of contract 119–120, 122
Breese, Sidney 53
Breguet, Louis 188, 199–202, 214, 218, 222–4, 235, 238, 243, 248
Brigham, Amariah 93, 162, 164
Bryant, William Cullen 200
Burbank, David 126, 131, 135–7, 144, 153, 161, 174, 179, 181
bureau knob insulator *see* pole insulators
Burr, Aaron 34
Butterfield, John 97, 181, 183, 206

cable-layer 109, 110–11, 118, 120–22, 125, 131, 143, 175–6; wrecking 127, 147, 157; patent 121–2, 138, 156, 161
Cambria 201
Castle Gardens 94, 103–4, 225, 227, 244–5; *see also* American Institute Fair
Chamber of Deputies (French) 200
Charleston 49, 68, 76; Cornell plough trip 93; *see also* South Carolina Academy of Fine Arts
Chase, Ebenezer 134–5; *see also* Bloomingdale ropewalk
circuit configuration 72–3, 87–8, 147, 203, 211, 234–5, 238, 240, 253; *see also* Quantity/Intensity circuit
civil engineering 105, 118, 132, 199
Civil Engineers and Architects Journal 161
Clark Company 218, 223, 243
Clay, Henry 75–9, 163; *see also* Corrupt Bargain
Clinton, DeWitt 39, 43, 64, 76
code 96, 263
Colt, Samuel 26, 94, 100–05, 117, 119, 122, 129, 134, 146, 181, 210, 227–8, 238–41, 244–6, 254, 262; *see also* lines; submarine battery
Columbia College 37, 53, 64, 254
commodities market 181, 245
compass experiment *see* Dana, James Freeman
congressional appropriation 97–8, 145–147, 149, 152, 169–70, 173; Colt's 146 first 94, 100, 101; second 105, 133–4; Washington-Baltimore line 175

Cooke, William 86–7, 94, 155–6, 161
copper 47, 72, 89, 154; *see also* wires
Corcoran, William 172
Cornell, Alonzo 160, 237, 246
Cornell, Alvin 93–4, 112–13, 158, 162, 164, 186, 257, 259, 261; murder case 93–4, 108, 112–13
Cornell, Edward 158, 256–61
Cornell, Elijah 108
Cornell, Ezra 66–7, 96, 108, 112–14, 135, 146, 152, 161–4, 172–86, 188–90, 193–95, 197, 201–6, 209–15, 217–19; cable-layer 107–11, 161; family life 160; glass pole insulators 81–2; health 174, 217, 257, 227; Hudson River, crossing 225–37, 246; letters to Mary Ann 158–61; madness in family 67, 93–4, 113, 164, 256–65; magnetic relay 85, 89–91, 221–30, 238–43, 246; Patent Office 132, 138–44; sales trip 92–4, 96, 106; Test Telegraph 117–127, 131, 136–8, 150; Washington 156–62, 249–54
Cornell, MaryAnn 93–4, 108, 112–13, 125, 141, 143, 156, 158–64, 173, 176, 185–86, 197, 256
Cornell, Rebecca 67, 257, 259
Cornell, Robert Comfort 108
Cornell, Thomas 257, 259
Cornell University 39, 85, 113
Corrupt Bargain 76–8
cotton 32; as insulation 73, 104, 122, 135, 143–4; shellacked 119, 122, 134, 192; *see also* varnish
cotton gin 26, 254
Court of Death 49
Crawford, William 76–7
cross cut *see* short circuits
cross-Hudson line 227; *see also* submarine cable
Cruikshank Battery (series) 87
current fall-off *see* resistance

Dana, James Freeman 45, 83–4, 192; *AJS* article 69–73
Davis, Daniel 205, 214, 222
Declaration of Independence 49
defective design (relay) 214; wiring 122, 125–7, 131–2, 137, 148, 150
DeWitt, Richard Varick 37, 108
DeWitt, Simeon 34, 38–9, 45
diplomatic status 196
Doolittle, Curtis 50
doubts on feasibility of telegraph 98–9, 105–6, 129
Draper, John 134, 192

Eastern Argus 98
Eaton, Amos 30, 43–4, 46, 57–60, 65, 68, 75; letters 31, 44; Rensselaer School 44, 46, 57–60
Eddy, James 228
Edwards, Ninian 79
Eights, Jonathan 37, 58
election: 1824, 51, 75–8; 1844, 163
electricity 72, 86–7
electromagnetic telegraph 146, 173, 207; concept 89; European 86–7; Henry's 87–89; *see also Paquebot Sully*
Electro-magnetic Telegraph Company *see* Magnetic Telegraph Company
Electromagnetism 29–30, 45, 65, 83–4, 87–8, 91, 143, 192, 199, 204, 208, 215, 239; electromagnets 60, 70–4; insulation 73
Ellsworth, Annie 138–9, 152, 163, 191
Ellsworth, Henry L. 29, 96, 101, 105, 132, 138, 140, 142, 147, 150–2
Ellsworth, Henry W. 198
English telegraph 72, 86–7, 94–5, 156, 187, 199; *see also* Barlow, Peter; Cooke, William; Sturgeon, William; Wheatstone, Charles
equipment fabricators *see* fabricators
Erie Canal 32, 40, 53, 57–9, 64–5, 68, 186, 210; Eaton's floating university 58; mineralogical wealth 46–7; survey 43–44; *see also* DeWitt. Simeon
Erskine, Robert 38
express services 174, 210, 226, 234, 239, 245; *see also* #10 Wall

fabricators *see* Clark Company; Davis, Daniel; Munger; Stokell Company
Faraday, Michael 72, 86–7, 143, 193
Faxton, Theodore 97, 181, 183, 203, 206, 211–12, 214, 218–19, 222–4, 235, 240, 242–3, 249
Federalists 19–22, 28, 30, 33–4
Fellenbergian System 59
Field, Cyrus 249
Fiji 141
fire alarm 174–5
Fisher, Alexander 68, 69
Fisher, James 103–5, 118–19, 126–8, 130–1, 133–4, 139–40, 146, 149–55, 184, 191, 254, 262; dispute with Morse over insulation 135–7, 150, 169, 177
Fletcher's rooming house 82, 141, 146, 157, 159, 162, 184
Fort Lee 211, 233–5, 237, 246; ferry 226, 236; letters from Kendall to Cornell on 226, 229
Foy, Alphonse 199–200

Franklin, Benjamin 18, 44; Franklin Institute 97; printing press 140
French Telegraph 199–201
Freneau, Phillip 34

Gale, Leonard 81, 84–5, 89–91, 94, 101, 104–5, 126, 127–133, 136–39, 143, 147–9, 153–4, 161, 170–2, 182–3; illness 130, 136, 152
galvanic cell 70, 72; galvanic electricity 134, 178
galvanic multiplier 70, 73, 86, 87
galvanometer 124, 137, 154; *see also* galvanic multiplier; Schweigger, Johann
Geological Survey 42–4, 58, 61
Girard College 161
Goell, A.C. 213–14, 221, 225–6
Goodyear, Charles 228–9; rubber 230, 236, 241
Governor's Island 104
Granite Farmer 106
Green, Norvin 249
Greenough, Charles 172, 201, 248
grist mill 33
ground return 161, 192
gum elastic 227

Hamilton, Alexander 20–6, 28, 31–4, 36, 244
Hammond, Jabez 58
Harrisburg, Pennsylvania 181, 211–13, 215
Henry, Joseph 20, 25, 28, 30, 39, 45–6, 58, 60–6, 69–71, 73–4, 80–5, 94–5, 100–102, 105–113, 117, 134, 148, 155, 157, 189, 191, 193–5, 200, 239, 243, 253, 256; *AJS* article 74, 208; deposition 269; fight with Vail 81–3, 204–5, 207–09, 211, 254–5; letters 62–3, 82, 137, 153–4; quantity vs. intensity 72, 86–91, 99, 176, 188, 190, 192, 210, 238, 269; *see also* Quantity/Intensity circuit
Henry, Wilkes 141
history of the telegraph: pamphlet 203, 206, 210–12; *see also* Vail, Alfred
hostile takeover 217, 238–40
House, Royal 194, 239, 251
House Committee on Commerce 97, 105, 149
House of Representatives 49–50, 53, 64, 75–6, 79, 102, 105
Hubbell, Horatio 83
Hudson: line across 180, 185, 211–12, 214, 218–19, 221, 224–30, 233–6, 242, 246–7; steamboat travel on 50–1, 53

independent circuit *see* marginal circuit
India rubber 135, 227–8, 241, 254

Index

induction 95, 143, 148, 153, 229; self-induction 88
Industrial Revolution 24
infringement 71, 80, 174–5, 201, 239–43
insanity 256–7, 261; legal defense 112–13, 162
insulation: problems with 140, 143–4, 153–5, 119–20, 127, 137, 150; *see also* asphaltum; cotton; cross cut; electromagnet; wires
insulator 82, 156–7, 160, 175, 211, 213–14, 241
intensity circuit *see* Quantity/Intensity circuit
Ithaca 92, 96, 107, 110, 121, 138, 141, 162–3, 173–4, 176, 179–80, 181–3, 185, 193, 197, 214, 217, 257–50, 264

Jackson, Charles 71, 171, 173, 191, 251
Jackson Andrew 76–7, 79, 160, 170
James E. Serrell Company 110–11, 118–127, 229–30, 132, 127, 140, 143, 148, 157
James Kent 53
Jefferson, Thomas 20–36, 41; *see also* Botany Tour
Johnson, Cave 170
Journal of Commerce 96

Kane, Elias Kent 53
Kendall, Amos 170–4, 178–181, 184–5, 190, 194–5, 197, 201–2, 210–17, 220–30, 233, 236, 238–245, 247–49, 251–2; dispute with Cornell 175, 177, 218–19, 229
Kennebec 118
Kent, James 18, 40, 51, 61, 64, 68–9
Kent, Moss 48, 52–3

Lancastrian School 59, 161, 163
LaTrobe, Benjamin, Jr. 132
lead pipe 104, 110, 119, 144, 148, 157, 161, 227–8, 230, 236; problems with 120, 127, 131, 134
L'Hommedieu, Ezra 34–5, 38
Livingston, Crawford 181, 226, 229, 242, 245; *see also* hostile takeover
Livingston, Robert 20, 22, 28–40, 58–9
Livingston and Wells 239
lobbyist 100
lyceum system 29, 35–6, 45–6, 60, 62

Madison, James 34
magnetic field 39; atmospheric 94–5
Magnetic Telegraph Company 170–2, 177–81, 190, 218–22, 239–42, 245
magnets 88, 207
Maine Farmer 92, 98, 107

mandril 120, 122–3
marginal circuit 176, 239, 252–3; *see also* relay
Mason, Charles 248, 251
Mayflower Compact 98–99
McClane, Louis 121, 132
McClure, William 19, 59
Medical Jurisprudence 64
Mercantile Stock Exchange 181, 245
Merchant's Exchange building 221, 234, 246
Mesmerism 170, 217
Mexico 54, 75, 76–9
Mineralogy 27, 44, 45, 58; Mineralogical Committee of SPUA 37–8; survey 42–3, 44
Minniesland *see* Audubon
Mirror 98–100
Mitchell, Samuel Latham 34–5, 38
Moll, Gerard 74, 207
Morse, Charles 50, 53
Morse, James Edward 200, 212
Morse, Jedidiah (Jedediah) 30, 48–52, 106; geographic text 64
Morse, Lucretia 48, 50, 52, 75, 78, 106; letters 51, 53
Morse, Richard 50
Morse, Samuel F.B. 70–2, 75–6, 78–86, 96–102, 104–7, 113–14, 176, 186, 191, 219–222; Albany 26, 29–30, 48, 52, 54, 61, 64, 66, 68; as artist 18, 39, 49–51, 53, 69; inventor 20–1, 88–91, 94–5, 110–11, 117, 118–119, 120–140, 143, 145, 48–64, 170–3, 211, 225, 229, 233–6, 238–47, 254–5, 257, 262–5; dispute over Vail's book 83–5, 205–6, 207–9; dispute with Cornell over relay 160, 174–5, 178–79, 184, 189–90, 193–5, 203–4, 214–16, 223–4, 242; European trip (1845) 177, 181, 185, 187–8, 190, 192, 196–202; health 104; letters from Faxton Saxton 218
Morse, Sidney 131, 136, 143, 151–2, 196, 199–200
Morton, Samuel George 27
Mount Clare Rail Station 121, 135, 163, 165
Munger 179, 183, 185, 193, 215
Munroe, Charles 141
Mutual Life Insurance Company 108

naptha 227
National Academy of Design 98, 187, 200
National Institute 100, 140–3, 145
National Intelligencer 131
New Harmony 19, 59
New Haven 30, 50–4, 74, 228–9
New Orleans 181
New York Athenaeum 29, 71, 83, 192
New York State Fair 183; *see also* Utica

Index

New York University 81, 84, 99–105, 118, 126, 133–4, 152, 192, 244
Newark 180, 226, 236; Kendall threat to send Cornell to 229, 235–6
Niernsee, John 132
#10 Wall Street 174, 178, 182–3, 219, 225–6, 228, 239–40

Observer 136, 151
Oersted, Hans 70
Ohm, Georg 74, 87, 91, 193, 204, 269
Ohm's Law 87–8, 210
ore separator 89
O'Reilly, Henry 71, 80–1, 83, 180, 181, 201, 203, 206, 209–16, 218, 239, 243, 249, 251, 254, 262
overhead pole solution 94–5, 131–2, 143, 154–6, 160–1, 179; superiority of 148
over-winding *see* overwound circuits
overwound circuits 70, 72–3, 84
Owen, Robert 19

Pacific Telegraph Act 249–50, 256
package service 186
Page, Charles Grafton 85, 138, 143, 147, 154–5, 157, 172, 179, 188–90, 192, 205, 216–17; manifesto 248, 251–2; relay 193–6, 200–1, 211–12; relay dispute with Cornell 182–5, 202, 215, 222–6
Palmyra 65
panic: of 1819, 220; of 1837, 92, 96
Paquebot Sully 29, 83, 191
Paris 32, 188, 196, 252; Morse trip 198–201
Passaic Falls 25, 244
Patent Arms Company 244–5
Patent Office 132, 138, 139, 141, 144–8, 152, 159, 172, 179, 184, 193–5, 201, 212–13, 227, 235, 243; library 140, 142–3, 157, 176; Page's rooftop experiment 154; *see also* submarine cable
patents 80, 83, 85, 89, 147, 161, 175, 194–5; cable layer 109, 134; Cornell 90–1, 96, 156, 178–9, 182, 184, 189, 201–3, 212–15, 217–18, 222, 227–8, 235, 240, 243; European 98, 187–8, 190, 199, 201, 202, 203, 244, 190; Henry, unwillingness to patent telegraph 89; Morse 97, 104–5, 164, 169–72, 174, 187, 222–4, 251–3, 267; Page 184, 190; renewal 85, 90, 248–53; Serrell 110
Paterson 21, 100, 120
Pattison, Catherine 97
Peale, Rembrandt 18, 26, 49–50
Peale, Titian 82, 142
peripheral circuit *see* relay
Pestalozzian education 59
Philadelphia 27, 30, 33, 93, 97, 112, 120, 122, 162, 181, 211, 213–5, 225, 235–6, 239, 242–3, 254
Pickering, Charles 142
plough venture 92–4, 108; *see also* Barnaby Mooer plough
Pocahontas 93
Poinsett, Joel 49, 54, 75–9, 145; *see also* National Institute
pole insulators 82–3, 156–7, 174–5
policy 22–3
Polynesian artifacts 141
port rule 89, 171, 191
Portland, Maine 92–3, 107, 110–13, 118, 121–2, 125, 160
Post Office Department 117, 145, 164; offer to take over telegraph 169–71
posts 179, 221, 226; *see also* poles
Princeton 74, 81–2, 85, 95, 100, 105; Morse visit to 99, 102, 155
printing press 140, 254

Quaker Day School 93
Quantity/Intensity circuits 72, 87–8, 238, 269; *see also* Henry, Joseph
Quimby, Aaron 64–5

rain gauge 256
receiving magnet *see* relay
register 187, 190, 196, 200, 202–3, 211–13, 234, 267–8
Reid, James 206
relay 85, 187, 190, 199, 203–8, 224, 235, 238, 248, 251, 264, 267; Breguet's 200–02, 235, 238, 240, 243; comparative test with Page's 183, 184–5; Cornell's 89–90, 176–180, 182–4, 188, 189, 193–4, 214–15, 217–19, 222–9, 234, 240, 242–3, 249–50; in-line; 129–30, 164, 176, 191, 233, 256; Page's version 184–5, 188, 195–6, 216, 227, 233, 252; Vail's book 210–13
Relay House, Maryland 125–6, 128–132, 137, 149, 163
remote switching 239
Rensselaer School *see* Eaton, Amos
Report on Manufactures 21, 26
resin *see* wires (insulation)
resistance 72–4, 87–8, 238; *see also* wires
rights of way 94, 121, 132, 180, 221
Robertson, Hiram 258–60
Rochester Democrat 206
Rochester Post Office 206
Rogers, Henry 169, 173, 206; plan for submarine cable 226–8, 241
rope walk *see* Bloomingdale rope walk
rotunda 49; Cornell visit to 141, 159; Morse commission for 96, 98–9

330　　　　　　　　　　　　　　　　　Index

Rouen 199, 200
Royal Society 70
rubber 219, 228, 236, 241, 254; *see also* Goodyear; gum elastic; India rubber
Ruhmkorff coil *see* transformer
Russia 188, 194, 198
Rutgers 38, 60

salaries 98, 118, 121, 132, 139–40, 147, 150, 169, 173, 204; Cornell 140, 159, 172–4, 178, 217, 241; Vail 132, 139
Scale of Chemical Logarithmic Equivalents 60, 89
Schweigger, Johann 70, 73, 86–7
scientific endeavor 17–19, 26–8
scientific organizations 20, 33–5, 39; telegraph 104–5
scientific racism 27
semaphores 199; optical telegraph 200, 227
sending key 171, 191, 241
SEUM (also SUM) 20–6, 28, 30, 33, 41, 100, 244
Shaffner's Telegraphic Journal 85
sheep 35
shellac *see* wires
short circuits 124, 127–8, 147–8, 153–7, 190, 193
Sibley, Hiram 249
silk 70, 73
Silliman, Benjamin 18, 27, 30, 36, 38, 43, 50–3, 64, 66, 68–9, 74–5, 81, 84, 192, 229; Yale electromagnet 88
singeing *see* insulation
Smith, Francis Osmond 26, 92–5, 97–9, 102, 104–7, 109–13, 118–22, 125, 129–31, 136, 139, 145–7, 159–60, 165, 170–5, 177–9, 181–5, 187–8, 191, 197–99, 209, 214, 217–18, 222–3, 225, 238–40, 242, 245, 253, 254, 262, 264
Smith, Joseph 65
Smithsonian 82, 84–5, 192, 209, 254; *see also* National Institute
solenoid 239
Somerville 211, 213, 221, 225, 233–4
South Carolina Academy of Fine Arts 49, 54, 75–6
SPAAM (SPUAM or SPUA) 20, 28–30, 33–42, 45, 51, 57–8, 75, 108, 145; journal 38
Spafford, Horatio Gates 46
Speed, J. J. 176
Speedwell Iron Works 97
Spencer, John C. 120, 135, 148–152, 164, 225; letters from 151; letters to 131, 139–40
Springfield, Albany and Buffalo Telegraph Company 219
state road project 58, 65
steam 32, 65; navigation 28, 35–6

Steinheil, Carl 95, 204; *see also* ground return
stock issuance (MTC) 171
stock subscription plan 171–2, 181, 217; *see also* investment
Stockholm 198
Stokell Company 223
Sturgeon, William 69–70, 72, 86–9
submarine battery 101, 104, 117
submarine cable 242; Cornell's 213–14, 218–19, 225–30; Colt's 103–4, 241, 245–6; Rogers and Page's 226–7, 236–7
substations 176, 203, 211, 234–5, 239
Superintendant of the Telegraph 117, 181, 186, 197, 209, 215, 221
Supreme Court 161, 163, 209, 251; infringement cases 80, 253

Tatham Brothers 122–137, 143, 147–54, 179, 246; *see also* defective design
telegraph lines 181, 239; Albany-Buffalo 195, 212, 215, 216, 219, 224, 235, 243; Baltimore-Philadelphia 172, 181, 195, 239; letter 230; New York-Albany 242; New York-Boston 182, 214, 224, 239, 242; New York and Offing 181, 241, 244–6, 254; New York-Philadelphia 170, 172, 199, 183–5, 197, 211, 216, 219, 221, 227; Philadelphia-Pittsburgh 210, 212–16, 282
telegraph stations 169, 173, 176, 190, 203, 205–6, 252, 268
Ten Eyck, Philip 60, 61, 73–4, 83, 88
Teschemacher, J. E. 120
Test Telegraph 105, 121–8, 204, 206, 221, 233, 253; completion 164; continuity 124, 140, 147–8, 154; current 86; lead pipe 119–20, 123–7, 136–7, 153, 157; Morse's deposition 267; overhead portion of 148, 162, 171; Page experiment 154; testing 101, 111, 134, 162–3, 183, 185, 206, 211, 216, 224, 227–30, 233–6, 252; work stoppage 130–2; *see also* Congressional appropriate; right of way; trenching
Thomas, Philip 249, 252–3; *see also* patents
Torrey, John 35, 65
transformer 254
Treasury Department 145; *see also* Spencer and Woodbury
trenching 105–6, 110–11, 125, 140, 151; in the Capitol 147–50
Troy, New York 46, 53, 57–9, 65, 97
Trumbull, John 18, 49–50
Tyler, John 163

underwater cable *see* submarine cable
Upshur, Abel 129

Index 331

Utica 53, 164, 181, 210, 212, 216, 218, 235; state fair 182–6, 205, 227

Vail, Alfred 105, 118–19, 123–7, 130–2, 152, 159, 161, 163, 165, 169, 170, 172–5, 182–4, 187, 194–5, 197–8, 203, 211–12, 214, 217–19, 221–3, 225–9, 233–6, 238–43, 252–4, 262–4; book 188, 204–209; NYU 97, 99–100; patent office 138–42; pole insulators 81–2, 156–7
Vail, George 204
van Buren, Martin 58, 76, 210, 220
Van Deussen, Edward 113, 258–9
van Rensselaer, Stephen III 30, 31, 32, 37, 39–43, 45–6, 48, 51–3, 59, 61, 64, 68, 76, 78, 81
van Rensselaer, Stephen IV 46, 51
Varick, Richard 60
varnish 70, 106, 164
Volta 103–4
Voltage 87; *see also* Quantity/Intensity circuit
voltaic potentials in long wire 87, 137, 148, 153–5; Henry letter on 154

Wagram Hotel 198, 200
Walsh, Robert 200, 201
Watson, Elkanah 38
Webster, Daniel 77, 110, 161–2
Wells, Henry 181
Wells Fargo 239
West Point 60, 65, 71; lyceum 35
Western Union 219, 249–50
Wheatstone, Charles 94–5, 155, 156, 187, 189, 199, 200, 204; letter to Henry 192, 207–8; Morse letter to Sidney about 199; reference to in Page's manifesto 89, 248
Whig Convention 162–3
Wickliffe, Charles 117
Wilkes, Charles 141–2, 145
Willard Hotel 250, 260
wire guides 174
wires 125, 131–2, 137, 140, 143–4, 148, 154, 164, 174, 178, 192, 203, 213–14, 222, 234; color coded 124, 127; diminution of current problem 72–4, 86–8, 269; ground wire 119, 155, 162; Henry's experiment 73; hot annealing 124; Hudson River 226–8, 230, 236; methods of insulation 101, 104, 106, 122, 134–5, 143, 192, 206; redundant wires 155, 176, 253, 267; relay magnet 193, 252; ten-mile spool 85, 97, 99–100; *see also* cotton; cross cut; defective design; resistance
wire-winding machine 143, 156, 175, 241
Wood, Orrin 173–5, 182–3, 185–6, 205–6, 215, 222, 228, 243
Woodbury, Levi 96–7; letter to 146

Yale University 29, 36, 49, 52, 69, 133, 192

zinc 72, 154

www.ingramcontent.com/pod-product-compliance
Lightning Source LLC
Chambersburg PA
CBHW020859020526
44116CB00029B/464